HAPPY ACCIDENTS

HAPPY ACCIDENTS
Serendipity in Modern Medical Breakthroughs

Morton A. Meyers, M.D.

ARCADE PUBLISHING • NEW YORK

FIRST EDITION

Library of Congress Cataloging-in-Publication Data

Meyers, Morton A.
  Happy accidents : serendipity in modern medical breakthroughs / Morton Meyers. —1st ed.
      p. cm.
  Includes bibliographical references and index.
  ISBN 978-1-55970-819-7 (alk. paper)
  1. Medicine—History. 2. Medical innovations—History. 3. Medicine—Research—History. 4. Discoveries in science. 5. Serendipity. I. Title.

  R149.M49 2007
  610.9—dc22   2006100551

Published in the United States by Arcade Publishing, Inc., New York
Distributed by Hachette Book Group USA

Visit our Web site at www.arcadepub.com

10 9 8 7 6 5 4 3 2 1

Designed by API

EB

To my wife, Bea,
my greatest serendipitous discovery

Contents

Preface   *xi*

Introduction: Serendipity, Science's Well-Guarded Secret   1

PART I:
THE DAWN OF A NEW ERA: INFECTIOUS DISEASES AND ANTIBIOTICS, THE MIRACLE DRUGS

1. How Antony's Little Animals Led to the Development of
   Germ Theory   29

2. The New Science of Bacteriology   34

3. Good Chemistry   38

4. The Art of Dyeing   48

5. Mold, Glorious Mold   59

6. Pay Dirt   82

7. The Mysterious Protein from Down Under   92

8. "This Ulcer 'Bugs' Me!"   99

PART II:
THE SMELL OF GARLIC LAUNCHES THE WAR ON CANCER

9. Tragedy at Bari   117

10. Antagonists to Cancer   127

11. Veni, Vidi, Vinca: The Healing Power of Periwinkle   131

12. A Heavy Metal Rocks: The Value of Platinum   135

13. Sex Hormones   138

14. Angiogenesis: The Birth of Blood Vessels   140

15. Aspirin Kills More than Pain   148

16. Thalidomide: From Tragedy to Hope   151

17. A Sick Chicken Leads to the Discovery of
    Cancer-Accelerating Genes   154

18. A Contaminated Vaccine Leads to Cancer-Braking Genes   159

19. From Where It All Stems   165

20. The Industrialization of Research and the War on Cancer   169

21. Lessons Learned   181

PART III:
A QUIVERING QUARTZ STRING PENETRATES
THE MYSTERY OF THE HEART

22. An Unexpected Phenomenon: It's Electric!   187

23. What a Catheter Can Do   195

24. "Dottering"   205

25. A Stitch in Time   208

26. The Nobel Committee Says Yes to NO   215

27. "It's Not You, Honey, It's NO"   220

28. What's Your Number?   225

29. Thinning the Blood   234

PART IV:
THE FLAW LIES IN THE CHEMISTRY, NOT THE CHARACTER:
MOOD-STABILIZING DRUGS, ANTIDEPRESSANTS,
AND OTHER PSYCHOTROPICS

30. It Began with a Dream   241

31. Mental Straitjackets: Shocking Approaches   246

32. Ice-Pick Psychiatry   252

33. Lithium   260

34. Thorazine   266

35. Your Town, My Town, Miltown!   270

36. Conquering the "Beast" of Depression   273

37. Librium and Valium   281

38. "That's Funny, I Have the Same Bug!"   284

39. LSD   287

Conclusion: Taking a Chance on Chance:
            Cultivating Serendipity   300

Acknowledgments   *321*

Notes   *323*

Selected Bibliography   *371*

Illustration Credits   *375*

Index   *377*

Preface

"My God, it moves!" I was astonished at what I saw on the X-ray screen. As an academic radiologist, I was trying several years ago to visualize anatomic structures and features never seen before. I was directing my research efforts to the abdominal cavity, the largest potential space in the body, which encloses complex organs and structures. To do this, I introduced liquid contrast material ("dye") in volunteers to fill its recesses and outline its contents on X-ray images. Surprisingly, I discovered that the contrast agent "spontaneously" flowed. Rather than being static and pooling, over time the fluid spread in a specific pattern. I would come to understand that this dynamic circulation was influenced not only by anatomy but also by factors such as gravity and pressures within the abdomen. This serendipitous epiphany provided the stepping-stone to understanding how cancers metastasize to specific remote body sites: cancer cells, shed into the fluid evoked, are carried by the circulating fluid to be deposited at distant sites of pooling. The malignant cells become attached at these points by adhesions and continue to divide to form what is referred to as a secondary deposit, or metastasis. It became clear that the spread of a disease throughout the body is not a random, irrational occurrence but rather follows a predictable pattern. Analysis of a large volume of patient data corroborated this conclusion. This insight regarding cancer was universally adopted and now serves as the basis of modern-day detection and management.

Radiology is a medical specialty in which the trained eye reaps enormous benefits in diagnosis. Every radiologist is certainly familiar with uncovering incidental findings in daily practice that may redirect the course of an investigation. Such a finding is sometimes called a "corner-of-the-film diagnosis." Based on a clinical suspicion, an X-ray is requested to search for a specific abnormality, but the results often reveal disease in the periphery of the original area of interest. The value of accidental discoveries is deeply rooted in diagnostic imaging. Indeed, it is the basis upon which the specialty was founded. When Wilhelm Röntgen was experimenting with a cathode-ray tube in 1895, he noticed a fluorescent glow in the darkened room of his laboratory and thought at first that the effect was caused by the sunlight beyond the wooden shutters. Röntgen had made an unexpected discovery: the X-ray. Equally unexpected was the discovery of radioactivity by Henri Becquerel in the following year.

My own serendipitous experience set me on a quest to understand the role of chance in scientific research and its contribution to medical advances in the past century. I was amazed at the findings.

Most people have had at least one experience in which an unintentional action or inadvertent observation, or perhaps even simple neglect, led to a happy outcome — to something they could not, or would not, have been able to accomplish even if they had tried. Surprising observations that led to the development of several commercial products have been well described, including champagne, synthetic sweeteners, nylon, the microwave oven, and Post-it notes. In scientific research, such incidents happen all the time, but they have generally been kept secret. In fact, they occur way more often than most researchers care to admit or are even aware of. Accidental discoveries have led to major breakthroughs that today save the lives of millions of people and to drugs and procedures whose names have become household words. Lithium, Viagra, Lipitor, antidepressants, chemotherapy drugs, penicillin and other antibiotics, Coumadin — all were discovered not because someone set out to find a specific drug that did a specific thing but because someone found something he or she wasn't even looking for. Similarly, the use of surgical gloves, the Pap smear, and catheterization of the heart's arteries leading to bypass surgery were all stumbled upon.

This is the essence of serendipity. Although the term has become popularized to serve as the synonym for almost any pleasant surprise, it actually refers to searching for something but stumbling upon an unexpected finding of even greater value — or, less commonly, finding what one is looking for in an unexpected way. Discovery requires serendipity. But serendipity is not a chance event alone. It is a process in which a chance event is seized upon by a creative person who chooses to pay attention to the event, unravel its mystery, and find a proper application for it.

Many of the most important breakthroughs in modern medicine have routinely come from unexpected sources in apparently unrelated fields, have often been the work of lone researchers or small close-knit teams operating with modest resources and funding, and have depended crucially on luck, accident, and error. With luck, the essential human factor is sagacity.

While serendipity is essential to discovery, it is nothing without the human beings who know an opportunity when they see one. Lucky accidents or happenstance that could point the way to great discoveries occur every day, but few people recognize them. Successful scientists may have the insight and creativity to recognize a "Eureka!" moment when it happens, see the potential, and know what to do to take it to the next step.

The scientific literature very rarely reflects this reality. The dominant convention of all scientific writing is to present discoveries as rationally driven and to let the facts discovered speak for themselves. This humble ideal has succeeded in making scientists look as if they never make errors, that they straightforwardly answer every question they investigate. It banishes any hint of blunders and surprises along the way.

Consequently, not only the general public but the scientific community itself is unaware of the vast role of serendipity in medical research. Typically, a discoverer may finally admit this only toward the end of his or her career, after the awards have been received. Memoirs, autobiographies, and Nobel Prize acceptance speeches may reveal the true nature of the discovery. From personal interviews with several Nobel laureates and winners of the prestigious Albert Lasker Award, I have come to understand the factors that have driven many of the critical medical advances of the twentieth century.

This book is intended to be a comprehensive account, for a general or scholarly readership, of the importance of serendipity in modern medicine. It reveals the crucial role of chance in each of the four major fields of medical advances in the past century: infectious disease, cancer, heart disease, and mental illness. These pivotal discoveries are part of our everyday culture; most of us are familiar with or directly benefit from the products and procedures that have resulted.

Casting a critical eye on the way in which our society spends its research dollars, *Happy Accidents* offers new benchmarks for deciding how to spend future research funds. We as a society need to take steps to foster the kind of creative, curiosity-driven research that will certainly result in more lifesaving medical breakthroughs. Fostering an openness to serendipity has the potential to accelerate medical discovery as never before.

HAPPY ACCIDENTS

Introduction

Serendipity, Science's Well-Guarded Secret

I exist
But only in you if you want me . . .
All things are meaningless accidents, works of chance
unless your marveling gaze,
as it probes, connects and orders,
makes them divine . . .

— WILHELM WILLMS, "GOD SPEAKS"[1]

Contemplating the genesis of the great medical breakthroughs of the last century, most people picture brilliant, well-trained scientists diligently pursuing a predetermined goal — laboriously experimenting with first this substance and then that substance, progressing step by step to a "Eureka!" moment when the sought-after cure is at last found. There in the mind's eye is Marie Curie stirring a vat of pitchblende over many years to recover minute amounts of radium, or Paul Ehrlich testing one arsenical compound after another until he finds Salvarsan, the "magic bullet" against syphilis, on his 606th attempt. In the contemporary setting, one looks to what might be called Big Science. Surely, we imagine, in the halls of ivy-draped universities and the gleaming labs of giant pharmaceutical companies, teams of researchers in smart white coats are working in harmony to cure cancer, banish the common cold, or otherwise produce the Next Big Thing in medicine.

For its own reasons, the medical establishment is happy to perpetuate these largely false images. By tradition and protocol, it presents science as a set of facts and strong beliefs that, like the Ten Commandments, have been set in stone by a distant all-knowing authority and, if followed, will lead inevitably through a linear process to the desired results. Furthermore, it portrays the history of scientific advances as a sequence of events that have led to more-or-less direct progress.

The reality is different. Progress has resulted only after many false starts and despite widespread misconceptions held over long periods of time. A large number of significant discoveries in medicine arose, and entirely new domains of knowledge and practice were opened up, not as a result of painstaking experimentation but rather from chance and even outright error. This is true for many of the common drugs and procedures that we rely on today, notably many antibiotics, anesthetics, chemotherapy drugs, anticoagulant drugs, and antidepressants.

Consider the following examples, all typical of how things happen in medical research:

- At the Johns Hopkins Hospital in 1947, two allergists gave a new antihistamine, Dramamine, to a patient suffering from hives. Some weeks later, she was pleased to report to her doctors that the car sickness she had suffered from all her life had disappeared. Drs. Leslie Gay and Paul Carliner tested the drug on other patients who suffered from travel sickness, and all were completely freed of discomfort, provided the drug was taken just before beginning the potentially nauseating journey. A large-scale clinical trial involving a troopship with more than 1,300 soldiers crossing the rough North Atlantic for twelve days (Operation Seasickness) decidedly proved the drug's value in preventing and relieving motion sickness. Dramamine is still used today, available over the counter.[2]

• A professor of biological chemistry and medicine at the Johns Hopkins University School of Medicine was studying a particular blood protein when he found another protein contaminating his sample. Rather than simply discarding it, Dr. Peter Agre realized that he had stumbled upon the structure of the channel — folded-up proteins piercing cell walls — that can control the flow of water molecules into and out of living cells. For making this basic discovery, which, he said, "really fell into our laps," he won the Nobel Prize in Chemistry in 2003.[3]

• A similar circumstance proved very beneficial to the neurobiologist David Anderson of the California Institute of Technology, who publicly announced his serendipitous breakthrough in the *New York Times* in July 2001. Researching neural stem cells, the cells that build the nervous system in the developing embryo, Anderson discovered the "magic fertilizer" that allowed some of them to bloom into neurons, sprouting axons and dendrites: "It was a very boring compound that we used to coat the plastic bottom of the Petri dish in order to afford the cells a stickier platform to which to attach. Never would we have predicted that such a prosaic change could exert such a powerful effect. Yet it turned out to be the key that unlocked the hidden neuronal potential of these stem cells."[4]

• An unanticipated variable seriously hampered the efforts of biochemist Edward Kendall to isolate the thyroid hormone thyroxine, which partly controls the rate of the body's metabolism. After four years of meticulous work on the gland, he finally extracted crystals of the thyroid hormone on Christmas morning 1914 at the Mayo Foundation in Rochester,

Minnesota. But when he moved to expand production, Kendall could no longer recover active material. Only after fourteen months of futile efforts was he able to trace the cause of this setback to the decomposition of the hormone by the use of large galvanized metal tanks in which the extraction from the gland was being done. The iron and copper in the metal tanks rendered the crystals ineffective. From then on, he used enamel vessels. By 1917, Kendall had collected about seven grams of crystals and was able to start clinical studies.[5]

THE NORMAL VERSUS THE REVOLUTIONARY

In his highly influential 1962 book *The Structure of Scientific Revolutions,* Thomas Kuhn contributed an idea that changed how we see the history of science.[6] Kuhn makes a distinction between "normal" and "revolutionary" science. In "normal" science, investigators work within current paradigms and apply accumulated knowledge to clearly defined problems. Guided by conventional wisdom, they tackle problems within the boundaries of the established framework of beliefs and approaches. They attempt to fit things into a pattern. This approach occupies virtually all working researchers. Such efforts, according to Nobel laureate Howard Florey, "add small points to what will eventually become a splendid picture much in the same way that the Pointillistes built up their extremely beautiful canvasses."[7]

Kuhn portrays such scientists as intolerant of dissenters and preoccupied with what he dismissively refers to as puzzle-solving. Nonetheless, a period of normal science is an essential phase of scientific progress. However, it is "revolutionary" science that brings creative leaps. Minds break with the conventional to see the world anew. How is this accomplished? The surprising answer may be "blindly"! Systematic research and happenstance are not mutually exclusive; rather they complement each other. Each leads nowhere without the other.

According to this view, chance is to scientific discovery as blind genetic mutation and natural selection are to biological evolution. The appearance of a variation is due not to some insight or foresight but rather to happenstance. In groping blindly for the "truth," scientists sometimes accidentally stumble upon an understanding that is ultimately selected to survive in preference to an older, poorer one.

As explained by Israeli philosophers of science Aharon Kantorovich and Yuval Ne'eman, "Blind discovery is a necessary condition for the scientific revolution; since the scientist is in general 'imprisoned' within the prevailing paradigm or world picture, he would not intentionally try to go beyond the boundaries of what is considered true or plausible. And even if he is aware of the limitations of the scientific world picture and desires to transcend it, he does not have a clue how to do it."[8]

An anecdote about Max Planck, the Nobel Prize–winning physicist, hammers home this reality. When a graduate student approached him for a topic of research for his Ph.D. thesis, asking him for a problem

"I'll be happy to give you innovative thinking. What are the guidelines?"

he could solve, Planck reportedly scoffed: "If there was a problem I knew could be solved, I would solve it myself!"

Induction and deduction only extend *existing* knowledge. A radically new conceptual system cannot be constructed by deduction. Rational thought can be applied only to what is known. All new ideas are generated with an irrational element in that there is no way to predict them. As Robert Root-Bernstein, physiology professor and author of *Discovering,* observed, "We invent by intention; we discover by surprise."[9] In other words, accidents will happen, and it's a blessing for us that they do.

THE RECEPTIVE SCIENTIFIC MIND

"Accident" is not really the best word to describe such fortuitous discoveries. Accident implies mindlessness. Christopher Columbus's discovery of the American continent was pure accident — he was looking for something else (the Orient) and stumbled upon this, and never knew, not even on his dying day, that he had discovered a new continent. A better name for the phenomenon we will be looking at in the pages to follow is "serendipity," a word that came into the English language in 1754 by way of the writer Horace Walpole. The key point of the phenomenon of serendipity is illustrated in Walpole's telling of an ancient Persian fairy tale, *The Three Princes of Serendip* (set in the land of Serendip, now known as Sri Lanka): "As their highnesses traveled, they were always making discoveries, *by accidents and sagacity,* of things they were not in quest of."[10]

Accidents *and* sagacity. Sagacity — defined as penetrating intelligence, keen perception, and sound judgment — is essential to serendipity. The men and women who seized on lucky accidents that happened to them were anything but mindless. In fact, their minds typically had special qualities that enabled them to break out of established paradigms, imagine new possibilities, and see that they had found a solution, often to some problem other than the one they were working on. Accidental discoveries would be nothing without keen, creative minds knowing what to do with them.

The term "serendipity" reached modern science by way of physiologist Walter B. Cannon, who introduced it to Americans in his 1945 book *The Way of an Investigator*.[11] Cannon thought the ability to seize on serendipity was the mark of a major scientist. The word is now loosely applied in the popular media to cover such circumstances as luck, coincidence, or a fortunate turn of events. This sadly distorts it. Serendipity means the attainment or discovery of something valuable that was not sought, the unexpected observation seized upon and turned to advantage by the prepared mind. The key factor of sagacity has been lost. Chance alone does not bring about discoveries. Chance with judgment can.

Serendipity implies chance only insofar as Louis Pasteur's famous dictum indicates: "In the field of observation, chance favors only the prepared mind." Salvador Luria, a Nobel laureate in medicine, deemed it "the chance observation falling on the receptive eye." *I have the answer. What is the question?* Turning an observation inside out, seeking the problem that fits the answer, is the essence of creative discovery. Such circumstances lead the astute investigator to solutions in search of problems and beyond established points of view.

The heroes of the stories told in this book are not scientists who merely plodded rationally from point A to point B, but rather those who came upon X in the course of looking for Y, and saw its potential usefulness, in some cases to a field other than their own. Chance is but one element, perhaps the catalyst for creativity in scientific research. And, yes, the process of discovery is indeed creative. It involves unconscious factors, intuition, the ability to recognize an important anomaly or to draw analogies that are not obvious. A creative mind is open and can go beyond linear reasoning to think outside the box, look beyond conventional wisdom, and seize on the unexpected. Most important, a creative scientific mind recognizes when it is time to start viewing something from a whole new perspective.

TURNING REALITY ON ITS SIDE

One day in 1910, the Russian painter Wassily Kandinsky returned to his studio at dusk and was confronted with an object of dazzling beauty on his easel. In the half-light, he could make out no subject but was profoundly moved by the shapes and colors in the picture. It was only then that he realized the painting was resting on its side. Like an epiphany, this experience confirmed his growing belief in the emotional powers of colors and in the ultimate redundancy of the traditional subject of a picture. Kandinsky, who broke through to what he called "nonobjective" painting, is widely acknowledged as the father of abstract art.

By dipping into the world of art, and especially into visual illusions, scientists can gain perspective on illusions of judgment, also known as cognitive illusions. Gestalt psychologists have elaborated on such things as the balances in visual perception between foreground and background, dark and light areas, and convex and concave contours. The gist of their message is that too-close attention to detail may obscure the view of the whole — a message with special meaning for those alert to serendipitous discovery.

To readily appreciate this phenomenon, consider the paintings of the contemporary artist Chuck Close. Viewed at the usual distance, they are seen as discrete squares of lozenges, blips, and teardrops.

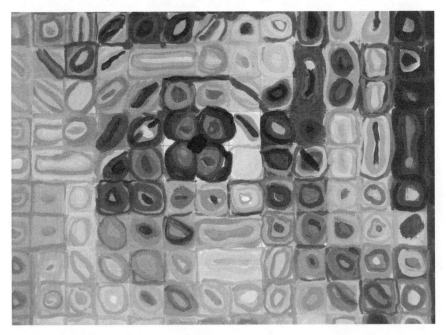

Detail from Chuck Close's *Kiki* — left eye.

Viewed from a much greater distance, they can be appreciated as large, lifelike portraits.

Chuck Close, *Kiki*, 1973. Oil on canvas, 100 x 84 in. Walker Art Center, Minneapolis.

In a Gestalt figure, such as the M. C. Escher drawing on the next page, one can see the devils or the angels, but not both at the same time. Even after you know that there is more than one inherent pattern, you see only one at a time; your perception excludes the others.

M. C. Escher, *Circle Limit IV*, 1960. Woodcut in black and ochre, printed from two blocks.

The same holds true for W. E. Hill's *My Wife and My Mother-in-Law*. It's easy to see his dainty wife, but you have to alter your whole way of making sense of the lines to see the big-nosed, pointy-chinned mother-in-law.

Certainly, if one's perspective is too tightly focused, gross distortion may result. This phenomenon has broad implications for medical research. So does the human tendency to believe that one's partial

view of an image — or, indeed, a view of the world — captures its en-
tirety. We often misjudge or misperceive what is logically implied or
actually present. In drama this may lead to farce, but in science it leads
to dead ends.

Illustrative of this phenomenon are poet John Godfrey Saxe's six
blind men (from his poem "The Blind Men and the Elephant")

W. E. Hill, *My Wife and My Mother-in-Law*, 1915.

observing different parts of an elephant and coming to very different but equally erroneous conclusions about it. The first fell against the elephant's side and concluded that it was a wall. The second felt the smooth, sharp tusk and mistook it for a spear. The third held the squirming trunk and knew it was a snake. The fourth took the knee to be a tree. The fifth touched the ear and declared it a fan. And the sixth seized the tail and thought he had a rope. One of the poem's lessons: "Each was partly in the right, And all were in the wrong!"[12]

Robert Park, a professor of physics at the University of Maryland and author of *Voodoo Science,* recounts an incident that showed how expectations can color perceptions. It happened in 1954 when he was a young air force lieutenant driving from Texas into New Mexico. Sightings of UFOs in the area of Roswell, New Mexico, were being reported frequently at the time.

> I was driving on a totally deserted stretch of highway. . . . It was a moonless night but very clear, and I could make out a range of ragged hills off to my left, silhouetted against the background of stars. . . . It was then that I saw the flying saucer. It was again off to my left between the highway and the distant hills, racing along just above the range land. It appeared to be a shiny metallic disk viewed on edge — thicker in the center — and it was traveling at almost the same speed I was. Was it following me? I stepped hard on the gas pedal of the Oldsmobile — and the saucer accelerated. I slammed on the brakes — and it stopped. Then I could see that it was only my headlights, reflecting off a single phone line strung parallel to the highway. Suddenly, it no longer looked like a flying saucer at all.[13]

People, even scientists, too often make assumptions about what they are "seeing," and seeing is often a matter of interpretation or perception. As Goethe said, "We see only what we know." As they seek causes in biology, researchers can become stuck in an established mode of inquiry when the answer might lie in a totally different direction that can be "seen" only when perception is altered. "Discovery consists of seeing what everybody has seen and thinking what nobody has thought," according to Nobelist Albert Szent-Györgyi.[14]

Another trap for scientists lurks in the common logical fallacy *post hoc, ergo propter hoc* — the faulty logic of attributing causation based solely on a chronological arrangement of events. We tend to attribute an occurrence to whatever event preceded it: "After it, therefore because of it."

Consider Frank Herbert's story from *Heretics of Dune*:

> There was a man who sat each day looking out through a narrow vertical opening where a single board had been removed from a tall wooden fence. Each day a wild ass of the desert passed outside the fence and across the narrow opening — first the nose, then the head, the forelegs, the long brown back, the hindleg and lastly the tail. One day, the man leaped to his feet with the light of discovery in his eyes and he shouted for all who could hear him: "It is obvious! The nose causes the tail!"[15]

A real-life example of this type of fallacy, famous in medical circles, occurred in the case of the Danish pathologist Johannes Fibiger, who won the Nobel Prize in Medicine in 1926 for making a "connection" that didn't exist. Fibiger discovered roundworm parasites in the stomach cancers of rats and was convinced that he had found a causal link. He believed that the larvae of the parasite in cockroaches eaten by the rats brought about the cancer, and presented experimental work in support of this theory. Cancer research at this time was inhibited by the lack of an animal model. The Nobel committee considered his work "the greatest contribution to experimental medicine in our generation." His results were subsequently never confirmed and are no longer accepted.

Another, less famous example of false causality occurred in New York in 1956. A young physicist, Chen Ning Yang, and his colleague, Tsung-Dao Lee, were in the habit of discussing apparent inconsistencies involving newly recognized particles coming out of accelerators while relaxing over a meal at a Chinese restaurant on 125th Street in

Manhattan frequented by faculty and students from Columbia University. One day the solution that explained one of the basic forces in the atom suddenly struck Yang, and within a year the two shared one of the quickest Nobel Prizes (in Physics) ever awarded. After the award was announced, the restaurant placed a notice in the window proclaiming "Eat here, get Nobel Prize."[16]

PATHWAYS OF CREATIVE THOUGHT

Researchers and creative thinkers themselves generally describe three pathways of thought that lead to creative insight: reason, intuition, and imagination.

Three Pathways of Creative Thought		
Reason	Intuition	Imagination
Logic	Informal patterns of expectation	Visual imagery born of experience

While reason governs most research endeavors, the most productive of the three pathways is intuition. Even many logicians admit that logic, concerned as it is with correctness and validity, does not foster productive thinking. Einstein said, "The really valuable factor is intuition. . . . There is no logical way to the discovery of these elemental laws. There is only the way of intuition, which is helped by a feeling for the order lying behind the appearance."[17]

The order lying behind the appearance: this is what so many of the great discoveries in medicine have in common. Such intuition requires asking questions that no one has asked before. Isidor Rabi, the Nobel Prize–winning physicist, told of an early influence on his sense of inquiry. When he returned home from grade school each day, his mother would ask not "Did you learn anything today?" but "Did you ask a good question today?"[18] Gerald Edelman, a Nobel laureate in medicine, affirms that "the asking of the question is the important thing. . . . The idea is: can you ask the question in such a way as to facilitate the answer? And I think really great scientists do that."[19]

Intuition is not a vague impulse, not just a "hunch." Rather, it is a cognitive skill, a capability that involves making judgments based on very little information. An understanding of the biological basis of intuition — one of the most important new fields in psychology — has been elaborated by recent brain-imaging studies. In young people who are in the early stages of acquiring a new cognitive skill, the right hemisphere of the brain is activated. But as efficient pattern-recognition synthesis is acquired with increasing age, activation shifts to the left hemisphere. Intuition, based upon long experience, results from the development in the brain of neural networks upon which efficient pattern recognition relies.[20] The experience may come from deep in what has been termed the "adaptive unconscious" and may be central to creative thinking.[21]

As for imagination, it incorporates, even within its linguistic root, the concept of visual imagery; indeed, such words and phrases as "insight" and "in the mind's eye" are derived from it. Paul Ehrlich, who won the Nobel Prize in Medicine in 1908 for his work on immunity, had a special gift for mentally visualizing the three-dimensional chemical structure of substances. "Benzene rings and structural formulae disport themselves in space before my eyes. . . . Sometimes I am able to foresee things recognized only much later by the disciples of systemic chemistry."[22] Other scientists have displayed a similar sort of talent leading to breakthroughs in understanding structures.

Creativity is a word that most people associate with the arts. But the scientific genius that leads to great discoveries is almost always rooted in creativity, and creativity in science shares with the arts many of the same impulses. Common to both are the search for self-expression, truth, and order; an aesthetic appreciation of the universe; a distinct viewpoint on reality; and a desire for others to see the world as the creator sees it. The novelist Vladimir Nabokov bridged the tension between the rational and the intuitive in his observation that "there is no science without fancy and no art without fact."

Among artists, the creative urge with its sometimes fevered

obsession has entered our folklore. Legend has it that one day, when Sir Walter Scott was out hunting, a sentence he had been trying to compose all morning suddenly leaped into his head. Before it could fade, he shot a crow, plucked off one of its feathers, sharpened the point, dipped it in the bird's own blood, and recorded the sentence.[23] In the twentieth century, Henri Matisse, bedridden in his villa near Nice during his recovery from abdominal surgery, could not restrain himself from using a bamboo stick with chalk at its tip to draw on his bedroom wall. Among scientists, the creative urge is no less compelling.

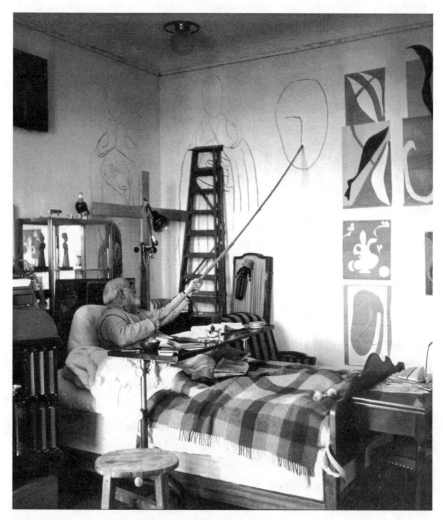

Henri Matisse, ca. 1950. Photo by Walter Carone.

Creative people are open-minded and flexible in the face of unusual experiences. They are alert to the oddity of unexpected juxtapositions and can recognize a possibility even when it is out of context. In his massive work *The Act of Creation,* Arthur Koestler proposes that bold insights are produced by juxtaposing items that normally reside in different intellectual compartments, a process he terms "bisociation." In many scientific discoveries, he asserts, "the real achievement is seeing an analogy where no one saw one before."[24]

In the late 1940s the biologist Aser Rothstein saw such an analogy. He was working at a unit of the then-secret Manhattan Project established at the University of Rochester. At that time, the cell membrane was basically an abstract notion, and the leading concept simply regarded diffusion across it as being of a passive nature. Rothstein was studying the toxic action of uranium salts on cells. Laboratory data were coming out well and reproducibly until, suddenly, everything went wrong. There was no progress, and no two results were the same.

One day, when Rothstein walked into his lab, he noticed a box of detergent that was used in the lab to clean glassware. On the box, surrounded by a flashy red star, were the words "New Improved Dreft." Comparing its label to that of an old box of Dreft, Rothstein saw that the new version contained an added ingredient — a water softener. As it turned out, this softener coated glass tenaciously and chemically bound the material Rothstein was studying (uranium ions) to the surface of the glass. His creative mind then made an extraordinary leap. He wondered about a possible analogy: If there is binding on the surface of glass, could there be binding on the surface of a cell?

Seizing upon this capability of the chemical in the water softener, he went on to prove that there are binding sites on the cell surface as well. Fortune had provided him with a contaminant similar to the natural enzymes involved in transport across the cell membrane. But Fortune might have come calling in vain if not for Rothstein's ability to draw the essential analogy. Some ten years before the cell membrane could actually be seen with the development of electron microscopy, Rothstein's "accidental" discovery enabled him to show that it was a metabolically *active* structure containing enzymes critical in transport mechanisms.[25]

Analogical thinking has certainly been a cornerstone of science. The seventeenth-century English physiologist William Harvey compared the heart to a pump. The physicists Ernest Rutherford and Niels Bohr pictured the atom as a tiny solar system. "Every concept we have," writes the cognitive scientist Douglas R. Hofstadter, "is essentially nothing but a tightly packaged bundle of analogies."[26]

Drawing analogies is one part of the creative discovery process, but an equally important one is seeing things that don't quite make sense. Thomas Kuhn introduced the idea that revolutions in science arise from the recognition of anomalies. Kuhn observed that the accumulation of anomalies — findings that cannot be assimilated into an accepted scientific framework, tradition, or paradigm — paves the way for scientific revolution. A single anomalous observation may stimulate an initial inquiry, but most productive to an alert mind is a special sort of anomaly, one that clearly falls into a *class* of anomalies. Resolving just one can provide insight into a whole category of more complicated ones. For example, the era of cancer chemotherapy was initiated by the recognition of never-before-seen symptoms in sailors saturated for long periods with liquid mustard gas during a military disaster in World War II. From this came the development of alkylating chemical agents, followed by a series of different categories of anticancer drugs.

In the early 1950s Nathan Kline, a psychiatrist at Rockland State Hospital in Orangeburg, New York, exploited an anomalous reaction in patients receiving the drug reserpine for hypertension. He noticed that it tranquilized agitated, restless patients. It was later shown that reserpine affected the levels of serotonin, dopamine, and adrenaline in the brain. This was truly a "Eureka!" finding because it steered psychiatry onto a whole new path that focused on brain chemistry. Kline's pioneer efforts in introducing the use of tranquilizers to the practice of psychiatry in the United States was followed by the development of a host of psychoactive drugs influencing the brain's neurotransmitters, culminating in today's multibillion-dollar mood-altering-drug industry.

Creative thinkers tend to take analogies and anomalies to higher levels. They have a gift for seeing *similar differences* and *different similarities* — phrases coined by the British theoretical physicist David

Bohm. True creation, Bohm argues, relies upon perceiving a new fundamental set of similar differences that constitutes a genuinely new order.[27] Indeed, it is the recognition of anomalies, discrepancies, inconsistencies, and exceptions that often leads to the uncovering of a truth, perhaps one of greater magnitude than the one originally pursued. Writing of Charles Darwin, his son said: "Everybody notices as a fact an exception when it is striking and frequent, but he had a special instinct for arresting an exception. A point apparently slight and unconnected with his present work is passed over by many a man almost unconsciously with some half-considered explanation, which is in fact no explanation. It was just those things that he seized on to make a start from."[28]

The ideal scientific mind comfortably incorporates unanticipated factors into an established body of work or, more likely, follows it in completely new directions. Such a mind handles error, inconsistencies, and accidents in a characteristic way that represents a special mark of creativity. In other words, the open mind embraces serendipity and converts a stumbling block into a stepping-stone. As Winston Churchill whimsically observed, "Men occasionally stumble across the truth, but most of them pick themselves up and hurry off as if nothing happened."

WHEN INSIGHT STRIKES

A perceptive breakthrough may be likened to grasping the "hidden" figure in a Gestalt diagram. In the 1950s Rosalyn Yalow, a biophysicist, and Solomon Berson, a physician, at the Bronx VA Hospital began using radioisotopes — radioactive forms of chemical elements — to study diseases. At that time, it was believed that the high levels of sugar in the blood of adult diabetics were due to insulin deficiency. Some researchers hypothesized that it was probably destroyed by a liver enzyme once it entered the bloodstream.

To their surprise, the researchers found that injected radioactive insulin remained longer — albeit uselessly — in diabetic patients who had received insulin than in people who had never received insulin before. Further studies led to an astonishing discovery: a large plasma protein, a gamma globulin antibody, was called forth as part of the body's immune system, inactivating the insulin and keeping it in the bloodstream.

Then a "Eureka!" moment occurred that would have delighted the Gestalt psychologists of perception. Because both natural insulin and injected radioactive insulin compete for sites on the antibody molecule, the amount of the natural hormone present in a patient's body can be measured. A curved line viewed from one side is convex but viewed from the other side is concave. As Yalow put it, "Once you saw it one way, you saw it the other way."[29] The inverse would measure the hormone itself. In this way an unexpected finding combined with a flash of insight, and the method of radioimmunoassay (RIA) was born.

The term was aptly chosen because the method used radioactively tagged substances to measure antibodies produced by the immune system. Circulating throughout the human body in solution in the blood are a multitude of hormones and other regulatory substances. They are each infinitesimal in quantity but exert profound effects. To understand bodily functions, it is necessary to determine the presence and amount of each substance.

Yalow and Berson discovered by accident a technique so sensitive that it can detect the equivalent of a sugar cube dissolved in Lake Erie. This revolutionized endocrinology and its application to virtually every system in the body. RIA is now routinely used to detect such things as hepatitis-associated antigen in the blood of patients and donors and the presence of steroids in the urine of athletes, and to ascertain blood levels of therapeutic drugs. Its discovery resulted from experiments initially designed to answer another question.[30]

Yalow acknowledged the role of serendipity: "It was luck . . . to discover that insulin disappears more slowly from one group of patients than from another. . . . That's what you mean by discovering something by accident. You make an observation. But it isn't by accident that you interpret the observation correctly. That's creativity."[31]

No one would expect much science to come out of a university dining hall experience. Nevertheless, Richard Feynman, as a twenty-eight-year-old at Cornell, was eating in the school cafeteria when someone tossed a dinner plate into the air. Its two simultaneous movements caught his attention. His eyes following the red medallion insignia of Cornell on one rim of the plate, he saw not only that it was spinning but also that it was wobbling. He noticed something amiss: the spinning rotation and the wobble were not precisely synchronous. Feynman turned his characteristic playfulness and unbridled curiosity to this trivial observation:

> I had nothing to do, so I start to figure out the motion of the rotating plate. I discover that when the angle is very slight, the medallion rotates twice as fast as the wobble rate — two to one. It came out of a complicated equation! Then I thought, "Is there some way I can see in a more fundamental way, by looking at the forces or the dynamics, why it's two to one?"

"There was no importance to what I was doing," he wrote later, "but ultimately there was. The diagrams and the whole business that I got the Nobel Prize for came from the piddling around with the wobbling path." That "whole business," as he charmingly called it, was the application of his observation about the Cornell plate to the spin of electrons, known as nuclear precession, and the reformulation of quantum electrodynamics, the strange rules that govern subatomic reality.[32]

DON'T ASK, DON'T TELL

Relatively few investigators have spontaneously acknowledged the contribution of chance and accident to their discoveries. Scientific papers in the main do not accurately reflect the way work was actually done. Researchers generally present their observations, data, and conclusions in a dry passive voice that perpetuates the notion that discoveries are the natural outcome of deliberative search. The result, in the words of Peter Medawar, winner of a Nobel Prize for his pioneering work in immunology, is to "conceal and misrepresent" the working reality.[33] Virtually without exception, scientific literature imposes a *post facto* logic on the sequence of reasoning and discovery. The role of chance would never be suspected from the logically rigorous sequence in which research is reported.

Too much is at risk for scientists early in their careers to admit that chance observations led to their achievements. Only years later, after reputations are solidly made, do they testify to the contributions of such mind-turning factors as unexpected results, fortuitous happenstances, or exceptions to a premise. The truth is aired in award-acceptance speeches, autobiographies, or personal interviews. Wilhelm Röntgen, who won the Nobel Prize in 1901 for his discovery of X-rays, readily acknowledged the accidental nature of his discovery in a lecture to his local physics society. However, it is typically not until the Nobel Prize acceptance lectures that the laureate will for the first time clearly acknowledge the role of chance, error, or accident — as happened with Charles Richet (immunology, 1913), Alexander Fleming (the first antibiotic, 1945), Baruch Blumberg (the hepatitis B virus, 1976), Rosalyn Yalow (radioimmunoassay, 1977), and Robert Furchgott (the signal molecule nitric oxide, 1998).

To his credit, the accounts of his experiments on nerve conduction by Alan Hodgkin — subsequently awarded the Nobel Prize in 1963 — are openly characterized by such phrases as "discovered by accident when trying to test something quite different," "to our surprise . . . ," "chance and good fortune," and "a great piece of luck."[34]

Thomas Starzl, a surgical pioneer in the field of liver transplantation, wrote about his early career in a personal letter to a colleague: "I have a very difficult confession to make. Practically every contribution I ever made in my professional life turned out to be exactly the opposite of my expectations. This means that all my hypotheses turned out to be wrong, and usually spectacularly so. Naturally, I would not admit this to anyone, but an old friend!"[35]

Based upon a series of serendipitous events in his own research, Aser Rothstein observed: "Many of our advances in biology are due to chance, combined with intelligent exploitation . . . It is for this reason that the image of the scientist is not a true one. He comes out as a cold, logical creature when in reality he can fumble around with as much uncertainty as the rest of humanity, buffeted by an unpredictable environment."[36]

Peter Medawar has asserted that "any scientist who is not a hypocrite will admit the important part that luck plays in scientific discovery."[37] Writing in 1984, after a distinguished career in immunology with the National Institute for Medical Research in England, J. H. Humphrey stated: "Most of [my experiments] that led to anything novel or interesting arose because of some unexpected or chance observation that I was fortunate in being able to follow up."[38] Humphrey felt obligated to make the point not only in his recollection but eventually in the *British Medical Journal,* where he wrote rather forcefully: "I am aware from personal experience or from acquaintance with the people concerned how little the original purpose of some important experiments had to do with the discoveries which emerged from them. This is rarely obvious from the published accounts. . . . By the time a paper is published the findings have usually been married with current ideas and made to look as though they were the logical outcome of an original hypothesis."[39] Some observers have euphemistically termed this process "retrospective falsification."[40] Others have baldly termed it "fraud."[41]

Those instances of rare and belated admissions underscore the deliberate omission of the creative act that originated the medical discovery. The general scientific paper simply does not accurately reflect the way the work was actually done.

At times, one can detect an inchoate longing among scientists themselves for a forum for recounting the distractions, obstructions, stumblings, and stepping-stones in the process. Richard Feynman, the plainspoken physicist, affirms in his 1966 Nobel lecture:

> We have a habit in writing articles published in scientific journals to make the work as finished as possible, to cover up all the tracks, to not worry about the blind alleys or describe how you had the wrong idea first, and so on. So there isn't any place to publish, in a dignified manner, what you actually did in order to get to do the work.

It is not hard to understand why most scientists remain circumspect. Embarrassment and fear of loss of stature may inhibit them from making full disclosure. They do not wish to jeopardize their chances to raise funds, win grants, earn publication, and advance their careers. It is unsettling for scientists to have to admit that so many discoveries came about purely by accident.

Reflecting on nonlogical factors in research, Rothstein concluded that "there is no body of literature to which one can turn . . . that reveals or collates the factor of chance and serendipity in research."[42] It is precisely this complaint that this book attempts to rectify.

A NEW SCIENTIFIC METHOD?

"Unless you expect the unexpected," warned the Greek philosopher Heraclitus, "you will never find truth, for it is hard to discover and hard to attain."[43]

Can a serendipitous discovery be predicted? Of course not. We cannot forecast that something — especially something valuable — will be found without specifically being sought. Does randomness play a role? Although chance implies unpredictability, it does not mean total randomness. In a random occurrence, there is complete absence of any explanation or cause. Randomness is generally seen as incompatible with creativity, as improbable as the writing of *Hamlet* by the legendary band of monkeys with typewriters in the basement of the British Museum.

Three things are certain about discovery: Discovery is unpredictable. Discovery requires serendipity. Discovery is a creative act. In the words of Peter Medawar:

> What we want to know about the science of the future is the content and character of future scientific theories and ideas. Unfortunately, it is impossible to predict new ideas — the ideas people are going to have in ten years' or ten minutes' time — and we are caught in a logical paradox the moment we try to do so. For to predict an

idea is to have an idea, and if we have an idea it can no longer be the subject of a prediction.[44]

Yet, despite the examples given, and all that follow, medical research stubbornly continues to assume that new drugs and other advances will follow exclusively from a predetermined research path. Many, in fact, will. Others, if history is any indication, will not. They will come not from a committee or a research team but rather from an individual, a maverick who views a problem with fresh eyes. Serendipity will strike and be seized upon by a well-trained scientist or clinician who also dares to rely upon intuition, imagination, and creativity. Unbound by traditional theory and willing to suspend the usual clinical set of beliefs, this outsider will persevere and lead the way to a dazzling breakthrough. Eventually, once the breakthrough becomes part of accepted medical wisdom, the insiders will pretend that the outsider was one of them all along.

So the great secret of science is how much of what is sought is not actually found, and how much of what has been found was not specifically sought. Serendipity matters, and it benefits us greatly to understand the true dynamics of the discovery process for many reasons: Because we are affected so directly by medical advances. Because directed research — in contrast to independent, curiosity-driven research that liberates serendipity — is often costly and unproductive. Because we need to be sound in our judgment of the allocation of funding and resources. Because profound benefits and consequences to society may be at stake. And — perhaps an equally compelling reason — because we thrill to hear and understand the many fascinating stories that lie at the intersection of science, creativity, and serendipity.

Part I

The Dawn of a New Era: Infectious Diseases and Antibiotics, the Miracle Drugs

Chance favors only the prepared mind.

— Louis Pasteur

1

How Antony's Little Animals Led to the Development of Germ Theory

In today's era of the electron microscope, the Hubble telescope, and satellite transmission of images from the surface of Mars, the observations of an unschooled shopkeeper in Delft in the 1670s of a hitherto unknown world prove even more astounding.

Antony van Leeuwenhoek (pronounced *Lay-ven-hook*) earned his living as a draper but surely ranks among the greatest self-taught geniuses in the history of science and medicine. Having become skilled in grinding and polishing lenses to inspect cloth fibers during a youthful apprenticeship in Amsterdam, this amateur scientist of limited education designed and built simple microscopes with astonishingly high magnification and resolution. With these he first observed microorganisms.[1]

Leeuwenhoek not only used his lenses to more closely inspect small structures that the naked eye could discern — duck feathers, seeds, mold, the parts of a bee — but he also had the curiosity to move beyond the understanding held by more learned contemporaries and peer into what had been an invisible world. Leeuwenhoek lived during the much-heralded age of exploration, which introduced to Europe, among many other products, spices from far-flung continents. Leeuwenhoek wanted to find out, by the microscopic examination of macerated peppercorns, why pepper is hot. (He thought the peppercorns might have spikes on their surface). Examining a suspension in water, he was surprised to see what he called "very little animalcules," which were without question bacteria.

His observations were detailed over the next fifty years in 375 letters, frequently illustrated, to the Royal Society of London for Improving Natural Knowledge and published in its *Philosophical Transactions.*[2] It was his famous Letter 18, dated October 9, 1676, that caused a sensation and earned him immortality for his discovery of protozoa and bacteria. His descriptions allow us to share his wonder at the microscopic appearance of a new world of "little creatures" in rainwater that had been left in a barrel for several days, well water, and seawater. He saw that what modern science knows as protozoa ("first animals") use tiny "legs" or "tails" to swim in the tiny drop of water that was their world. We can share his sense of awe as he observed that "the motion of most of these animalcules in the water was so swift, and so various upwards, downwards, and round about, that 'twas wonderful to see." He observed protozoa entangled in a filament expand and contract their shape, "struggle by strongly stretching themselves," to extricate themselves. And he marveled when an "animalcule" brought on a dry place "burst asunder." Observing no skeletal parts or obvious skin in an animal whose body consisted of soft "protoplasm" was a considerable novelty at this date.

These observations would not be followed up until Letter 39, dated September 17, 1683, when he examined the plaque from his own teeth and was stunned at the number of bacteria he found. He estimated that "there are more animals living in the scum on the teeth in a man's mouth than there are men in a whole kingdom." From this microscopic menagerie, he clearly described and illustrated all the morphological types known today: round (cocci), rod-shaped (bacilli), and spiral-shaped (spirochetes).[3]

Leeuwenhoek came tantalizingly close to grasping the germ theory of disease, when he found animalcules swarming in the decaying roots of one of his teeth and in the plaque from his own and other people's mouths. He noted that people who cleaned their mouths regularly had much less plaque than those who did not. And coming within hailing distance of heat pasteurization, he saw that the animal-cules in plaque "could not endure the heat of my coffee."

It took two hundred years from Leeuwenhoek's first observations until the germ theory of disease and the first effective germ-fighting treatments were established. The delay was in large part due to the fact that all biological processes are chemically based and mediated, and thus progress in medicine was intertwined with progress in chemistry.[4]

Frenchman Louis Pasteur discovered the role of bacteria in the causation of disease. A chemist and not a physician, Pasteur was work-ing on problems besetting burgeoning French industries. They quickly led him from chemistry to reveal the workings of biology.

Looking at wine under a microscope, he unraveled the fermenta-tion roles played by living yeast organisms and bacteria in the delicate balance between wine and vinegar. But Pasteur wondered: If these or-ganic changes were caused by tiny living microbes, where did they come from? Were they in the air, waiting for favorable conditions to multiply, or were they generated spontaneously by the lifeless matter itself? By 1864, in a series of ingenious experiments, he proved that living organisms do not spontaneously arise but are present in any material because they are introduced, then reproduce. He showed that the air is never free from living organisms.

The first disease Pasteur attributed to a living organism was one that was devastating the silkworm industry. By 1870 he showed that it was due to a protozoan infesting the grain the silkworms were fed.

Pasteur was elected to the Académie de Médecine in 1873. On February 19, 1878, before the academy, he presented his germ theory of infection. He laid out his conviction regarding the causal relation-ship between microorganisms and disease: that specific organisms pro-duce specific conditions; that destroying these microorganisms halts transmission of communicable diseases; and that vaccines might be prepared for prevention. A revolutionary dictum now illuminated the way toward productive research, practice, and therapeutics. However, this was not universally greeted with acclaim. Some doctors called it "microbial madness" and disdainfully asked Pasteur, "Monsieur, where is your M.D.?"

Are Your Hands Clean?

Meanwhile, an English surgeon named Joseph Lister, inspired by Pasteur's work, inaugurated the era of modern surgery. In 1865 he began a program to prevent sepsis by using carbolic acid as a disinfectant, markedly reducing the incidence of postoperative infections. As the remonstrations of Ignaz Semmelweis in the late 1840s and 1850s against the unwashed hands of obstetricians received the scorn of his colleagues, so did Lister's technique meet the rigid objections of the surgical establishment before it was generally accepted two decades later.

Another advance in reducing surgical infections was serendipitously introduced by William Stewart Halsted, professor of surgery at the Johns Hopkins University Hospital and Medical School in Baltimore. Halsted's OR nurse, Caroline Hampton, whom he would later marry, suffered dermatitis of her hands from repeated exposure to the sterilizing solution (mercuric chloride). In 1891 Halsted asked the Goodyear Company to make thin rubber gloves, which he then had her use to protect her hands.

Before long it became clear that the wearing of gloves — and later surgical gowns and masks, and the heat sterilization of instruments — by operating room staff prevented infection in surgical patients.

Pasteur himself took the next giant step as he turned his attention to a disease destructive to French poultry farms, chicken cholera. An unplanned discovery provided the first useful model for the preparation of a vaccine. Pasteur knew the value of cultivating a mind receptive to surprising occurrences. In his inaugural address at the age of thirty-two as professor and dean of the new Faculty of Sciences in Lille in 1854, he proclaimed a maxim that resonates to this day: "Where observation is concerned, chance favors only the prepared mind." Repeatedly, his own activities proved to be striking illustrations of his statement.

If healthy chickens were injected with a culture of cholera microbes, they invariably died within twenty-four hours. But one day in 1879, upon returning from a three-month-long summer vacation, he tried to restart his experiments, using a culture he had prepared before leaving. He was surprised to discover that nothing happened: the injected chickens remained quite healthy and lively. With the genius for exploiting what looked like an experiment gone wrong, he followed with the next logical step. Injections of fresh virulent cultures into the same hens now failed to produce the disease. Pasteur immediately recognized the significance of what he had blundered into. He had found a way of attenuating the cholera microorganisms artificially. He had succeeded in immunizing the chickens with the weak, old bacterial culture. A new truth was discovered: attenuated microbes make a good vaccine by imparting immunity without actually producing the disease. Vaccines could now be produced in the laboratory.

Two years later Pasteur produced a vaccine against anthrax, a highly contagious disease that was killing large numbers of cattle in Europe. It was also known as "wool sorters' disease" because people contracted it from their sheep. In 1885 Pasteur had a triumphant success with the introduction of a vaccine against the terrifying disease rabies. Three years later the Pasteur Institute was established in Paris, becoming, in time, one of the most prestigious biological research institutions in the world.

2

The New Science of Bacteriology

The meteoric rise of Robert Koch (pronounced "coke") from obscure country doctor to international celebrity was based on his talent for developing techniques for the isolation and identification of microbes. When he was a young physician in a small town near the German-Polish border, his wife bought him a microscope for his birthday, and he started looking at microbes as a pastime in a makeshift laboratory partitioned from his living room by a dropped sheet. By 1876, at the age of thirty-three, he had discovered the bacterium that causes anthrax. Within two years he published his monumental paper *Investigations Concerning the Etiology of Wound Infections,* scientifically proving the germ theory beyond doubt.[1]

Koch's landmark papers generated much excitement because he transformed the concept of what caused so many diseases. He would go on to establish with the clarity and purity of Euclidean logic the essential steps ("Koch's postulates") required to prove that an organism is the cause of a disease: that the organism could be discoverable in every instance of the disease; that, extracted from the body, the germ could be produced in a pure culture, maintainable over several microbial generations; that the disease could be reproduced in experimental animals through a pure culture removed by numerous generations from the organisms initially isolated; and that the organism could be retrieved from the inoculated animal and cultured anew.

His painstaking work transformed bacteriology into a scientifically based medical discipline. In 1891, three years after the Pasteur Institute was established in Paris, the German government founded the Koch Institute for Infectious Diseases in Berlin under his directorship. In this era of European imperialist expansion into Africa, Asia, and the Indian subcontinent, he discovered the bacillus that causes cholera and studied the disease known as sleeping sickness in East Africa. Within a few years, through his pioneering methods, the bacterial causes of a host of other diseases — diphtheria, typhoid, pneumonia, gonorrhea, meningitis, undulant fever, leprosy, plague, tetanus, syphilis, whooping cough, and various streptococcal and staphylococcal infections — were uncovered, largely by his students.

An earlier chance observation by Koch had resulted in a critical breakthrough in culturing microorganisms. Up to this point, scientists had grown bacteria in flasks of nutrient broth. In the lab Koch just happened to notice that a slice of old potato left on a bench was covered in spots of different colors. He placed a spot from the potato slice on a slide and saw that all the microorganisms were identical and clearly different from those in another spot. He realized that each spot was a colony of a specific microorganism. The importance of this observation was that in broth all types of microbes are randomly mixed together and only with great difficulty can be selectively cultured. The potato allowed discrete colonies of separate bugs to grow, enabling distinction and selection from culturing for identification and testing. Serendipity pointed the way to obtain pure cultures.

The next step was to develop a more usable culture medium. Adding gelatin to the liquid broth resulted in a solid medium. A tiny loop of platinum wire was used to capture a droplet from a broth containing various species of bacteria, and it was streaked across the surface of the solid broth plate. But much to Koch's disappointment, the gelatin liquefied when placed in an incubator. His colleague, Richard Julius Petri, designed a shallow flat round dish with a cover, and agar, a jelling compound derived from Japanese seaweed, was used as the solid growth medium. Koch came upon this in an indirect way. Japanese seaweed was suggested by the wife of a colleague who had been posted in the Dutch East Indies; she had used it for making jam. In this way was born the Petri dish or "agar plate," the mainstay of a bacteriology laboratory. Clinical specimens such as throat swabs, sputum, or blood are streaked over the surface of the agar and then incubated at body temperature. Colonies containing millions of microbes shortly grow. The technique was revolutionary. Pure cultures of bacteria could now be obtained, enabling their isolation and identification.

Once he had a mechanism with which to isolate microorganisms, Koch was able to identify organisms that caused specific diseases. At this time he began focusing his attention on tuberculosis. Classically referred to as "consumption," human tuberculosis was then responsible for one in seven of all European deaths. Identifying the organism was challenging work over a four-year period. The tiny rod-shaped organism was difficult to recognize with the staining techniques available at the time. Fortunately, a new advance came to Koch's attention. Paul Ehrlich, a young physician with a passion for chemistry, had developed a tissue stain called methylene blue. It was this stain that Koch used to detect the tiny rod-shaped bacillus in the tissues infected with tuberculosis. Because the bacillus grows slowly, it required the addition of blood serum to the agar as a nutrient and incubation for several weeks before colonies became apparent.[2]

In 1882, in an evening address to the Berlin Physiological Society, Koch — now employed by the Imperial Health Office — thrilled the audience with the news that he had discovered the bacillus that causes tuberculosis, *Mycobacterium tuberculosis*. Koch's singular discovery led to his winning the Nobel Prize in 1905.

Inspired by Koch's success with the methylene blue stain, Ehrlich went on to devise and develop, over the years 1878–88, the technique of counterstaining, whereby the washing of a stained specimen with a second, acidic chemical removes the color from only specific cells or parts of cells and thus permits greater differentiation. He experimented unsuccessfully with a number of dyes to stain the TB bacillus until, after a few months, chance intervened. Finishing up his work late one night, Ehrlich found the small iron stove in his home laboratory a handy place to leave his stained preparations to dry overnight. The next morning, before the scientist was up, his housekeeper lit the stove without noticing the glass slides lying atop it. Upon entering the laboratory, Ehrlich was aghast at the sight of the fire in the stove. He rushed to pick up his slides and inspect them through his microscope. What he saw was astonishing. The tubercle bacilli stood out wonderfully in bold color. The accidental heating had fixed the stain to the waxy-coated TB bacteria, allowing ready microscopic identification. This "acid-fast" staining technique is still used today.

3

Good Chemistry

By the middle of the nineteenth century, chemists had started synthesizing new substances with particular properties. The contributions of Friedrich Kekulé regarding the ring structure of the benzene molecule and Dmitri Mendeleev's periodic table in the 1860s provided a sound theoretical basis. An element's chemical properties were shown to be largely determined by two factors: the weight of its atoms, and the number of electrons each atom has in its outermost electron shell. With the understanding, toward the end of the nineteenth century, of the arrangement of electrons around a nucleus, atomic linkages could be constructed. Chemistry thus became a true science, capable of making predictions based on theory.

In this period, a new source of energy for illumination, coal gas, became widely employed. It was discovered that when coal is heated to high temperatures in the absence of air, it yields an inflammable gas. Coal gas became a popular replacement for candles and sperm-oil lamps. But perhaps more important, its ill-smelling waste product, coal tar, was quickly recognized as a gold mine. Coal tar contains aniline, an organic base, and azo compounds with nitrogen linkages attached to the benzene group. Benzene is a hydrocarbon compound, perhaps the most important of the organic compounds. It is the parent substance in the manufacture of thousands of other substances. Coal tar yields such varied products as dyes, drugs, perfumes, and plastics. In Germany synthetic dye production became a major industry, with thousands of coal-tar dyes of varying colors and properties.[1]

Many commercial products were inventions rather than discoveries. By tweaking a chemical structure — adding a side chain here, modifying one there — the property of a compound could be programmed. In the German chemical industry, this activity led to what Georg Meyer-Thurow, a historian of science, calls the "industrialization of invention."[2] Research was tightly managed within a bureaucratic structure. In an organization of industrial scientific laboratories, under a research administrator, invention was planned for and channeled through a series of steps based on past experience. But no one expected coal tar and its derived dyes to be the source of breakthrough drugs.

As a Jew, Paul Ehrlich was not eligible for appointments in academia or government research institutions in Germany. Koch, as director of the Institute for Infectious Diseases, reached out to him but, given the official constraints, could offer him only an unpaid position. Ehrlich gratefully accepted. Within a few years, he was working with Emil von Behring on a treatment for diphtheria. The name of this dreaded childhood disease is derived from the Greek for "leather," *diphthera*, reflecting the leathery false membrane that coats the throat and palate, blocking the airways. The two men developed a diphtheria serum by repeatedly injecting the deadly toxin into a horse. The serum was used effectively during an epidemic in Germany. Ehrlich skillfully transformed diphtheria antitoxin into a clinically effective preparation, the first achievement that brought him world renown.

However, he was cheated out of both recognition and reward in a bitter experience with his colleague, von Behring. A chemical company preparing to undertake commercial production and marketing of the diphtheria serum offered a contract to both men, but von Behring found a devious way to claim all the considerable financial rewards for himself. To add insult to injury, only von Behring received the first Nobel Prize in Medicine, in 1901, for his contributions.

The experience with diphtheria stimulated Ehrlich to think about the way toxins and antitoxins work, and he visualized groups of atoms fitting together like a lock and key. He proposed that antigens — toxins or other pathogens — lead to the generation of antibodies of reciprocal molecular shapes. This formulation is inherent in the term "antigen" — that is, leading to the *gen*eration of *anti*bodies. Although this hypothesis was proved false, Ehrlich's contribution was to conceptualize the process as essentially chemical responses and to establish the vocabulary of immunology. This would lead to a Nobel Prize in 1908 for Ehrlich and Ilya Metchnikoff of the Pasteur Institute in Paris, who shared the award for their work on immunity.

Ehrlich was a seminal thinker who had the unique ability to mentally visualize the three-dimensional chemical structure of substances: "Mine is a kind of visual 3-dimensional chemistry. Benzene rings and structural formulae disport themselves in space before my eyes. It is this faculty that has been of supreme value to me. . . . Sometimes I am able to foresee things recognized only much later by the disciples of systematic chemistry."[3] His mind raced from natural antibodies to the range of dyes manufactured by the German chemical industry. These were very promising because, as histological staining made clear, their action was specific, staining some tissues and not others.

His earlier studies had indicated that different dyes react specifically with different cellular components. This led to the fundamental concept underlying his future work: the idea that chemical affinities govern all biological processes.[4]

Shortly thereafter, his staining of the malarial parasite provided him with an opportunity. For centuries, malaria had been a disease encountered in Europe and rampant in Africa. It had long been believed to result from poisonous mists and evil vapors drifting off low-lying marshes. In time, with the suspicion that transmission of germs was due to mosquitoes, large parts of the Pontine swamps, breeding grounds for malaria, were drained. In the latter decades of the nineteenth century, the agent of the disease was identified, the parasite transmitted by female anopheles mosquitoes.

Bad Air

The term "malaria" — a misnomer derived from the Italian *mal aria* ("bad air") — was first used in the 1740s by Horace Walpole (the same person who coined the term "serendipity" after reading the fable *The Three Princes of Serendip*). He described "a horrid thing called the mal'aria that comes to Rome every summer and kills one."

Ehrlich began using methylene blue, the first aniline dye, to stain bacteria in 1880. By 1885, upon injecting it into a living frog, he found that it avidly stained nerve fibers. Even more startling was another experiment Ehrlich conducted with the dye, on a frog with a parasitic urinary infection. Studying the frog's excised bladder, he found that not only was the parasite still to be seen sucking blood from the frog tissue, but its nerve fibers were stained blue. Could the chemical dye, he reasoned, affect biological function to interfere with nervous transmission and exert an analgesic, or pain-killing, action in people?

Trials in humans with neuritic and arthritic conditions did not work. In 1891 he pursued to its logical conclusion the finding that malaria parasites stain well with methylene blue, and he administered the dye to a German sailor with a mild case of the disease. The patient recovered, but the dye was later proved to be of no value against the more severe manifestations of malaria experienced in the tropics. The dye also turned an individual blue. Nevertheless, this represented the first instance of a synthetic drug being successfully used against a specific disease.

Ehrlich had arrived at a totally new and revolutionary way to treat infections caused by microorganisms. He realized that such coal-tar-derived dyes bonded with substances like nucleic acids, sugars, and amino acids to form a definable chemical reaction. If, he reasoned, there were dye receptors — structures that received dyes — there might be drug receptors, substances fixed by microbes but not by the human host.

In this way, the concept of "magic bullets" was born. His basic idea, or hypothesis, was that since some dyes selectively stained bacteria and protozoa, substances might be found that would be selectively absorbed by the parasites and would kill them without damaging the host. Ehrlich termed fighting diseases with chemicals "chemotherapy" (a term that came to be used exclusively for cancer treatments) and later allowed that "initially, chemotherapy was chromotherapy," meaning treatment with dyes.

Ehrlich combined some of the endearing characteristics of an absentminded professor with a mercurial mind. A small, wiry, unprepossessing man, he spent long hours absorbed in the laboratory, during which he typically smoked up to twenty-five strong cigars a day and frequently neglected to take his meals. Struck with an idea or an observation, he would scribble notes on his detachable shirt cuffs. To remind himself of an important forthcoming task or something he must not forget, he would on occasion send himself a postcard. He never claimed credit for the work of his assistants and always listed them as coauthors in the publications describing their joint work.

In 1899 Ehrlich was appointed director of the new government-supported Royal Prussian Institute for Experimental Medicine in the industrial city of Frankfurt. The government's desire to support research against infectious diseases overcame prevailing religious biases. Chemical companies, especially Hoechst and Bayer, were turning their attention to diseases as targets of industrial innovation, and hopes were high that Ehrlich could lead the industry into the burgeoning field of pharmaceuticals. A wealthy Jewish banker, and subsequently his widow, funded the building of Georg-Speyer-Haus, a laboratory with ample resources, to serve as the Research Institute for Chemotherapy.

In 1906 Ehrlich turned his attention to the age-old, worldwide scourge of syphilis. By the turn of the twentieth century, this disease was of epidemic proportions throughout Europe, involving probably 15 percent of the adult population in major cities like Berlin, Vienna, Paris, and London. Generally referred to as "the pox," syphilis was by this time known to be transmitted sexually.

A Shepherd Named Syphilus

The name "syphilis" comes from the poem *Syphilis, or the French Disease*, written in Latin by the sixteenth-century Italian philosopher and physician Girolamo Fracastoro, about a shepherd named Syphilus who offended the god Apollo and was punished with the world's first case of the disease. The poem was frequently translated and eventually was published in more than a hundred editions.

The general course of syphilis occurs in three stages. Within a few weeks of the initial exposure, the victim develops an ulcerous sore at the point of contact, which disappears after a few weeks. Fever, an extensive rash, and fatigue typically follow. These symptoms may recur over the course of a few years. A long, uneventful interval lasting for decades then ensues before the ravages of the systemic infection, known as tertiary syphilis, become apparent. It results in destructive lesions of the nervous system, heart, and bones. Both the long interval and the plethora of symptoms that are also common to other ailments often made diagnosis difficult. Treatment with oral mercury compounds was an ordeal and only marginally effective.

During the course of Ehrlich's research, two particular aspects demonstrated the remarkable vagaries of fortune. One had to do with a wrong hypothesis that nevertheless set him upon the right path, and the other was a laboratory misadventure that preserved the therapeutic miracle of the famous compound 606.

The causative bacterium of syphilis, a pale, threadlike, undulating spirochete later identified as *Treponema pallidum*, was discovered in 1905 by a German protozoologist and microbiologist, Fritz Schaudinn, with the dermatologist Erich Hoffmann. On the basis of what he took to be similarity of shape and movement, Schaudinn erroneously said that the spirochete was closely related to the trypanosomes. These comprise a group of protozoa, one form of which causes African sleeping sickness when transmitted to humans by tsetse flies.

Hoffmann personally emphasized the spirochete's similarity to try-panosomes when he visited Ehrlich's laboratory the following year. Trypanosomes have an elongated, somewhat spindly shape and a whip-like tail. This misidentification set Ehrlich upon the right path for an unsound reason. He was determined to find a chemical that would kill the spiral-shaped microorganism, and he had considerable experience with a form of arsenic in the treatment of trypanosome infections.

Ehrlich had read that an arsenical compound, optimistically named Atoxyl because it was thought to be free of toxic effects, had been successfully used in mice infected with trypanosomiasis at the Liverpool School of Tropical Medicine; however, it proved to cause optic nerve damage and blindness in up to 2 percent of humans. Ehrlich and his team worked unceasingly to prepare compound after compound that incorporated arsenic and, with methodical diligence, started testing them one by one. By 1907, compound no. 418 proved effective in humans but was unacceptable because a small number of patients exhibited severe and often fatal hypersensitivity. Chemical analogs continued to be produced in the search for a potential agent against trypanosomes.

That same year, the progenitor of compound no. 606 was pre-pared by one of Ehrlich's assistants. A piece of equipment, a vacuum apparatus necessary for reducing the compound, was lacking, so it stood on a shelf in the laboratory for more than a year before being re-examined. This proved to be a lucky lapse. If the apparatus had been available and the tests carried out, the compound might well have proved ineffective in the mice infected with trypanosomes and have been discarded, never to be tried against syphilis.

In the spring of 1909 Ehrlich was joined by Sahachiro Hata, a bacteriologist from Tokyo who had developed a method of infecting rabbits with syphilis. Under Ehrlich's direction, he undertook to retest, using the new animal model, every arsenical compound that the laboratory had synthesized over the preceding three years. The first 605 compounds they tried didn't have any effect. On August 31 of that year, they approached yet another cage with an infected rabbit exhibiting large syphilitic sores on its scrotum. Compound no. 606, which had been sitting on the shelf for almost two years, was injected into the rabbit's ear vein. By the next day, its blood had been cleared of syphilis spirochetes and the sores were drying up and healing. With restrained glee, Ehrlich wrote, "It is evident from these experiments that, if a large enough dose is given, the spirochetes can be destroyed absolutely and immediately with a single injection!"[5]

Ehrlich's secretary, Martha Marquardt, commented on the feverish activity behind this crowning achievement:

> No outsider can ever realise the amount of work involved in these long hours of animal experiments, with treatments that had to be repeated and repeated for months on end. No one can grasp what meticulous care, what expenditure and amount of time were involved. To get some idea of it we must bear in mind that [the first hopeful compound] had the number 418, "Salvarsan" the number 606. This means that these two substances were the 418th and 606th of the preparations which Ehrlich worked out. People often, when writing or speaking about Ehrlich's work, refer to 606 as the 606th experiment that Ehrlich made. This is not correct, for 606 is the number of the substance with which, as with all the previous ones, very numerous animal experiments were made. The amount of detailed work which all this involved is beyond imagination.[6]

The medical historian W. I. B. Beveridge wrote that Ehrlich's discovery of the cure for syphilis "is perhaps the best example in the history of the study of disease of faith in a hypothesis triumphing over seemingly insuperable difficulties"[7] — the hypothesis being the belief that a substance could be found that would be absorbed by the microbe and then be the source of its undoing.

Two young laboratory assistants at Ehrlich's institute injected themselves with the yellow crystalline powder dissolved in water to find out the right doses and test the toleration of 606. The preparation proved successful. Its chemical makeup directed the arsenic poison to the spirochete rather than to the patient. When 606 (generic name arsphenamine) was licensed to Farbwerke Hoechst near Frankfurt in a contract with Georg-Speyer-Haus, it was marketed under the proprietary name Salvarsan — an arsenical salvation. Ehrlich chose not to make the contract with himself. He asked for only a modest interest in it, which was laid down in an agreement with the company.

The drug was eagerly sought by both physicians and patients, and large clinical trials in patients with syphilis were promptly undertaken. About 65,000 doses were provided free by Speyer-Haus from June to December 1910, until Hoechst Chemical Works could produce vast commercial quantities. The press throughout the world trumpeted the success of the treatments. A more easily used form, named Neosalvarsan, was soon brought to market. But even here, the intensity of the effort is evident in that Neosalvarsan was compound no. 914, more than three hundred compounds after 606.

Ehrlich found a great source of happiness in the first postcard he received from a cured patient. He kept this always in his wallet in the breast pocket of his coat.

Sometime after the outbreak of World War I in 1914, freed from patent protection, British and French scientists began collaborating to synthesize the drug, and 94,762 injections were administered by the French Military Medical Services alone. While some might find the title of Fracastoro's poem of 1530, *Syphilis sive morbus Gallicus* (*Syphilis, or the French Disease*) vindicated, this large figure provides an index of the widespread extent of venereal disease at the time. Neosalvarsan remained the only serious treatment for syphilis until the advent of penicillin in the 1940s.

Nazi Gratitude

When the Nazis came to power in 1933, they confiscated all books about Ehrlich and burned them in an attempt to expunge his name from German history. In 1938 his widow and family fled to the United States.

After Salvarsan was discovered, there was widespread hope that other chemical "salvations" would be rapidly uncovered. Optimism ran high that an arsenal of "magic bullets" could be directed against man's ills. In 1913 Ehrlich enthused, "In the next five years we shall have advances of the highest importance to record in this field of research." But this was not to be the case. Many compounds, including some new synthetic dyes, were tried by many researchers against the common bacterial diseases. Over and over, each failed. Dashed hopes and pessimism reigned over the next two decades.

Toward the end of his career, Ehrlich took stock of his experiences. He had earlier prophesied, "It will be a caprice of chance, or fortune, or of intuition which decides which investigator gets into his hands the substances which turn out to be the very best materials for fighting the disease, or the basal substances for the discovery of such."[8] And, in fact, it was a "caprice of chance or fortune" that yielded the next triumph of the new art of chemotherapy, the sulfonamides — not five but twenty-five years later.

4

The Art of Dyeing

By the time World War I began, Germany's chemical industry had achieved world leadership. It had evolved in a series of steps begun in the 1880s that led from synthetic dyestuffs through pharmaceuticals to "chemotherapy." The synthetic dye industry had originated with a stunning circumstance of serendipity in 1856 by an eighteen-year-old chemist, William Henry Perkin, trained at the Royal College of Chemistry in England.

Using a tiny laboratory that he had set up in his home, Perkin tried to synthesize the antimalarial drug quinine. The British Empire and other European nations were expanding their colonies into malaria-ridden areas of India, Southeast Asia, and Africa. Quinine, the only known cure and preventive for malaria, was commercially extracted from the bark of the South American cinchona tree. Quinine was in short supply, and its chemical structure unknown. In searching for a synthetic form, Perkin stumbled upon the first coal-tar dye, named mauve, or aniline purple.

The Origin of Ex-Lax

Spurred by Perkin's discovery, the Germans too began to synthesize dyes. One of these, with the tongue-twisting name phenolphthalein, became a worldwide medical remedy by sheer accident. At the turn of the twentieth century in Hungary, many vineyards had been destroyed by a pestilence, and vintners turned to marketing artificial wine colored by natural dyes. Phenolphthalein, which had been used as an indicator of pH for more than thirty years, turns a purple-red color in an alkaline solution. When the Hungarian government began using phenolphthalein as an additive to identify adulterated white wines, an epidemic of diarrhea quickly broke out. This purgative action was a complete surprise,[1] and the chemical's use in the wine industry was discontinued. Subsequently, a few commercial laxatives based on phenolphthalein (Ex-Lax, Feen-A-Mint) were distributed worldwide. Preparations were sold in gum and candy forms palatable to children, and the products became as well known as aspirin. In time, the active ingredient was replaced by another laxative, senna.

In 1918, after World War I, Germany was a depleted nation, virtually bankrupt, stripped of its colonies and thereby its textile empire. With neither the colonial sources of fibers nor the money to pay for imported wool and cotton, Germany feared that England would soon assume ascendancy in the world textile market. Germany harnessed its innovative minds and technological resources to emerge, within a decade, as the world's leading manufacturer of textiles, especially synthetic fibers. But textiles need to come in colors, and part of Germany's quest for market share in this burgeoning industry involved a search for a wide spectrum of dependable synthetic dyes.

Originally, independent firms with names like Farbenfabriken Bayer and Farbenwerke Hoechst developed the field of synthetic textile coloring. *Farben* is the German word for "colors." In 1904 Hoechst merged with Farbwerke Cassella. Bayer joined forces with BASF and AGFA (AG für Anilinfabrikation) dye works, forming a group known as the little IG (*Interessengemeinschaft*, a coalition of shared interests). Bayer made a fortune in 1899 when it began marketing aspirin, the world's best-selling pain reliever, which it had made from the dyestuff intermediate salicylic acid. The gigantic conglomerate formed in 1925 that controlled all the German pharmaceutical houses, as well as virtually all other branches of the German chemical industry, was named IG Farben. In this way, Germany's unrivaled eminence in the pharmaceutical, and particularly the "chemotherapeutic," sector took root from developments in synthetic dyeing and staining. In 1924 the synthesis of a drug against malaria was announced, and within a few years success was achieved against several other tropical protozoal diseases, such as relapsing fever, sleeping sickness, yaws, leprosy, and filaria. But one fact engendered gloom: the utter failure of antibacterial chemotherapies over the two decades since Ehrlich's breakthrough.

With entrepreneurial vision, IG Farben now prepared to launch a systematic search, in combination with the development of new dye products, for antibacterial drugs along the lines laid down by Ehrlich. In 1927 it appointed a thirty-two-year-old physician and bacteriologist, Gerhard Domagk, to be director of its research laboratory at Elberfeld. Under his direction a laborious process of screening new compounds was undertaken, beginning with tests of effectiveness against various microbes in culture, followed by studies in animals, first to determine what doses could be tolerated and then their effects against infections; finally clinical tests in humans would be considered. He quickly concentrated on azo dyes, compounds in which two nitrogen atoms were linked by a double bond. First developed in 1909, these dyes were very resistant to fading from washing and light since they bond so effectively with the proteins in wool and silk. Might these agents react as well with the proteins of a bacterial cell?

Domagk was utterly single-minded in his approach. He soon focused his interest on a single bacterium, the streptococcus. Domagk was elegantly economical in his choice, aiming to treat a host of diseases due to one germ. Streptococcus was the cause of devastating infections that were both acutely life-threatening and chronically debilitating with frequent lethal effects. It was the cause of severe tonsillitis, infected glands in the neck, bacterial infections of the large joints (septic arthritis), puerperal fever, epidemics of scarlet fever, and rheumatic fever, which damaged heart valves and resulted in chronic kidney damage. Infection of the middle ear (otitis media) could result in permanent deafness. Physicians also dreaded its skin manifestations when it spread into the tissues from an infected site: the scarlet wave of erysipelas (streptococcal cellulitis) and the sharp red lines indicating lymphatic spread. In such cases, amputation of an infected limb was often the only recourse. Most ominous was its tendency to cause blood poisoning, or streptococcal sepsis, which, along with its toxic products, could infect almost any site in the body. This invariably constituted a death warrant.

In 1932 Domagk was looking at a new brick-red azo dye that had a sulfonamide group attached to its molecule. It was being used to dye leather and was commercially marketed as Prontosil Rubrum. Tests in the laboratory showed it to be virtually inactive against bacteria in a Petri dish. Undaunted, Domagk went on to test it against streptococcal infections in laboratory mice.[2] The results astonished him. Despite having been injected with a deadly species of streptococcus, every mouse inoculated with this new crystalline chemical was alive and frisky, running about in its cage. All those in the control group given that same bacterium but no red dye were dead.

Domagk was an intense researcher who was driven by an independence of thought. His close-shaven head and pale blue eyes gave him the appearance of an unemotional man. But witnessing those results in the laboratory cages, five days before Christmas, moved him to exaltation: "We stood there astounded at a whole new field of vision, as if we had suffered an electric shock."[3] IG Farben wasted no time in claiming its priority. On Christmas Day, it submitted an application for a German patent on Prontosil.

Over the next two years, as Domagk quietly investigated the new compound with the help of selected local physicians treating patients with streptococcal infections, several dramatic cures in humans testified to the new drug's effectiveness. The first patient was a ten-month-old infant severely ill with staphylococcal septicemia (blood poisoning), a condition from which no one had ever been known to recover. Treating the baby with Prontosil was a daring gamble on Domagk's part. If the child had not survived, it would not have been clear whether the drug or the disease had killed him. The child's skin turned red, and his physicians were able to calm his excited nurse only by explaining that the drug was basically a dye. Within days, the child miraculously recovered. Prontosil was distributed for clinical trial in 1933.

In a dramatic twist of fate, a few months later Domagk was forced to use Prontosil to cure his own six-year-old daughter, Hildegard, who was facing a similar life-threatening infection. She had pricked her hand with an embroidery needle in the web of tissue between her thumb and index finger. The wound became infected with streptococcus, and the infection spread to the arm and then into the bloodstream. Swollen lymph nodes in the armpit were lanced of pus fourteen times, but the infection was relentless, and the surgeon recommended amputation of the child's arm. Domagk refused. In desperation, he took one of his precious experimental drug samples and treated Hildegard with it. Within two days her fever broke, and after repeated courses of the drug, she went on to recover fully — with, however, one unfortunate and permanent side effect: her skin turned a light reddish, lobsterlike color. This was not true of all patients who used Prontosil.

Early reports that "something was brewing in the Rhineland" regarding chemotherapy cures of streptococcal infections reached Britain by the summer of 1935. The brew was Prontosil. Used in an outbreak of puerperal sepsis (childbed fever) in 1936 in Queen Charlotte Maternity Hospital in London, the new "miracle drug" slashed mortality from 25 to 4.7 percent.

In February 1935 Domagk finally published his classic paper, which was modestly titled "A contribution to the chemotherapy of bacterial infections."[4] He reported that "in the course of our studies, we chanced upon" the effective cure.

Domagk had waited more than two years before announcing his revolutionary discovery to the world. His report came out just one month after the German patent for Prontosil was finally approved, by which time clinicians had spread word of the unprecedented improvement in streptococcal infections. His article clearly stated that Prontosil "works like a true chemotherapeutic agent only in the living organism," not in the test tube. The delay in publishing the findings was most likely due to the desire to keep a valuable industrial secret from

competitors. IG Farben would have recognized that the active principle, sulfanilamide, was not patentable.

No One Ever Thought of Testing It

Sulfanilamide itself had been synthesized and described long before, in Vienna in 1908 by Paul Gelmo in the course of working
on his doctoral thesis. Like many other synthetic chemicals of no
use, it had been gathering dust on some laboratory shelves. It had
been produced in huge quantities as a by-product in the dye industry, and nobody had ever thought of testing it as an antibiotic.
Had its properties been known, sulfanilamide might have saved
750,000 lives in World War I alone. After the war, Gelmo was
found eking out a precarious existence as an analytical chemist for
a firm of printing ink manufacturers.

Domagk and his team of chemists may have been in a frantic race
for a patentable analog. This concern for secrecy would explain why
the IG Farbenindustrie did not respond to researchers at the Pasteur
Institute in Paris who requested samples of Prontosil to investigate its
inactivity in the laboratory. The Germans, of course, were pursuing an
exclusive commercial product with huge potential profits. Not only
scientific but also nationalistic rivalry was undoubtedly at stake.

In a breathtaking series of events, the French scientists Jacques
Tréfouël and his wife Thérèse, together with Frédéric Nitti and Daniel
Bovet, quickly succeeded in synthesizing similar compounds on their
own. Within nine months of Domagk's publication, their experiments
not only verified his results but, in a compelling instance of fortune,
uncovered what would prove to be the essential "wonder drug."

They intended to test the curative powers of their chemical dyes
on eight groups of mice injected with a highly virulent culture of
streptococcus. Only seven dyes were readily available. For the eighth
group, they reached for a long-neglected dust-laden bottle on their laboratory shelf containing a colorless mixture. This would turn out to
be sulfanilamide, the chemical common to all the compounds in the

experiment. The next morning, the control group of mice — those not treated with a synthesized dye — were all dead. Those treated with Prontosil and analogs had been protected — as had, most surprisingly, the eighth group of mice treated with the colorless chemical. A "Eureka!" moment had been occasioned by chance.

Why Prontosil was inactive in vitro (in the test tube) but effective in vivo (in the body) became apparent and was soon proved. In the body, Prontosil is broken down to liberate the active principle, sulfanilamide, which alone is the agent against streptococcal growth. Furthermore, Tréfouël and his team demonstrated that sulfanilamide in contrast to Prontosil is active not only in vivo but also in vitro. They were amazed to find that its action was in no way related to its being a dye but rather was due to its containing sulfanilamide, which is not a dye. "From that moment on," writes Bovet, "the patent of the German chemists would carry no more weight."[5] French patent law recognized Prontosil as a dye and not a medicine.

As reported in a French journal, the team's findings were presented as though they had been reached via deductive reasoning.[6] But recalling the experiences fifty-one years later in a book, Bovet clearly describes them as a fortunate happenstance.[7]

IG Farben's euphoric hopes for huge profits from the marketing of Prontosil had been shattered.[8]

Sulfanilamide's derivatives were patentable, but all these products were cheap and easy to make. They had the further advantage, over Prontosil, of not turning the patient as red as a boiled lobster. These circumstances greatly helped the new antibacterial therapy to receive widespread acceptance.

Within two years of the publication of Domagk's paper, his cure made a splash on the other side of the Atlantic. In December 1936 twenty-two-year-old Franklin D. Roosevelt Jr., son of the president of the United States, developed a severe streptococcal tonsillitis that soon involved his sinuses. Blood-borne spread appeared imminent, and the prognosis was feared grave. But he was given sulfanilamide and made a rapid recovery. Newspaper accounts spread the miraculous potential of this new wonder drug. In America, such reports stimulated rapid growth of the pharmaceutical industry.

A new era dawned. An individual with a severe infection need not start making funeral arrangements. For the first time in history, patients suffering from overwhelming infections could be cured without surgery, simply by taking a course of tablets or injections.

The Nazis versus the Nobelists

In October 1939 the forty-four-year-old Domagk was awarded the Nobel Prize in Physiology or Medicine.[9] In announcing their selection one month after Germany's invasion of Poland, the Nobel Committee in Sweden displayed not only its integrity but also its bravery. The National Socialist government had been hostile to the Nobel Committee since 1936 when the Peace Prize was awarded to the German radical pacifist writer Carl von Ossietzky, who had been imprisoned in 1932 for exposing German rearmament. At the time, he suffered from pulmonary tuberculosis and, following the award, he was transferred — in a public relations gesture — to a hospital, where he soon died. After this incident, the Hitler regime established a Nazi Party Prize that could be won only by a German of impeccable Aryan ancestry and decreed that acceptance of a Nobel Prize was forbidden.

Domagk sought advice from the authorities on whether it would be possible to accept the prize. Two weeks later he was arrested by the Gestapo and forced to send a letter drafted for him by the Nazi government refusing the prize. After being released from jail, he confided in his diary: "My attitude to life and its ideals had been shattered."[10] When he was arrested a second time while traveling to Berlin for an international medical conference, he realized that he was under constant surveillance and thereafter acted cautiously to protect himself and his family. These experiences plunged him into years of depression. Only after the war, in 1947, was he able to travel to Stockholm to receive his Nobel Prize medal — but not the prize money, which had been redistributed. By this time, his breakthrough was eclipsed by a new class of drugs: antibiotics.

THE BREW THAT WAS TRUE

Once sulfanilamide was in the public domain, pharmaceutical houses undertook a frenzied search for even more effective analogs. Foremost among these was the British firm of May and Baker, a subsidiary of the French chemical firm Rhône-Poulenc. Over the best part of three years, its chemists tweaked Prontosil's formula in a wide variety of ways and tested no fewer than 692 separate compounds. Their system of methodical trial-and-error yielded no good results until it was blessed by serendipity.

The true brew turned out to be a compound in a dusty bottle prepared seven years previously without any antimicrobial action in mind. It would shortly evolve into a form known as sulfapyridine. The historian of science John Lesch uncovered the event:

> One day in October 1937, someone in the laboratory . . . noticed an old bottle sitting on the shelf. It turned out to be a sample of the base aminopyridine, prepared by a chemist named Eric Baines back in 1930 for another chemist who had since left the company. The sample . . . had stood on the shelf collecting dust ever since. Looking back two and a half decades later, company chemists . . . agreed that it would have been very unlikely that this compound would have been prepared for use in the sulfanilamide derivatives program. But since it was there, "at the front of the third shelf on the left hand side of the cupboard," as another chemist, L.E. Hart, recalled, it was used. "Well, you know what would happen very much in those days," Baines remarked to company colleagues in 1961, "you would set out to try and make something, and you would find something on the shelf, and you would try it."[11]

Sulfapyridine not only proved more potent than sulfanilamide but also had a wider spectrum of antibacterial activity. It was effective against pneumococci, meningococci, gonococci, and other organisms.

It reduced the mortality rate among patients with the dreaded lobar pneumonia from 1 in 4 to only 1 in 25.

By the late 1930s and early 1940s the initiative for research pursuits passed to the British and the Americans.[12] May and Baker began marketing sulfapyridine in 1938. It and the other new sulfa drugs were prescribed in huge quantities. By 1941 some 1,700 tons of them were given to 10 million Americans. Sulfadiazine, prepared by the American Cyanamid Company in 1940, was used extensively during World War II. The mortality rate among American soldiers in epidemics of meningitis was cut from 39 percent to 3.8 percent. (However, the course of gas gangrene, a bacterial infection following injury to muscles, characterized by bubbles of gas in the tissues, could not be successfully altered.) Every American soldier carried a packet of sulfa powder in his first-aid kit during World War II.

In late 1943 a singular incident involving Winston Churchill created a major impact on the popular imagination. The sixty-nine-year-old Churchill, in Tunisia for the planning of the invasion of Italy, developed life-threatening pneumonia. A course of sulfapyridine saved him, and, in typical Churchillian prose, he proclaimed that "the intruders were repulsed."[13] The "miracle drug" became a household term. The medical profession, impressed by reports of highly favorable clinical results, was now awakened to the new field of bacterial chemotherapy. A profusion of different "sulfa drugs," all white crystalline powders, followed in quick succession.

The sulfa drugs were a breakthrough in the treatment of a host of deadly infections, such as pneumonia, meningitis, childbed fever, wound infections, erysipelas, mastoiditis, bacillary dysentery, gonorrhea, and most urinary infections.[14] For the first time, the majority of these patients survived and recovered without lasting damage. These drugs would later be superseded by antibiotics, but the hope and faith in medical science engendered by Gerhard Domagk's discovery reverberate to this day.

The FDA Is Born

In 1937 shortly after young FDR Jr. recovered from his serious infection, a small Tennessee firm named Massengill and Company, which made pharmaceuticals for animals, began marketing a sulfa drug for people. To make it more easily administered to children in a sweet liquid form, they dissolved the drug in diethyl glycol, a commercial solvent used to make antifreeze, and sold it widely throughout the South as "Elixir of Sulfanilamide." The company tested neither the solvent nor the final product for toxicity. Within weeks, more than a hundred people died, most of them children. The company's president refused to take responsibility and came to be convicted only on a technicality: the fact that the word *elixir* means a medicine containing alcohol, and there was none in the product sold.

Massengill's chief chemist committed suicide. The incident outraged the public, and Congress, and on June 15, 1938, President Franklin Delano Roosevelt signed into law the Food, Drug and Cosmetic Act, providing for safety tests on drugs before they could be marketed. This milestone legislation twenty years later spared the United States from the thalidomide tragedy.

5

Mold, Glorious Mold

In 1928 a quiet Scottish bacteriologist, Alexander Fleming, made a discovery that would change the course of history. But before Fleming came Sir Almroth Wright.

In charge of the bacteriology department at St. Mary's Hospital, a famous but run-down old hospital near the Paddington railway station far from the center of London, Wright was a brilliant larger-than-life physician and researcher who directed the hospital laboratory in the spirit of enlightened despotism. Contemporaries described him, with his large head, sandy hair, and moustache, as looking like the John Tenniel illustration of the Lion in Lewis Carroll's *Through the Looking Glass*. He was a strong-minded individual convinced of his own infallibility. His early scientific training was at several of the leading institutions on the Continent.[1]

Early in his career Wright became convinced that the only means of defeating a bacterial infection was through immunization, which was achieved in one of three ways: by having the disease, by getting vaccinated, or by using immune serum. The most remarkable advances at the turn of the century — Emil von Behring's diphtheria serum therapy, the first vaccines, and Ilya Metchnikoff's studies of phagocytosis (the immune response by white blood cells) — were in the realm of immunotherapy and exploited the natural defensive capabilities of the body to fight disease. Wright became single-minded in this pursuit.

Wright's earliest experiences were with the British military in India and then in South Africa, involving the prevention of typhoid fever in the Boer War (1899–1902). Although the incidence of typhoid fever had fallen in cities with high standards of sanitation, Wright realized that these measures would not work with large armies in war, owing to the near impossibility of dealing effectively with waste. More than 60 percent of German deaths during the Franco-Prussian War were attributed to typhoid. Wright knew that typhoid fever was not just an infection of the bowel, but that death occurred when the bacilli invaded the bloodstream. He developed a culture of the typhoid germs killed by heat to serve as a vaccine.

During the Boer War, Wright was grudgingly given permission from the War Office to inoculate "such men as should voluntarily present themselves." However, the army medical authorities were more worried by the body's reaction to vaccination — which often rendered a soldier unfit for several days — and therefore ordered many troopships departing for the war to throw caseloads of Wright's vaccines overboard. With only 4 percent volunteering to be vaccinated, 13,000 soldiers were lost to typhoid on the South African veld as against 8,000 battle deaths. (When giving evidence before a military tribunal, Wright was asked if he had anything more to say. His response was typically blunt: "No, sir. I have given you the facts. I can't give you the brains.")[2]

Both Wright and the military were glad when the opportunity for him to go to St. Mary's Hospital arose in 1902. Here, with a group of dedicated young physicians, he undertook a long program of research on immunity and on techniques to stimulate immunity by vaccination in order to treat as well as prevent disease. A special component of this was the Inoculation Department, set up in 1907 for the production and sale of vaccines for a variety of illnesses ranging from acne to pneumonia. Independent of the hospital and medical school, it was, in effect, the first private clinical research institute in England.

An early recruit to this program was a recent prize-winning medical graduate of St. Mary's, Alexander Fleming. Fleming was born to a sheep-farming family in southwest Scotland in 1881. He excelled in school, and when he entered St. Mary's Hospital in London to study medicine, he fell under Wright's sway.

Fleming was just under five feet six inches in height, and slim. A boyhood accident that had flattened the bridge of his nose may have given him the pugnacious appearance of a bantamweight boxer, but such an impression would have been very misleading. Invariably referred to as Flem, he was easygoing, good-tempered, and likeable, though very much the quiet type. (Fleming was generally so laconic that one of his colleagues claimed that trying to have a conversation with him was like playing tennis with a man who, when he received a serve, put the ball in his pocket.) He was respected for his common sense and ingenuity in experimental techniques.

Although Fleming's habitual silence contrasted sharply with Wright's flamboyance, the two men established a working relationship that would last until Wright's retirement at the age of eighty-five in 1946 (a year after Fleming won the Nobel Prize in Medicine). Through Wright's friendship with Paul Ehrlich, Fleming became one of the first to use Salvarsan to treat syphilis in Great Britain.

With the outbreak of World War I, both Wright and Fleming joined the Royal Army Medical Corps and were posted to Boulogne on the north coast of France. Here they studied the wounds and infection-causing bacteria of men who were brought straight from the battlefields of Mons, Ypres, and the Marne. Many of these men died from septicemia or the dreaded gas gangrene. It became evident that gas gangrene and tetanus were due to wounds contaminated by organisms found in horse manure in the many farmers' fields that war had turned into killing grounds.

Wright and Fleming made two significant observations regarding the ineffectiveness of the traditional use of antiseptic methods in curing established infections. Not only were antiseptics, such as carbolic acid, not reaching the many hidden crevasses of the deep, jagged war wounds typically caused by shrapnel, where bacteria could flourish, but the antiseptics themselves were destroying the white blood cells that were part of the body's natural immune system. Consequently, the infections spread rapidly. These findings reinforced Wright's contention that the enhancement of natural immunity with vaccines was superior to chemotherapy or antiseptic methods.

About half of the 10 million soldiers killed in World War I died not directly from explosives, bullets, shrapnel, or poison gases but from infections in often relatively mild wounds. For his part, Fleming, then thirty-three, was forever affected by the suffering and dying he saw during the war, and he decided to focus his efforts on the search for safe antibacterial substances. From France in 1918 he wrote: "I was consumed by the desire to discover . . . something which would kill these microbes." He would, he resolved, find "some chemical substance which can be injected without danger into the blood stream for the purpose of destroying the bacilli of infection."[3]

Within three years, Fleming would surprise himself with a chance revelation. In November 1921 he had a cold. While he was working at his laboratory bench at St. Mary's, a drop from his runny nose fell on a Petri dish and "lysed" (dissolved) some colonies of bacteria. These common, harmless airborne microbes became translucent, glassy, and lifeless in appearance. Excited, Fleming prepared a cloudy solution of the bacteria and added some fresh nasal mucus to it. The young bacteriologist V. D. Allison, who was working with Fleming at the time, described what happened next: "To our surprise the opaque suspension became in the space of less than two minutes as clear as water. . . . It was an astonishing and thrilling moment."[4]

Fleming had discovered lysozyme, a naturally occurring antiseptic substance present in tears, nasal mucus, and saliva. In order to gather such secretions to extend his investigations, he made colleagues, technicians, and even visitors weep batches of tears by putting drops of lemon juice into their eyes. Peering through his microscope, he marveled that bacteria in the presence of tears became swollen and transparent, then simply disappeared before his eyes. Of equal importance, he found that lysozymes were present in many animal and plant tissues, including blood, milk, and egg whites.

Fleming's accidental illumination revealed that human fluids have some bacteria-fighting properties and that these are parts of the body's defense system essential to life. Unfortunately, lysozyme has relatively little medical use, being most effective against bacteria that do not cause illness. The discovery, however, put Fleming on the lookout for other nontoxic antibacterial substances.

Fleming had made an important step toward discovering how the perfect antiseptic would work. The next month, he reported his findings as a work in progress before a group of colleagues interested in research. But he was such a bad lecturer — desultory and self-deprecating — that his listeners greeted the presentation with stony silence and asked no questions. All that the audience could gather was that some rare and harmless germ was soluble in tears or saliva. According to Fleming's biographer, "What would later be regarded as an historic event in medical history seemed to have sunk without a ripple."[5]

Fleming published five papers on the subject between 1922 and 1927, and these too were met with indifference. He suspected that the "factor" might be an enzyme but did nothing to prove it. Lysozyme came to be regarded as an interesting oddity. Nevertheless, Fleming, not one to depend on outside encouragement, prepared and incubated thousands of slides to observe the effects of bacteria over the next few years, his keen eye alert to the minutest changes.

And then, in the summer of 1928, there ensued an incredible chain of fortunate circumstances that even a scientist might characterize as having been blessed by angels. As is true of events that profoundly changed the lives of many people, Fleming's discovery has become enveloped in myth. The image of the penicillin-bearing mold floating through Fleming's open window from the polluted London air to land on his open culture plate is appealing but apocryphal. Moreover, most people assume that penicillin's potential usefulness was apparent immediately. In fact, it took many years for its miraculous powers to be unveiled.

In his 1970 book *The Birth of Penicillin,* a remarkable feat of medical history detection, Ronald Hare, a bacteriologist and former Fleming colleague, detailed the astonishing sequence of chance events that made the discovery possible. Confusion, ambiguity, and mythology had long clouded the occurrence because of the lack of contemporary accounts and the absence on Fleming's part of notebook entries, journals, and relevant letters. Indeed, the conventional account was not published until 1944, sixteen years after the discovery, once penicillin's usefulness had been demonstrated by Florey and his colleagues at Oxford and the drug had received worldwide interest. The true account is every bit as extraordinary as the myth.

The odds against the discovery were astronomical. Fleming was working with cultures of *Staphylococcus aureus* from boils, abscesses, and nose, throat, and skin infections. He piled the plates on a bench and looked at them every few days to see what was happening. Each colony, about the size of a letter on this page, was typically golden-yellow. Some bacteria characteristically develop colonies of various bright colors. Fleming set out to investigate whether a change in color of the colonies from different environmental factors might indicate a different virulence.

Fleming had a discerning eye and liked to "play with bacteria." The artist in him developed a unique palette of colors. For instance, *Staphylococcus aureus* produced golden-yellow colonies, whereas *Serratia* produced a vivid red color, *B. violaceous* violet, and so on. He had collected a whole range of these colorful bacteria and would streak agar plates with them in a carefully planned sequence and placement. In this way, he grew pictures and designs within a four-inch circle — rock gardens ablaze with color, a red, white, and blue Union Jack, a mother feeding a baby — that would appear after incubation, as if by magic, twenty-four hours later.

At the end of July, it was time for Fleming to go on his summer holiday. He left the staphylococcal plates on his bench in a pile of about forty to fifty.[6] Upon his return to work on September 3, he started clearing his workbench of old plates, first looking at each one to see whether anything interesting had developed before placing it into a shallow tray of disinfectant to be washed for reuse. (Today's culture plates are plastic and disposable.) This illustrates one of his compelling traits: he always needed, and got, a second chance to notice the interesting fact. When an assistant stopped by, Fleming wanted to show him some of the plates he had put into the tray for disinfection. He picked one Petri dish at random, one which miraculously rested on others and was just clear of the disinfectant that would have destroyed the bacteria cultures on it. It had been contaminated by a mold. This was not unusual, but Fleming was surprised to see that for some distance around the patch of mold, there was a zone cleared of bacteria, presumably due to some substance manufactured by the mold. At one edge was a blob of yellow-green mold, with a feathery raised surface, and on the other side of the dish were colonies of staphylococcus bacteria.

But in a circular zone around the mold, the bacteria had been lysed, dissolved.

Molds are living creatures of the same broad group as fungi. They can be found on jam, bread, wet carpets, damp basement walls. They grow as a tangle of very fine threads that periodically produce fruiting bodies. The fruiting bodies of molds shed millions of microscopic spores, which are the reproductive seeds of the organism. These are carried in the air, and when a spore lands, by chance, on a hospitable medium, it germinates and starts producing a new mold. During its metabolic process of absorbing nutrients, it produces by-products. In the instance that caught Fleming's attention, this by-product was killing bacteria!

Fleming had seen bacteria lyse years earlier when the drop from his nose had fallen on a Petri dish. But he had never seen disease-causing bacteria lyse because they were near a fungal colony. This mold wasn't just preventing bacteria from growing, it was killing all that were present!

Chance had thus alighted on a prepared mind.

Fleming later noted: "But for the previous experience [with lysozyme], I would have thrown the plate away, as many bacteriologists must have done before. . . . It is also probable that some bacteriologists have noticed similar things . . . but in the absence of any interest in naturally occurring antibacterial substances, the cultures have simply been discarded."[7]

V. D. Allison later paid tribute to Fleming's work patterns: "Early on, Fleming began to tease me about my excessive tidiness in the laboratory. At the end of each day's work, I cleaned my bench, put it in order for the next day and discarded tubes and culture plates for which I had no further use. He for his part kept his cultures . . . for two or three weeks until his bench was overcrowded with 40 or 50 cultures. He would then discard them, first of all looking at them individually to see whether anything interesting or unusual had developed. I took his teasing in the spirit in which it was given. However, the sequel was to prove how right he was, for if he had been as tidy as he thought I was, he would never have made his two great discoveries."[8]

What can be taken as Fleming's untidiness should not be misun-

Alexander Fleming at his lab bench.

derstood. In truth, it was his way of cultivating the unexpected. He maintained a curiosity about the seemingly trivial or unusual.

Other elements contributed to the combination of chance circumstances improbable almost beyond belief. The mold that contaminated the culture was a very rare organism, *Penicillium notatum*, ultimately traced to a mycology laboratory on the floor below, where molds from the homes of asthma sufferers were being grown and extracts of them made for desensitization. (*Penicillium* is named from the Latin, meaning "brushlike," since the fruiting organs have a brushlike form.) Its spore wafted up the stairwell to settle on one of Fleming's dishes at a particularly critical instant, at precisely the time he implanted the agar with staphylococci. If the mold spores had been deposited later, the flourishing bacterial growth would have prevented multiplication of the penicillium spores.

Penicillium notatum acts on microbes only while they are young and actively multiplying. A typical colony of staphylococci on a bacteriologist's culture plate consists mainly of microbes that are already dead or well past the stage of growth at which the mold could affect them; the destruction of the relatively few susceptible microbes would not produce any visible change.

In chronicling the story for his book, Hare consulted old meteorological records and found out that an intense heat wave, which had been smothering London and which would have prevented the growth of the penicillium spore on the Petri dish, broke on the day Fleming opened the dish, thereby allowing the penicillium spore to thrive. The following cool spell in London created conditions in which the mold grew first, followed by the bacteria when the weather turned warm again. These were the only conditions, Hare later found, under which the discovery could have been made.

No wonder that Gwyn Macfarlane, Fleming's biographer, saluted this episode as "a series of chance events of almost incredible improbability." Listing the sequence of events leading to Fleming's discovery, he likened it to drawing the winners of seven consecutive horse races in the right order from a hat containing all the runners.[9]

Fleming showed the plate to his coworkers, who were unimpressed. They assumed that the phenomenon was an example of a lysozyme that had been produced by the mold. But Fleming was undaunted. He confirmed that when the bright yellow fluid extracted from the mold was dropped on staphylococcal colonies, after several hours the bacteria died, disappearing before his very eyes. He began to dilute the extract, but even after further and further dilutions up to 1:800, it still retained its killing power, not only against staphylococci but against most other gram-positive bacteria, including streptococcus, pneumococcus, meningococcus, gonococcus, and the diphtheria bacillus. Unfortunately, it had no effect on gram-negative bacteria like those responsible for cholera and bubonic plague.[10]

The extract, however, was unstable, losing its efficacy over a period of weeks, and he could not isolate and purify the active principle from the mold filtrate. To his detriment, he was not a chemist and had little background and limited resources for studies on the "mold juice," as he called it.

Unaccountably, Fleming failed to pursue two critical paths of investigation. Despite the fact that he was a leading physician who was treating hundreds of patients with syphilis, he never tested his mold extract on the syphilis bacterium. Nor did he think to try it on animals infected with streptococci or any other pathogens. If he had, he would have been astonished at its overwhelming effectiveness. He may have been misled by the fact that when he added the penicillin extract to blood in a test tube, it seemed to be inactivated, suggesting that it would be useless in the human body. Disappointed, Fleming concluded that penicillin did not seem clinically promising, only that it might be helpful applied to superficial local infections and that it might be used in the laboratory to isolate certain microbes.

He addressed the Medical Research Council at a meeting chaired by the physiologist Henry Dale, who would go on to become one of the most distinguished British scientists of his time. Again, Fleming's delivery was so mumbling, monotonous, and uninspired that Dale, along with most of the audience, could not fathom the vast potential medical uses of the discovery. Fleming's stoicism was shaken. Looking back nearly twenty-five years later, he described the silence he faced after his presentation as "that frightful moment."[11] In his landmark paper, published in the *British Journal of Experimental Pathology* in 1929, Fleming placed more emphasis on the potential use of penicillin as a kind of lab trick to isolate other microbes that did not respond to it than on its potential use as a medical drug.[12] Indeed, for the next ten years or so, isolating other microbes was the principal purpose for which penicillin was used. The scientific community otherwise paid no heed.

Yet Fleming, perhaps out of mere stubbornness, preserved the strain of mold and kept it available, which turned out to be to his credit and mankind's great benefit. Over the following years, he made many attempts to reproduce his original discovery but never succeeded. He searched among scores of different molds — samples from cheese, jam, breads, boots and shoes, old books, and dust and dirt generally — but uncovered no other with antibacterial activity. Without the continued preservation of Fleming's strain of mold, the discovery would have remained only a scientific anecdote.[13] Although he gave subcultures of the precious penicillium mold to several colleagues in other laboratories, he never mentioned penicillin in any of his twenty-seven papers and lectures published between 1930 and 1940, even when his subject was germicides.[14]

Fleming would have remained an obscure figure in medical history but for the accomplishments several years later of dedicated researchers a few miles away at Oxford University. They transformed penicillin from a laboratory curiosity to keep unwanted bacteria out of culture dishes into a miraculous medicine. When it was finally recognized for what it was — the most effective lifesaving drug in the world — penicillin would alter forever the treatment of bacterial infections.

A Profoundly Ignorant Occupation

Lewis Thomas, author of *The Youngest Science,* commented on his experience as a medical student in the early 1930s: "It gradually dawned on us that we didn't know much that was really useful, that we could do nothing to change the course of the great majority of the diseases we were so busy analyzing, that medicine, for all its façade as a learned profession, was in real life, a profoundly ignorant occupation."[15]

All feared the scourge of infectious diseases. Before the discovery of penicillin, a doctor's black bag contained digitalis, insulin, and some powerful plant-based painkillers and sedatives, such as morphine and cocaine, but not much else of scientific value. Doctors could do little else than offer palliative treatment and reassurance. Such a scene is captured in Sir Luke Fildes's 1891 painting *The Doctor,* in which a physician keeps watch at the bedside of a seriously ill child as the anxious parents look on. Ironically, the painting was reproduced on a postage stamp in 1947 to commemorate the centennial of the American Medical Association.

A NUGGET OF GOLD TRANSFORMED INTO A GOLD MINE

Howard Florey, a pathologist and physiologist, was appointed at the age of thirty-seven as director of Oxford's large new Sir William Dunn School of Pathology in 1935. With forceful vision and superb organizational skills, he set out to recruit to his department researchers with interdisciplinary skills. It was a radical concept at the time. He was joined by a talented and supremely self-confident biochemist, Ernst Chain, a twenty-nine-year-old Jewish refugee from Nazi Germany. The differences in personality and temperament between the two were sharp, but they had an amicable relationship with highly productive complementary talents for several years before conflicts arose.

Florey, handsome and square-jawed, had a laconic, undemonstrative, frequently prickly and dismissive manner. His driving enthusiasm for research was concealed by a tendency toward understatement: "We don't seem to be going backwards" meant real progress.[16] Chain, short, dark-haired, with a large bushy moustache and an unfashionably long mane of hair, likened by many in appearance to Albert Einstein, was excitable and contentious but radiated infectious enthusiasm. In contrast to his starchy British colleagues, Chain half-mockingly described himself in a phrase Cole Porter would have admired as a "temperamental Continental";[17] he was nicknamed Mickey Mouse for his bouncing, smiling energy.

Another biochemist, Norman Heatley, who was painfully shy but extremely ingenious and skillful, with a remarkable flair for designing and constructing laboratory equipment to overcome analytic and production problems, joined the team.

All these individuals were certainly astute, but their talents would likely have come to naught without a series of lucky circumstances. Florey and Chain's journey of discovery was directed by the muse of serendipity. Florey, long involved in a systematic investigation of the mucus of the digestive tract, was interested in its biochemistry and saw lysozyme as a possible lead. After discovering the presence of lysozyme in intestinal secretions, he wondered if he had found the cause of intestinal ulcers. Chain's primary interest was the chemistry of cell walls. Florey asked Chain to undertake a study to determine how lysozyme broke down the cell walls of the intestinal lining. Chain based his experiments on Fleming's original observations on lysozyme's effect on bacteria. He found that it was a protein that destroyed the cell walls of bacteria, in effect killing them. This seminal work laid the foundation for understanding the chemical structure of the bacterial cell wall. It did not produce any direct advance in therapeutics, but was valuable for showing that powerful antibacterial agents were not incompatible with living human tissues, and that an understanding of chemistry was essential to an understanding of biology.

In 1938 Florey and Chain decided to extend their investigations to other natural antibacterial compounds. They happened across Fleming's original description of penicillin published nine years earlier and wondered if his mold extract might be a sort of mold lysozyme. Chain believed that a detailed study of how penicillin lysed the bacterial cell wall would afford much useful information on the structure of this wall, just as his studies on lysozyme had done.

Chain was surprised and delighted to find that he needed look no further for a source of *Penicillium notatum* than the hallways of his own laboratory. "I was astounded at my luck in finding the very mould about which I had been reading, here, in the same building, right under our very noses."[18] He encountered a lab assistant carrying germ culture flasks on the surface of which a mold had grown and asked her what the mold was. She told him it was the mold Fleming had described in 1929. The Oxford laboratory had been culturing *Penicillium notatum* for years simply to keep Petri dishes free of contamination! He requested a sample of the mold and started experiments with it right away.[19] "We happened to go for penicillin," he later said, "because we had a culture growing at the school."

Chain accepted the challenge abandoned by Fleming. The idea was to crystallize penicillin by cooling it in order to obtain it in stable form. Neither he nor Florey contemplated possible medical uses. "So," in 1938, "we started our work on the isolation and purification, not in the hope of finding some new antibacterial chemotherapeutic drug, but to isolate an enzyme which we hoped would [inactivate a chemical] common on the surface of many pathogenic bacteria."[20] Years later, Chain reaffirmed the serendipity of the devious, unintended paths leading to this: "That penicillin could have a practical use in medicine did not enter our minds when we started work on it."[21] For his part, Florey admitted, "I don't think it ever crossed our minds about suffering humanity; this was an interesting scientific exercise."[22]

Meanwhile, Chain and Heatley's task of extracting enough material of sufficient purity from the penicillium cultures was enormous. And with the start of World War II in 1939, resources for research became extremely limited. The mold grows only in the presence of oxygen. Large surface areas seemed essential so that every bit of mold would come into contact with air. Heatley borrowed sixteen sterilized bedpans from the Radcliffe Infirmary and filled them with nutrient meat broth so the mold could grow on the surface. By early 1940 a small sample of brown powder, no more than 0.02 percent pure, was obtained. Its injection into two mice produced no ill effect.

What happened next involved luck on three different counts. First, they chose mice rather than guinea pigs, the other common lab animal in use at the time. Penicillin is nontoxic to mice but quite toxic to guinea pigs. Second, there was an extremely high content of impurities in the powder they injected into the mice; had the impurities been toxic to the mice, the scientists would not have realized how safe penicillin actually is and might have terminated the study. Third, Chain and Florey noted that the urine of the injected mice had a deep brown color, signaling to them that penicillin is mostly excreted unchanged in the urine. This meant not only that it is not destroyed in the body but also that penicillin could seep through any fluids to fight infections wherever they existed in any tissues.

The team now fully appreciated the extraordinary antibacterial activity of penicillin for the first time. Furthermore, they found that penicillin would stop bacteria from growing even at a dilution of one part in a million. They began to grasp the therapeutic potential of the substance and pressed Florey to begin full-scale animal experiments. With refinements in technique, Heatley was able to increase the purity to 3 percent.

On May 25, 1940, Florey, Chain, and Heatley carried out a groundbreaking experiment to demonstrate penicillin's effects on eight white mice infected with lethal doses of streptococci. Four were injected with penicillin, while the other four, the controls, were not. Heatley, excited, stayed in the laboratory all night. By morning, all the controls were dead. Three of the four mice who had received penicillin survived. While Chain broke into an exuberant dance, Florey cautiously allowed that "it looks like a miracle."

These crucial experiments were taking place as Britain was plunged into war and the British Expeditionary Force was being evacuated from Dunkirk. Florey and Chain's first paper, "Penicillin as a chemotherapeutic agent," appeared in August 1940 in the *Lancet*, the most widely read British medical journal.[23] The journal's editors, recognizing the landmark results, had rushed it into print, two weeks before the Battle of Britain began. It was a brief account of only two pages, but it said everything that could be said. For the next two years, Florey applied his time and energy to persuading people to take notice of what he was convinced was the medical discovery of the century.[24]

Sulfa drugs had proved their usefulness, but they had drawbacks. In time of war, more powerful antibacterial drugs would be immensely valuable. The experiments on mice had used up the team's entire stock of penicillin, and a mouse is 3,000 times smaller than a human being. It was clear they were going to have to expand the operation on a massive scale. Needing more containers, Florey decided to turn his department into a penicillin factory to produce enough for a clinical trial on human patients and commissioned a pottery to produce six hundred ceramic vessels modeled after bedpans. Between February and June 1941, these tests spectacularly confirmed the results of the animal experiments: penicillin was far more effective and nontoxic than any known antibiotic.

The first patient was a forty-three-year-old Oxford policeman, Albert Alexander, who was near death from a mixed staphylococcal-streptococcal septicemia that developed at the site of a scratch at the corner of his mouth that he had sustained while pruning his roses. He had not responded to sulfapyridine, had undergone removal of his infected left eye, and had numerous abscesses on his face, scalp, and one arm. Only about a teaspoon of the unrefined penicillin extract was available. After being treated with it, he improved dramatically, only to suffer a fatal relapse when the minute supply ran out. The last three days of his treatment were maintained by collecting all his urine and extracting and recycling as much of the drug as possible. Florey likened the situation to "trying to fill the bath with the plug out."[25] While shaken by the tragedy, Florey nevertheless knew that only logistical problems remained to be solved before penicillin could begin to save lives.

During the summer of 1941 four children seriously ill with streptococcal or staphylococcal infections were treated with penicillin at Oxford. Children were chosen in the hope that they would need less of the substance than an adult. All were cured. Penicillin not only was active against a range of bacteria but it also had no harsh side effects, such as the kidney toxicity that had been reported with sulfanilamides, or their tendency to destroy the ability of the bone marrow to produce white blood cells. The results were promptly published in the *Lancet*.[26] Modestly titled "Further observations on penicillin," the article served to announce the dawn of the antibiotic era.

Events now took on the character of an Eric Ambler thriller. Florey was determined to keep penicillin from the Germans. When a German invasion of the British Isles under Hitler's Operation Sea Lion seemed likely, the Oxford team smeared the lining of their clothes with the penicillin mold in the hope that spores of the precious discovery could be smuggled to safety.

As the war dragged on, the need for penicillin to treat Allied soldiers became more crucial. Realizing that an adequate supply could not be produced in wartime England, Florey and Heatley left for the United States at the end of June 1941. They flew via neutral Lisbon in a blacked-out Pan Am Clipper to New York City, storing their precious freeze-dried penicillin mold in the plane's refrigerator. Arriving several days later, in 90-degree heat, they rushed across Manhattan with their precious sample by taxi to a midtown hotel where it could be refrigerated. The American government was easily persuaded to undertake large-scale production of the drug, and events progressed from there at lightning speed.

The two men were referred to the regional research center of the Department of Agriculture in Peoria, Illinois, located in the heart of the Midwest.[27] This facility had a large fermentation research laboratory for the production of useful chemicals from agricultural by-products.

Heatley set to work with Robert Coghill, head of the fermentation division, and Andrew J. Moyer for the large-scale cultivation of the mold.

As luck would have it, Coghill's and Moyer's experiences had been with corn steep liquor, a cheap syrupy by-product of the manufacture of cornstarch. They found that the mold thrived in this valuable nutrient, and with the deep fermentation technique in 15,000-gallon tanks, the yield of penicillin was greatly increased. They had yields of 900 units per milliliter of brew, whereas Florey had gotten only 2 at Oxford. The serendipitous nature of this turn of events, which resulted in the large-scale production of penicillin, was emphasized by Coghill: "One of the least understood miracles connected with it [penicillin] is that Florey and Heatley were directed to our laboratory at Peoria — the *only* laboratory where the corn steep liquor magic would have been discovered."[28]

Even further success was then ensured by another stroke of good luck. America's entry into the war in December 1941 guaranteed a total dedication to a project that would ultimately benefit battlefield casualties. Military personnel were ordered to gather handfuls of soil from around the world in the hope of tracking down a fungus that produced high quantities of penicillin. Mold from soil samples flown in by the Army Transport Command from Cape Town, Bombay, and Chungking were the front-runners. In the end, the army was beaten by Mary Hunt, a laboratory aide who one day brought in a yellow mold she had discovered growing on a rotten cantaloupe at a fruit market right in Peoria. This proved to be *Penicillium chrysogenum*, a strain that produced 3,000 times more penicillin than Fleming's original mold![29] This made commercial production of penicillin feasible. The laboratory assistant was called Moldy Mary for the rest of her life.

In its first trial in the United States, penicillin proved miraculous. In early 1942 Mrs. Ogden Miller, the thirty-three-year-old wife of Yale's athletic director, hospitalized for a month at the university's medical center, was near death from hemolytic streptococcal septicemia (blood poisoning) after a miscarriage. She was often delirious, with her temperature spiking to nearly 107 degrees. Her doctors tried everything available, including sulfa drugs, blood transfusions, and surgery. All failed. In desperation, she was given a precious small supply of penicillin made available by Merck in Rahway, New Jersey. The clinical response was dramatic, and she rapidly recovered. The record of this singular event, her hospital chart, is preserved at the Smithsonian Institution. (Mrs. Miller went on to live to the age of ninety.)

By the spring of 1943 some two hundred cases had been treated with the drug. The results were so impressive that the surgeon general of the U.S. Army authorized trial of the antibiotic in a military hospital. Soon thereafter, penicillin was adopted throughout the medical services of the U.S. Armed Forces and became a decisive factor in World War II. By the time of the D-Day invasion of Normandy in June 1944, the American drug companies Merck, Squibb, and particularly Pfizer were producing 130 billion units of penicillin per month, enough to treat all 40,000 wounded soldiers.

British pharmaceutical companies were able to produce only tens of millions of units of penicillin per month. This paled almost into insignificance in contrast to the large-scale American effort. Considerable friction between the two countries arose over legal protection of manufacturing methods. Chain, trained in the German tradition of collaboration between academic research and industry, vigorously advocated patenting penicillin for "protecting the people in this country against exploitation" (that is, he did not want to see penicillin used for personal gain). British law did not cover patenting a natural product, but the production process used various innovations that could be protected by patents. Chain was opposed by the head of the Medical Research Council and the president of the Royal Society on the grounds that patenting lifesaving drugs was unethical. In Britain there was a prevailing attitude that universities should keep a gentlemanly distance from collaboration with industry. This stemmed from the social distance the landed gentry had been anxious to maintain from the entrepreneurs during the Industrial Revolution. Florey acquiesced. Chain was furious.

The Americans had no such ethical qualms regarding patents and worked out a vast commercial enterprise. The money generated by U.S. pharmaceutical companies from penicillin production has underpinned the industry ever since. By 1979 sales topped $1 billion. Britain would be required to pay many millions of dollars to the United States to use, under license, the culture methods developed in the United States in 1941.[30]

The Cocoanut Grove Miracle

The American public was largely unaware of the momentous co-operative effort among academia, government, and industry regarding penicillin. Little news had leaked out about it. In a disastrous fire on the night of November 28, 1942, at the Cocoanut Grove nightclub in Boston, 492 people perished. Penicillin was successfully used to treat 220 badly burned casualties. But the public remained ignorant of this "miracle," as penicillin was then classified as a military secret.

With the end of the war, the necessity for secrecy came to an end, and in March 1945 commercial sales began. Pneumonia, syphilis, gonorrhea, diphtheria, scarlet fever, and many wounds and childbirth infections that had once been typically fatal suddenly became treatable. Deaths caused by bacterial infections plummeted.

And so began the modern era of antibiotics. The term chemotherapy, referring to drugs manufactured by laboratories, would in time be replaced by the term antibiotics for substances produced by microbes that can be used to treat infections. Antibiotics are chemical substances produced by various species of microorganisms (bacteria, fungi, actinomycetes) that suppress the growth of other microorganisms and may eventually destroy them. Unlike the therapeutic agents pioneered by Ehrlich and Domagk from synthetic chemicals, antibiotics have always existed in the biological world as natural defenders. They are chemicals produced by one microorganism in order to prevent another organism from competing against it.[31]

Out with Gout

Unexpectedly, attempts to counteract the flushing of penicillin into the urine led to development of a major drug for gout. Merck spent years searching for a compound that would conserve penicillin blood levels by inhibiting its excretion by the kidneys. By the time it came upon one, probenecid, in 1950, it was no longer needed, since penicillin was being produced in large quantities at low cost. But the drug was found to increase the excretion of uric acid and became an effective treatment for the crystallized uric acid that causes the crippling destruction of joints in gouty arthritis. Probenecid was developed for a specific purpose, but was found to be more useful than originally intended. The muse of serendipity had smiled again.

THE FLEMING MYTH BECOMES ENSHRINED

Another fascinating element of the penicillin story involved a "return engagment" in the midst of the drama by its original discoverer, Alexander Fleming. In August 1942 a friend of Fleming's was dying of streptococcal meningitis at St. Mary's Hospital. Fleming telephoned Florey to ask if there was any chance of using some of his penicillin. Florey traveled down to London and handed over all the available stock. Fleming injected the drug into his friend intravenously, but then, when he discovered that the penicillin was not gaining entry into the fluid around the brain to kill the germs there, he courageously injected it through a spinal tap. This route of administration had never been used on humans before. The man's life was saved.

The story of this success was reported in the *Times,* without naming the researchers but referring to work in Oxford. In response, Sir Almroth Wright, Fleming's old mentor, wrote an effusive letter to the *Times* in which he credited the discovery to Fleming (and the Inoculation Department of St. Mary's). When interviews with Fleming appeared in the press, an Oxford professor informed the *Times* that

Florey's group deserved much of the credit. But Florey shunned the press, believing that publicity debased scientists and their work. In addition, he felt it would be a public disservice to raise hopes before he knew how to mass-produce the drug.

Thus Fleming alone was seen in the papers, and so was born the Fleming Myth, as the press and public, longing for good news during the grim war years, seized happily on the humble story of the moldy culture plate. Perhaps even more important was the fact that it was easier for the admiring public to comprehend the deductive insight of a single individual than the technical feats of a team of scientists. At first startled by and rather shy in response to the adulation, Fleming soon began to revel in it, as he went on to be lionized by the world. Nobody raised the question of what Fleming had done with his discovery between 1928 and the time it was evaluated by the Oxford group in 1940.

At war's end, the contributions of the Oxford group were finally recognized, and Fleming, Florey, and Chain shared the 1945 Nobel Prize for Physiology or Medicine. As for Norman Heatley, his official recognition was delayed until 1990 when Oxford University gave him an honorary M.D. degree.

In accepting the Nobel award in 1945, Fleming said: "In my first publication I might have claimed that I had come to the conclusion as a result of serious study of the literature and deep thought, that valuable antibacterial substances were made by moulds and that I set out to investigate the problem. That would have been untrue and I preferred to tell the truth that penicillin started as a chance observation."[32] Some years later, he reminisced: "Penicillin happened. . . . It came out of the blue."[33]

In September 1945 Fleming was honored by the Académie de Médecine. In his speech, he asked his Paris audience, "Have you ever given it a thought how decisively hazard — chance, fate, destiny, call it what you please — governs our lives?" And he went on to hail the freedom of a scientist to pursue a serendipitous observation: "Had I been a member of a research team, engaged in solving a definite problem at the moment of this happy accident that led me to penicillin, what should I have done? I would have had to continue with the work of the team and ignore this side entrance. The result would have been either that someone else would have discovered penicillin or that it would still remain to be discovered."[34]

Fleming reportedly confessed to a scientific contemporary in 1945 that he didn't deserve the Nobel Prize but added disarmingly that he nevertheless couldn't help enjoying his fame. Perhaps this explains why to the end of his life he felt that his greatest achievement was the discovery of lysozyme. He spent his last ten years collecting twenty-five honorary degrees, twenty-six medals, eighteen prizes, thirteen decorations, and honorary membership in eighty-nine scientific academies and societies. There is even a crater on the moon named after Fleming. When he died of a heart attack in 1955, he was mourned by the world and buried as a national hero in the crypt of St. Paul's Cathedral in London, beside Nelson, Wellington, and Wren.

Fleming's discovery was a stroke of brilliant observation, but he simply lacked the ability to convince. It was not in his nature to develop his discoveries, as Florey and the Oxford group did, with perseverance, ingenuity, and organization. Florey's biographer (and later also Fleming's) writes: "Fleming was like a man who stumbles on a nugget of gold, shows it to a few friends, and then goes off to look for something else. Florey was like a man who goes back to the same spot and creates a gold mine."[35]

Ernst Chain was generous in his appraisal of Fleming: "We ourselves at Oxford . . . had plenty of luck. Therefore it is petty and irrelevant to try to detract from the importance of Fleming's discovery by ascribing it entirely to good luck, and there is no doubt that his discovery, which has changed the history of medicine, has justly earned him a position of immortality."[36]

Florey, known for his scientific integrity, also acknowledged the indirect, often stumbling nature of the Oxford accomplishment. But, as is typical of many researchers, this did not happen until long after the work was published and high awards bestowed. In his Dunham Lectures at the Harvard Medical School in 1965, Florey said: "We all know that when we compose a paper setting out . . . discoveries we write it in such a way that the planning and unfolding of the experiments appear to be a beautiful and logical sequence, but we all know that the facts are that we usually blunder from one lot of dubious observations to another and only at the end do we see how we should have set about our problems."[37]

Chain had observed with a keen eye how well government, industry, and academic research collaborated in a unique combination of resources to bring this wonder drug to the world. Yet he understood the futility of any "mass attack" on diseases where knowledge of fundamental biologic facts is lacking. In 1967 he shared this wisdom with the world:

> But do not let us fall victims of the naive illusion that problems like cancer, mental illness, degeneration or old age . . . can be solved by bulldozer organizational methods, such as were used in the Manhattan Project. In the latter, we had the geniuses whose basic discoveries made its development possible, the Curies, the Rutherfords, the Einsteins, the Niels Bohrs and many others; in the biologic field . . . these geniuses have not yet appeared. . . . No mass attack will replace them. . . . When they do appear, it is our job to recognize them and give them the opportunities to develop their talents, which is not an easy task, for they are bound to be lone wolves, awkward individualists, nonconformists, and they will not very well fit into any established organization.[38]

6

Pay Dirt

One day in 1943 a New Jersey poultry farmer noticed that one of his chickens seemed to be wheezing. None of his other birds exhibited this symptom, but he was fearful that it might be the harbinger of an infectious disease that could afflict his entire flock. He rushed the chicken to a state laboratory for testing, which found that a clod of dirt with a bit of mold on its surface was stuck in the bird's throat. The lab's pathologist, knowing of the interests of Selman Waksman, the distinguished soil microbiologist at the New Jersey Agricultural Experiment Station of Rutgers University in the city of New Brunswick, took a throat swab and sent it to him.

Waksman had established himself as the world authority in his science in 1927 with a 900-page textbook, *Principles of Soil Microbiology,* which gave the field the stature of a scientific discipline. He attracted dozens of graduate students to his department to work and study under him, helping to transform Rutgers from a small agricultural college into a world-class institution.

Soil microbiology — the study of its microscopic inhabitants — had been a passion of Waksman's for many years. A Russian Jew who had emigrated to the United States from the Ukraine in 1910 and earned a Ph.D. in biochemistry from the University of California at Berkeley in 1918, he would later recall the "earthy" fragrance of his rich native dirt. He came to understand that the smell was not of the earth itself but rather of a group of harmless funguslike bacteria called actinomycetes. Waksman's aim was to find practical applications of his science, particularly in improving soil fertility and crop production, and he knew that the organic chemical content of soil was a crucial factor. He analyzed soil, humus, clay, loam, and sand for their microbes to see how they grew, how they multiplied, what nutrients they took in, and what waste products they exuded. Such microbes exist through a delicate ecological competition, some producing chemicals to kill others. But still unknown was their potential value to modern medicine.

Waksman missed several opportunities to make the great discovery earlier in his career, but his single-mindedness did not allow for, in Salvador Luria's phrase, "the chance observation falling on the receptive eye." In 1975 Waksman recalled that he first brushed past an antibiotic as early as 1923 when he observed that "certain actinomycetes produce substances toxic to bacteria" since it can be noted at times that "around an actinomycetes colony upon a plate a zone is formed free from fungous and bacterial growth."[1] In 1935 Chester Rhines, a graduate student of Waksman's, noticed that tubercle bacilli would not grow in the presence of a soil organism, but Waksman did not think that this lead was worth pursuing: "In the scientific climate of the time, the result did not suggest any practical application for treatment of tuberculosis."[2] The same year, Waksman's friend Fred Beaudette, the poultry pathologist at Rutgers, brought him an agar tube with a culture of tubercle bacilli killed by a contaminant fungus growing on top of them. Again, Waksman was not interested: "I was not moved to jump to the logical conclusion and direct my efforts accordingly. . . . My major interest at that time was the subject of organic matter decomposition and the interrelationships among soil microorganisms responsible for this process."[3]

Although Waksman knew that tubercle bacilli are rapidly destroyed in soil, his fixed paradigm of thinking simply kept him from carrying the problem to a practical conclusion. From that same year onward, Merck made grants to Waksman for research into antibiotics. In 1942 he was urged by his son, then a medical student, to isolate strains of actinomycetes active against human tubercle bacilli, but he replied that the time had not come yet. The general lack of interest in antibiotics at this time was no doubt discouraging, but it did not stop a former pupil of Waksman's, René Dubos, who was working at the Rockefeller Institute in New York.

In 1939 Dubos, a tall Frenchman with a robust personality and piercing intelligence, isolated a crystalline antibiotic, tyrothricin, from a harmless soil organism. It proved active against a range of disease-causing bacteria but was too toxic for use in humans. Several years earlier, working with Oswald T. Avery at the Rockefeller Institute, he had found that pneumonia microbes could be killed by dissolution of their protective capsules by other microorganisms. With the success of the sulfa drugs, however, this line of research had been abandoned. Now Dubos tried to interest other researchers in tyrothricin's promising properties, but the response from most researchers was lukewarm. Waksman was the exception. Inspired by Dubos's achievements, he began a dedicated screening of microbes. Meanwhile, also in 1939 came news from the other side of the Atlantic that Florey and his team of researchers at Oxford University were feverishly pursuing the antibiotic properties not of a bacterium but of a mold.

Over the next four years, Waksman's intensive research program involved isolating some ten thousand different microbes from soil and other natural materials, such as dung, and methodically screening them for their killing abilities. To cultivate them on various media and test their ability to inhibit the growth of pathogenic bacteria was a painstaking process. Waksman's main focus was on actinomycetes.[4]

Only about a thousand of the cultures had antibiotic properties, and only a hundred excreted the antibiotic to the growth medium. (Antibiotic potential is not enough. The microbe has to yield something like "penicillin juice.") Of these, ten were studied in detail, and by 1942 Waksman had isolated two different antibiotics: actinomycin and streptothricin.[5] They turned out not to be clinically useful because they were too toxic. But he knew he was on the right trail.

SOIL YIELDS A BOUNTY

Among Waksman's graduate students was twenty-three-year-old Albert Schatz. He had been discharged early from army service because of a back problem and sought to complete his Ph.D. at Rutgers. Waksman regarded him as unusually bright — a "star" among his student researchers. For more than three months, Schatz analyzed soil samples from the multitudes collected until, on October 19, 1943, in the words of fellow student Doris Jones, "Al hit paydirt!"[6]

The throat swab taken from the wheezing chicken's throat had by now made its way to Schatz. It grew greenish gray colonies of an actinomycete on an agar plate. Because of their color they were called *Streptomyces griseus.* By a remarkable twist of fate, this same organism had been found by Waksman himself twenty-eight years earlier while he was working on his doctorate, but he had not pursued it.

Tests quickly showed that the organism not only was active against staphylococci but, most exciting, also had dramatic killing effects on the gram-negatives, bacteria that cause a variety of diseases that had been unaffected by penicillin, such as typhoid, bacillary dysentery, bubonic plague, brucellosis, and tularemia. Clearly, the organism generated a powerful antibiotic. On Waksman's instruction, Doris Jones tested it against salmonella infections in chick embryos and confirmed its ability to cure. Furthermore, she found it had no toxicity in laboratory animals.

At this point, Schatz took it upon himself to move to the next stage of research on the potential of the antibiotic now called *Streptomyces griseus.* It had been a fundamental aim of his research to seek an antibiotic that would cure tuberculosis. At the time, the only treatments for TB were prolonged bed rest and nutritious food. He had almost a religious zeal in this pursuit and could not be dissuaded by his colleagues' insistence that the heavy wax capsule of the bacilli would resist drug action. Penicillin was useless against it, but Schatz held to the premise that if nutrients could enter the cell and waste products leave, so could antibiotics pass through the capsule. With great purpose, he tested streptomycin against tuberculosis and patiently waited several weeks for the slow-growing colonies to appear on the culture medium. The results were clear. Where streptomycin was added, not a single colony appeared, whereas in the controls dense growths were apparent. Schatz had shown that the new antibiotic was dramatically effective in inhibiting the growth of tuberculosis germs.[7] As directed by Waksman, Elizabeth Bugie, the lab's lead technician, verified the results.

A short report published by Schatz, Bugie, and Waksman in January 1944 emphasized the antibiotic activity of streptomycin against gram-positive and gram-negative bacteria, but curiously included only marginal notation regarding its effectiveness in tuberculosis. Nevertheless, researchers at the Mayo Clinic, long interested in the treatment of the disease, committed to undertaking animal experiments and then human trials. Schatz spent eighteen-hour days working in his small basement laboratory to concentrate the antibiotic for the Mayo Clinic trials, sleeping with two old, torn blankets on the floor to monitor the continuous running of laboratory apparatus.

Later in the year, a second paper by Schatz and Waksman based on test tube work announced the effectiveness of streptomycin against tuberculosis.[8] By April, Schatz was able to provide a meager supply of impure streptomycin to be used at the Mayo Clinic.

Favorable results in guinea pigs led eight pharmaceutical companies to provide an estimated $1 million worth of streptomycin for the largest clinical study of a drug ever undertaken, involving several thousand tuberculosis patients. This constituted the first privately financed, nationally coordinated clinical drug evaluation in history. Medicine had taken a giant step in embracing the value of controlled experiments repeated thousands of times in the evaluation of a new drug. Once again, the drug worked wonders. Side effects, including impairment of the sense of balance and deafness, proved to be transitory and could be minimized by controlling the dosage. The drug's use was also limited by its tendency to produce resistant strains of the tubercle bacillus. In 1947 streptomycin was released to the public.

The Scourge of TB

Tuberculosis is one of the oldest infectious diseases, having afflicted humans since Neolithic times. In the previous century and a half, it claimed an estimated billion lives worldwide. The "white plague of Europe" that raged in the seventeenth century was due to growing urban populations.

Mycobacterium tuberculosis was identified by Robert Koch in

1882. Because it has the ability to enclose itself in nodules in the body, the tubercle bacillus can remain dormant for years, then strike any part of the body with deadly results, although it most commonly affects the lungs. For centuries, "consumption," as it was called, was believed to physically stimulate intellectual and artistic genius. (Some of the greats who were so afflicted include Molière, Voltaire, Spinoza, Schiller, Goethe, Kafka, Gorki, Chekhov, Paganini, Chopin, Dr. Johnson, Scott, Keats, the Brontë sisters, D. H. Lawrence, Thoreau, Emerson, Poe, and O'Neill.) The pasteurization of milk, establishment of sanitariums, and posting of "No Spitting" signs were all public health measures that were eventually taken to control tuberculosis.

Waksman was hailed as a medical hero — the discoverer of the world's newest "miracle drug" — whose victory over TB resonated with symbolic value in the wake of World War II. In the resulting avalanche of public acclaim, he toured the world, gave lectures, and took tours of medical facilities. Although Schatz, the graduate student, had actually made the discovery, Waksman, as head of the department, was in a position to arrange for commercial development and get the glory. He won the highly prestigious Albert Lasker Award, often a presager of the Nobel Prize in Medicine, and was featured on the cover of the November 7, 1949, issue of *Time* magazine.

Other "Down and Dirty" Drugs

With the success of streptomycin, the pharmaceutical industry embarked upon a massive program of screening soil samples from every corner of the globe. Other soil microorganisms soon yielded their secrets, and it was the actinomycetes that generated a cornucopia of new antibiotics.[9]

Chloramphenicol (chloromycetin) — so named because it contains chlorine — was isolated in 1947 from a strain found in a sample of Venezuelan soil. The tetracyclines came from various streptomyces and proved of great benefit as broad-spectrum antibiotics, useful against a wide array of infections: Aureomycin was introduced in 1948; Terramycin, so named because it was cultured from soil samples collected near Pfizer's Terre Haute factory, in 1950; tetracycline (Achromycin) in 1952; and Declomycin in 1959. Erythromycin was cultured in 1952 from a soil sample collected in the Philippines. And vancomycin, one of the most potent antibiotics ever found, comes from an actinomycete isolated in 1956 from a clump of Indonesian mud.[10]

But the success of streptomycin also brought problems in its wake, involving who was to reap the rewards of its discovery. Rutgers was able to maintain the patent and the royalties from the drug's sales, based on nonexclusive licensing, yielding a huge financial windfall for the university and Waksman personally. Rutgers was then receiving two cents for each gram of the drug sold, and by 1950 its royalties amounted to almost $2.5 million. Waksman received some $350,000 in royalties.

Then something happened that thoroughly rocked the scientific community. Albert Schatz filed a legal claim demanding formal recognition as codiscoverer of streptomycin and a share of the royalties. This widely publicized lawsuit by a former doctoral student against a distinguished, internationally recognized professor was unprecedented.

Schatz's name had been listed first on the journal articles announcing streptomycin to the medical world, in accord with Waksman's policy of encouraging discoveries by his students or assistants. Furthermore, Schatz as well as Waksman was listed on the patent application filed by Rutgers in January 1945. Streptomycin was the subject of Schatz's Ph.D. thesis, which he defended in 1945. Waksman had always considered Schatz a brilliant star among his students, but the twenty-six-year-old, feeling that he was unfairly being shut out by his mentor, who was now enjoying worldwide acclaim over streptomycin, left Rutgers resentfully in 1946.

A Rutgers lawyer dismissed Schatz as "a carefully supervised laboratory assistant," and "a small cog in a large wheel."[11] Waksman felt that the discovery of streptomycin was the last inevitable step in a path he had personally paved, with the prior discoveries of actinomycin and streptothricin, and that Schatz was lucky to be "in at the finish." It was Waksman, after all, who had established the entire program of antibiotic research.

Shocked and embarrassed, Waksman reluctantly agreed to settle. He acknowledged in court that Schatz was "entitled to credit legally and scientifically as co-discoverer of streptomycin." Schatz received an immediate lump sum of $125,000, a public statement by Rutgers officially acknowledging him as codiscoverer of streptomycin, and 3 percent of the royalties. Waksman shared 7 percent of his 17 percent royalty with the other laboratory workers at Rutgers. With half of the rest of the money he started the philanthropic Foundation for Microbiology, thus reducing his own portion to 5 percent. Waksman took pains to include twelve laboratory assistants, clerks, and even the man who washed out the laboratory glassware. The settlement was broadcast on page one of the *New York Times* and in *Time* magazine.

Schatz felt vindicated and hoped to use the money to continue in microbiology research. But, following the notoriety, the doors of top-grade laboratories were closed to him. No other department head wanted to hire an aggrieved "whistle-blower," as such a contentious, publicly aired episode had been hitherto virtually unknown within the scientific community. The establishment closed ranks. He was, in effect, blacklisted.

In 1952, two years after the settlement, Waksman became the sole recipient of the Nobel Prize in Physiology or Medicine, even though Nobel regulations allow up to three people to share it. Schatz waged an unsuccessful campaign to obtain a share of the Nobel Prize. Burton Feldman, a historian of Nobel awards, uncharitably refers to Waksman as getting the Nobel "for not discovering streptomycin."[12]

The public dispute was tragic for both men. Waksman, heartbroken, regarded the year 1950 as "the darkest one in my whole life" because of the perceived betrayal by Schatz and the horrible publicity brought about by the whole affair.[13] Schatz, a penniless student, was at least financially recompensed but was deprived of what he felt to be his rightful share of the Nobel Prize and of career possibilities. Years later, Waksman would quote Dubos: "In science the credit goes to the man who convinces the world, not to the man to whom the idea first occurs."[14]

Schatz had to wait fifty years to be awarded Rutgers' highest honor, the Rutgers Medal, in 1994 as codiscoverer of streptomycin.

Meanwhile, in the same year that Waksman received the Nobel Prize, 1952, his discovery was eclipsed by the introduction of the chemical isoniazid (often called INH), a significantly more effective treatment for tuberculosis, which could be taken orally. With the further discovery that drugs given in combination increased efficacy, the annual death rate from TB per 100,000 population in the United States fell from 50 to 7.1 by 1958.

Serendipity and Cyclosporin

The search for new antibiotics and the microbes that produce them led to the serendipitous discovery of another breakthrough drug in 1978. The Sandoz laboratory in Basel, Switzerland, tested a fungus found in a soil sample from Norway for antibiotic activity. None was found. But one of their enterprising biochemists, Dr. Jean-François Borel, found that it could suppress immunity in cell cultures. An immunosuppresant, a drug that interferes with the body's rejection of foreign tissues, had been unearthed: cyclosporin. This led to the explosive development of human-organ transplant surgery, initially in kidney transplantation, and then in more difficult surgery, such as liver, heart-lung, pancreas, and bone marrow transplants.

The problem with antibiotics is that bacteria reproduce at an astonishing speed and, like all organisms, mutate. As a result, new strains emerge that are resistant to the drugs. The more frequently antibiotics are used, the more quickly bacteria seem to outwit them. In the United States by the year 2005, physicians in practice wrote 130 million prescriptions per year for antibiotics, as many as half of which were unnecessary. Beyond this, at least 30 percent of all hospitalized patients received one or more courses of therapy with antibiotics. One result of this widespread use is the continuous emergence of antibiotic-resistant pathogens, which in turn fuels an ever-increasing need for new drugs.

In 1993 the World Health Organization (WHO) declared tuberculosis, which claimed about 2 million lives each year, a global emergency. WHO estimates that by 2020, one billion people will be newly infected.

7

The Mysterious Protein
from Down Under

Hepatitis was the farthest thing from Baruch Blumberg's mind as he pursued his research. Armed with an M.D. from Columbia University and a Ph.D. in biochemistry from Oxford University, he decided to study how and why people of different backgrounds react differently to disease. How do genetically determined biochemical and immuno-logical variants among different populations determine differences in susceptibility and resistance to disease? This concept, known as poly-morphism, became the focus of his work at the National Institutes of Health in Bethesda, Maryland.

In the 1950s his field trips over a seven-year period, which fo-cused on drawing blood from diverse peoples, took him all over the world: West and East Africa, Spain, Scandinavia, India, China, Tai-wan, the Philippines, Australia, Micronesia, Canada, South America. In high boots and a broad-brimmed hat, Blumberg looked more like Indiana Jones than a genetic researcher. In the search for genetically determined proteins, he assembled an increasingly diverse collection of blood samples.

"By the Miracle of Chance . . ."

In 1960 a leap in his thinking occurred. His premise was straightfor-ward and based on the classic antigen-antibody response. If the blood of one individual — say, a blood donor — contains a protein that is not natural to another individual, the foreign protein (antigen) may excite an immune response and elicit a defensive globulin (antibody) in the recipient. As he put it, "We decided to test the hypothesis that patients who received large numbers of transfusions might develop antibodies against one or more of the . . . serum proteins (either known or unknown) which they themselves had not inherited but which the blood donors had."[1]

Transfused patients are exposed to blood from many different donors and therefore to whatever varying substances exist in that blood. Their immune system reacts to these foreign substances and makes antibodies, which are stored. Blumberg's aim was to use this store of antibodies to find hitherto unknown blood proteins.

He developed a simple technique that could readily identify antigen-antibody reactions. In a central depression in a thin sheet of gel, Blumberg would place the blood of, say, a multiply transfused hemophiliac patient. Presumably exposed many times over to proteins in donors' blood not common to his own, it would contain a host of antibodies. Blumberg then placed blood samples from various subjects in several other punched holes arrayed around the central well. If any of these contained antigens recognized by antibodies in the central well sample, the resulting chemical reaction would form a dark line in the region of diffusion between them.

In 1963 he came across a surprising finding. Of twenty-four different tests in the panel from individuals all over the world, the blood of a multiply transfused American hemophiliac had an antibody that reacted with only one. That specimen was from a decidedly unlikely source, an Australian aborigine. It had been sent to Blumberg by a colleague from the University of Western Australia in Perth. Blumberg dubbed the mysterious protein "Australia antigen." What could possibly have been the connection between a hemophiliac patient from New York and an aborigine from Australia?

The following year Blumberg joined the staff of the Fox Chase Cancer Center in Philadelphia and, with colleagues there, tested the unusual New York blood sample against thousands of other samples, trying to find out why the precipitating antibody had occurred. They observed a similar biochemical fingerprint in some patients with Down's syndrome and certain kinds of leukemia. The presence or absence of the Australia antigen seemed to be a constant feature of an individual.

But then fortune smiled, and a single exception turned the course of the investigation. In 1966 a patient with Down's syndrome, originally negative for the Australian antigen, turned positive. Laboratory tests and liver biopsy confirmed that he had developed hepatitis, without jaundice, in the interval. Blumberg had a true "Eureka!" moment as he came to suspect a link between the Australia antigen and acute viral hepatitis. The following year, another fortuitous event elated the researchers in Blumberg's laboratory. A technician who had previously tested negative for Australia antigen developed symptoms of hepatitis and tested her own serum for liver enzymes and the antigen. Both proved positive, enabling Blumberg to truly connect the dots for the first time.[2]

In the previous two centuries, epidemics of what was termed "infectious jaundice" had swept through military troops in many countries. Transmission studies by Japanese and German researchers in 1941–42 established the viral origin of the disease, which came to be known as hepatitis. Then, based on other experiences during World War II, with the advent of transfusions of blood and blood products in 1943, the terms "infectious hepatitis" and "serum hepatitis" appeared, later designated more simply as hepatitis A and hepatitis B. Researchers had not yet identified the actual viruses and didn't know where to look for them.

Hepatitis B attacks liver function. In its more severe form, the viral liver infection produces a loss of appetite, vomiting, and fatigue, but the most characteristic clinical sign is the vivid yellow color — jaundice — it imparts to the whites of the eyes and, often, the entire body. It is transmitted through sex or through contact with blood, via transfusions or dirty needles. About 1.5 million Americans are chronic carriers who may show no signs of the disease but can transmit the virus to others. Far more infectious than the virus that causes AIDS, the hepatitis B virus can live for a week outside the body on a dry surface. Most cases of hepatitis are acute and self-limiting, but many progress to a chronic form that can be deadly. Over 350 million people worldwide are chronically infected with this virus. It kills more than 1.5 million people a year and is the leading cause of liver cancer.

Blumberg conducted clinical and laboratory studies that tested and supported the hypothesis that the Australia antigen was a hepatitis virus or was located on it. Blumberg then briefly encountered hostility from what he characterized as "the establishment" when his report was rejected for publication. He tried to understand the dynamics of the resistance: "We found the hepatitis virus while we were looking at quite different things. We were outsiders not known to the main body of hepatitis investigators, some of whom had been pursuing their field of interest for decades."[3] We have seen in several other medical discoveries that the basis of a conceptual reformulation is often ignored initially or faces delayed consideration and acceptance. Only after other groups corroborated Blumberg's unexpected finding was his initial paper published two years later.[4]

Within a few years, intensive investigations would identify the Australia antigen as the protein that envelops the hepatitis B virus. The connection between the American hemophiliac and the aborigine eventually became clear. Subjected to multiple blood transfusions, the hemophiliac had, at some point, been exposed to blood that contained the hepatitis B virus and had developed antibodies against it. These reacted against the virus's protein in the blood of the aborigine, who happened to be a virus carrier. The inclusion of the aborigine's sample had been a most fortunate happenstance. If Blumberg had examined only the blood of Americans, he might not have stumbled upon his discovery, as fewer than half a percent of Americans are carriers and would yield a positive reaction. In contrast, 20 percent of Australian aborigines have a positive reaction. Even so, Blumberg was lucky to get the reaction at all, but his odds were much improved by the Australian sample. It represented a turning point that, in Blumberg's words, "by the miracle of chance, led us directly to the hepatitis B virus."[5]

By 1970 a group of British scientists visualized the whole virus itself by electron microscopy. By the mid-1970s, infection with hepatitis after transfusions essentially disappeared from the United States and the other countries where compulsory testing in blood banks was instituted.

Baruch Blumberg, with his colleague Irving Millman, next turned his attention to developing a vaccine using the sera of patients with the antigen, to elicit antibody production and prevent hepatitis B infection. What was needed was a carefully controlled trial on a population at high risk for hepatitis B. He filed his patent application at the end of 1969.

The perfect individual to conduct such a trial was Wolf Szmuness, an epidemiologist at the New York Blood Center. Szmuness had himself once been part of a population "at high risk": in the 1960s he had fled with his family to the United States from a series of pogroms in his native Poland against the remaining Jews that had survived World War II. Under his direction, the vaccine was tested on more than a thousand volunteers from the city's gay male community, representing a population at particularly high risk for hepatitis B because of their sexual practices. Almost 70 percent were carriers, and as many as a third of those previously uninfected would become so in a year.

The vaccine was proved highly effective and safe and has been in general use since 1982 when the Fox Chase Cancer Center licensed it to the Merck pharmaceutical firm. More than a billion doses of vaccine have been administered, mainly via national campaigns in Africa, Asia, and the Pacific to reduce the risk of primary liver cancer. New infections in the United States dropped to 80,000 a year in 2006 from more than 200,000 in 1982 when the vaccine was first used, according to the Centers for Disease Control and Prevention in Atlanta. However, the disease is still fairly common among drug abusers who inject themselves.

THE WILLOWBROOK EXPERIMENTS

At the same time Blumberg was doing his work, ninety-two miles away, a pediatrician interested in communicable diseases, Saul Krugman of the New York University School of Medicine, was conducting a series of experiments at the Willowbrook State School, an institution for the care of mentally handicapped children on Staten Island, New York. In the face of pitiful overcrowding, with soiled beds jammed one next to another in the ward and in the hallways, most of the newly admitted children acquired viral hepatitis in the first six to twelve months. The ward was essentially one giant culture dish inhabited by children.

Krugman divided the children into groups to test their responses to food containing the suspected virus and the effectiveness of injections of gamma globulin (the part of the blood rich in antibodies) in protecting against hepatitis. During the course of the study, a chance observation provided an illuminating insight. Second attacks of hepatitis occurred in about 8 percent, raising the possibility that there was more than one type of hepatitis virus.

By 1967 Krugman's conclusions could not be challenged. Hepatitis is caused by at least two distinct viruses: hepatitis A, of short incubation and highly contagious, and hepatitis B, of long incubation, up to six months, and of lower contagion.[6] Krugman also demonstrated for the first time that hepatitis B can be transmitted not only by blood transfusions or poorly sterilized instruments but also by intimate physical contact. His studies were crucial in providing clinical evidence that the Australia antigen was a specific marker for the hepatitis B virus.[7]

Wanted: Ethical Standards

The studies undertaken by Saul Krugman at the Willowbrook State School involved giving children food containing the live hepatitis virus. In keeping with the accepted standards of the era, Krugman received permission from the parents of the children involved as well as the approval of administrative authorities.

It is hard to believe, but at this time individual scientists and their financial backers could decide for themselves what constituted ethical research. Most of the time, their judgment was sound, but society became incensed at such appalling exceptions.[8] These led to the National Research Act of 1974, which required institutional review boards to approve and monitor all federally funded research.

Soon after the identification of the hepatitis B virus, the hepatitis A virus was found by electron microscopy in the waste of patients with hepatitis A and characterized as a small RNA virus. HAV is spread primarily by fecal oral contamination, contaminated water supplies, and food. It has been transmitted rarely by blood transfusion. In developing countries with substandard hygiene and sanitation, the disease is endemic.[9]

In 1976 the Nobel Prize ceremony in Stockholm openly acknowledged the serendipitous nature of Blumberg's discovery: "Like the princes of Serendip, [he] had found something completely different from the types of substances he was looking for. The protein he had discovered was not a part of normal body constituents but instead a virus causing jaundice."[10] In his presentation, Blumberg underscored that what he had stumbled upon was a compelling example of curiosity-driven research. Blumberg, whose friends and colleagues call him Barry, had always fondly attributed his elementary education in a yeshiva in Brooklyn, with its practice of fact-based argumentation, as conducive to modern scientific thought. It involves comparison of different interpretations, analysis of all possible and impossible aspects of the given problem, in order to arrive at an original synthesis that has never been offered before. Beyond that, Blumberg learned an invaluable lesson: "In research, it is often essential to spot the exceptions to the rule — those cases that do not fit what you perceive as the emerging picture. . . . Frequently the most interesting findings grow out of the 'chance' or unanticipated results."[11]

Baruch Blumberg did not set out to discover the hepatitis B virus. As occurred in many of the greatest medical advances of the past hundred years, he stumbled upon the answer to a question he never intended to ask.[12] His unexpected "Eureka!" moment, which was favored, in his words, "by the miracle of chance," illuminated a potentially devastating disease affecting millions worldwide.

8

"This Ulcer 'Bugs' Me!"

Nothing is so firmly believed
as that which we least know.

— Michel de Montaigne

"Oysters," my grandfather, a Russian immigrant to America, called them. "Calm down, relax, or you'll get oysters in the stomach," or "Don't eat so much spicy food — it'll give you oysters!" In the small inland town in which I grew up, oysters were an unknown delicacy, but to an immigrant's ear, the nuances of the English language often were lost. Pronunciation issues aside, my grandfather nevertheless displayed enough familiarity with the prevailing medical opinion of the time regarding the relationship between stress or spicy foods and stomach ulcers to admonish the members of his family.

"No Acid, No Ulcer"

An ulcer is an open sore on the mucous membrane lining the stomach or the duodenum, the portion of the intestine extending immediately beyond the stomach. Since ulcers affect an estimated 5 million Americans, with about 400,000 new cases reported each year, the condition — with the complications of pain, bleeding, and perforation — is no small matter.

For many decades, Schwartz's dictum of 1910 — "No acid, no

ulcer" — governed treatment of peptic ulcer disease.[1] The inside of the stomach is bathed every day in about half a gallon of gastric juice, which is composed of digestive enzymes and concentrated hydrochloric acid. Doctors typically treated ulcers by initially ordering changes in a patient's diet, in an attempt to protect the stomach walls from its acid. The Sippy diet, introduced by Chicago physician Bertram W. Sippy in 1915, was practiced well into the 1970s. Sippy called for three ounces of a milk-and-cream mixture every hour from 7:00 A.M. until 7:00 P.M., and one soft egg and three ounces of cereal three times a day. Cream soups of various kinds and other soft foods could be substituted now and then, as desired. Accompanied by large doses of magnesia powder and sodium bicarbonate powder, such "feedings," as meals were known, would continue for years — if not for life.[2]

Unfortunately, the intake of food brings about not only increased saliva but also the production of stomach acid in preparation for digestion. The Sippy method was thus rarely successful because, though doctors didn't realize it, the diet actually increased levels of stomach acid and therefore aggravated the symptoms of ulcers. In contemplating this well-intended but entirely inappropriate approach, one is reminded of the all-purpose huckster advertising slogan: "Successful except in intractable cases." Physicians treating their ulcer patients with the Sippy diet must have been amazed at how many of their cases proved intractable. The fact that this falsely based approach nonetheless persisted for six decades illustrates the unfortunate fact that conventional wisdom once adopted remains stuck in place even when it flies in the face of reality.

Ulcers often progressed to the point of chronic distress with the potential of life-threatening bleeding. In such cases, the answer was surgery to remove the acid-secreting portion of the stomach.

Besides blaming stress, diet, and tobacco as factors in the development of ulcers, many physicians believed there was also a psychosomatic aspect. Building on Freud's insights and Adler's idea of "organ inferiority," the concept of psychosomatic medicine was furthered by the experience of World War I, in which as many as 80,000 "shell-shocked" soldiers suffered from various severe somatic symptoms that seemed to have emotional origins. In his 1950 book *Psychoso-*

matic Medicine, Franz Alexander defined the cause of ulcers as internal conflict caused by a regressive wish to be dependent on others, the feelings stimulating increased gastric motility and secretion in anticipation of being fed by idealized parents. When the wish cannot be gratified through normal adult relationships, it was concluded, an ulcer results. In *Stress and Disease,* an influential medical textbook published in 1968, Harold G. Wolff, a well-known psychiatrist, described excess gastric secretion as a reaction to rampant "competitive striving" and other pressures of contemporary society, despite the fact that stressful life events had not been shown to be more common in patients with ulcers than in the general population.

Nevertheless, the stress-acid theory of ulcers gained further credibility when safe and effective agents to reduce gastric acid, known as H2 blockers, were introduced in the 1970s. Acid inhibitors, such as cimetidine (Tagamet) and ranitidine (Zantac), became so popular and were taken over such extended periods that they became the world's biggest-selling prescription drugs. By 1992 the worldwide market for prescription ulcer medications amounted to $6 billion a year. Ulcers affect one in ten adults in the Western world. These acid inhibitors are effective at easing symptoms, but they fail to prevent ulcers from recurring.

SWIMMING AGAINST THE TIDE OF "CERTAINTY"

No one ever even considered a bacterial cause for ulcers. Gastroenterologists invariably thought the stomach a sterile environment. Gastric juices are so acidic, a tooth immersed in a container of the fluid overnight would have its enamel dissolved. It was taken as doctrine that bacteria could not survive and flourish in such a harsh, inhospitable environment. It is nevertheless a curious historical fact that, over the years, spiral bacteria were glimpsed in specimens of the human stomach but were dismissed as either contaminants or opportunists that had colonized the tissue near ulcers. In 1954 a widely respected gastroenterologist reported finding no spiral bacteria in more than a thousand biopsy specimens of the stomach, and this report provided what was considered "conclusive" evidence that gastric bacteria, if present, were incidental.[3]

The overthrow of long-held but false concepts frequently is initiated not by those at the top of their fields but rather by those at the fringes. And technological advances in the 1970s opened a new door. Flexible fiberoptic gastroscopes enabled doctors to take targeted biopsy samples from a larger area of the stomach. In 1979 J. Robin Warren, a staff pathologist at the Royal Perth Hospital in Western Australia, was examining biopsy samples from patients with gastritis when he made a puzzling observation: there were unexpectedly large numbers of curved and spiral-shaped bacteria. He then used a special stain to make their number and shape more evident and saw that stomach cells near the bacteria were damaged.

Warren persisted in his systematic observations over the next two years. The bacteria were present in many gastric biopsies, usually in association with persistent stomach inflammation, termed chronic superficial gastritis. They were sited, often in colonies, beneath the mucus layer, indicating that they were not contaminants from the mouth, and were thriving within the mucus. Furthermore, Warren noted that the bacteria were all the same, not varied, as would be expected of secondary invaders.

Slightly Crackers

Later, reflecting on his discovery of what would be identified as *H. pylori* in the stomach, Warren commented: "It was something that came out of the blue. I happened to be there at the right time, because of the improvements in gastroenterology in the seventies. . . . Anyone who said there were bacteria in the stomach was thought to be slightly crackers."[4]

The fact that the bacteria settle particularly in the gastric antrum (the lower part of the stomach), as well as local variations in bacterial density, may explain why previous biopsies had failed to detect it. More to the point, stomach bacteria may have been overlooked simply because pathologists were not looking for it.

Despite his intriguing findings and repeated attempts to interest his hospital's gastroenterologists in the bacteria, Warren consistently encountered indifference. At this time, an unlikely deus ex machina appeared on the scene.

Barry Marshall, a lanky twenty-nine-year-old resident in internal medicine at Warren's hospital, was assigned to gastroenterology for six months as part of his training and was looking for a research project. The eldest son of a welder and a nurse, Marshall grew up in a remote area of Western Australia where self-sufficiency and common sense were essential characteristics. His personal qualities of intelligence, tenacity, open-mindedness, and self-confidence would serve him and Warren well in bringing about a conceptual revolution. Relatively new to gastroenterology, he did not hold a set of well-entrenched beliefs. Marshall could maintain a healthy skepticism toward accepted wisdom. Indeed, the concept that bacteria caused stomach inflammation, and even ulcers, was less alien to him than to most gastroenterologists.

Is There Such a Thing as Stagnant Knowledge?

Thus in subjects in which knowledge is still growing, or where the particular problem is a new one, or a new version of one already solved, all the advantage is with the expert, but where knowledge is no longer growing and the field has been worked out, a revolutionary new approach is required and this is more likely to come from the outsider. The skepticism with which the experts nearly always greet these revolutionary ideas confirms that the available knowledge has been a handicap.[5]

— W. I. B. Beveridge, medical historian, Cambridge University

Marshall's first indication that the bacteria were clinically relevant occurred in September 1981, when he treated a patient with severe abdominal discomfort caused by gastritis with tetracycline, an antibiotic. After fourteen days of treatment, the gastritis cleared up. Clearly, the next step was to culture the bacteria for identification and testing.

LEFT BEHIND ON A LONG EASTER WEEKEND

Culturing a microorganism in a sterile medium with the proper nutrients is essential for its further investigation, including establishing its identification, characterization, and degree of susceptibility to antibiotics. This is generally accomplished by streaking a bacteria-laden site upon an agar plate or in nutrient broth.

Over the latter months of 1981, Marshall repeatedly tried to grow the bacteria in both oxygen-rich and oxygen-depleted environments but was unsuccessful. He thought the spiral bacteria were of the *Campylobacter* genus, bacteria that require only two days to grow in incubation, and on this reasoning the clinical laboratory discarded the agar plates after forty-eight hours if no growth was visible. In April 1982, after many months of failed attempts at culturing the bacteria, Marshall left for the four-day Easter weekend feeling dejected.

Marshall's agar plates were inadvertently left in the dark, humid incubator for five days. No one had ever thought to give the culture this much time to grow. But five days did the trick. Returning to the laboratory, Marshall was elated to find thriving colonies on some of the plates. Peering through a microscope at a culture smeared on a slide, Marshall saw dozens of corkscrew-shaped organisms. He and Warren had grown the bacteria! It was now evident that longer culture time was necessary. Within a few months, it became clear exactly what the fastidious organism required: an enriched culture medium, such as sheep blood or chocolate agar, incubated under humid, microaerobic conditions for five to seven days.

MARSHALL-ING THE FACTS

Marshall then designed a clinical study in one hundred patients to look for the occurrence of the bacteria and the presence of stomach disorders. He found that bacteria were commonly associated with gastritis and were present in 80 percent of patients with gastric (stomach) ulcers and in 100 percent of patients with duodenal (intestinal) ulcers, in contrast to 50 percent of patients with a normal stomach.[6]

The literature strongly indicated not only that chronic gastritis was associated with peptic ulcers but also that acid-reducing drugs, such as cimetidine, merely healed rather than cured duodenal ulcers. Clearly, stomach acid was not the sole cause.

Then Marshall had an idea. Early in the century, Paul Ehrlich had used arsenical compounds against the spirochete of syphilis. Bismuth, an element in the same chemical group, had long been used as a remedy for abdominal discomfort and peptic ulceration. As recently as 1980 it had been reported that duodenal ulcers treated with a bismuth compound — marketed over the counter as Pepto-Bismol — had a diminished relapse rate. "This suggested to me," reasoned Marshall, "that bismuth compounds might inhibit the spiral bacteria" and thereby heal ulcers and gastritis.

The proper treatment regimen revealed itself in an unexpected circumstantial way. Marshall had earlier determined that, within a day of applying bismuth to a colony growing on an agar plate, all the bacteria died. In clinical trials, chewable bismuth tablets eradicated all signs of ulcers or gastritis, but to Marshall's disappointment, after discontinuing the treatment, nearly every patient suffered a relapse. Why, he wondered, would bismuth kill bacteria in a Petri dish but not in a patient's stomach? Then he noticed something strange. One patient, months after treatment with chewable bismuth tablets, continued to show no signs of recurrence. Reviewing the patient's record, Marshall saw that the man had suffered from a gum infection and, to treat it, had been given an antibiotic. Marshall hypothesized that bismuth alone did not reach all of the bacteria enshrouded in the stomach's thick mucus layer and had to be used in combination with an antibiotic.

The next development in the saga occurred due to another technological advance — the development of electron microscopy. A detailed morphology using the new equipment revealed that the stomach bacterium was not campylobacter. Clearly, this misidentification had been the cause of Marshall and Warren's earlier failures in growing the organism, as campylobacter required only a couple of days to get established. It would take another seven years, after RNA analysis and other studies, for these bacteria to be assigned to a new genus,

Helicobacter, and receive a proper name, *H. pylori. Helicobacter* refers to its helical shape and *pylori* refers to the pylorus ("gateway"), the exit from the stomach to the duodenum.

The *H. pylori* Lifestyle

That *Helicobacter* flourishes in the human stomach is a testament to evolutionary ingenuity. Both its shape and its biochemistry favor the adaptation of the bacterium to its harsh acidic environment. The lining of the stomach is protected from the acid by a thick viscous layer of mucus, in which *H. pylori* can take refuge. Four whiplike flagella at one pole of each bacterium provide rapid motility through the gastric juice, and the helical structure of the bacteria facilitates a burrowing corkscrew path from the stomach cavity through the thick mucus to establish colonies next to the lining cells.

Once *H. pylori* is safely sheltered in the mucus, it is able to fight the stomach acid that does reach it with an enzyme it possesses called urease. Urease converts urea, which is abundantly present in the gastric juices, into bicarbonate (a natural Alka-Seltzer) and ammonia. These are both strong alkalis. In this fashion, the *Helicobacter* cocoons itself in a protective acid-neutralizing mist. Furthermore, the bacteria thrive best in 5 percent oxygen — exactly the level found in the stomach's mucus layer.

Helicobacter colonies may thrive for a long time before symptoms appear — indeed, for decades, as recent studies have shown. Eventually the bacteria cause injury to the stomach lining through their toxic waste products and release of destructive enzymes, causing chronic inflammation. Some *H. pylori* even use a structure similar to a hypodermic needle to inject a particularly potent toxin into the stomach lining, boosting the associated inflammatory response. On top of all this, *H. pylori* infection stimulates the secretion of gastric acid. At that point, the bacteria's human host experiences symptoms of gastritis, and a peptic ulcer may eventually result from the inflammation of the stomach lining.

The crafty invader gets its supplies from the unfortunate be-sieged victim itself — specifically, the stomach tissue — using the inflammatory response as a means of obtaining a constant and re-liable source of nutrients. Thus, micronutrients from the body's bloodstream seep across to feed and sustain the *H. pylori* popula-tion for years or decades. The bacteria possess yet another sur-vival mechanism: In unfavorable environments, such as nutrient deprivation and antibiotic exposure, *H. pylori* may ball itself up and assume a rounded (coccoid) form for survival in a dormant state.

The gastritis induced directly results in structural and func-tional changes in the stomach and intestinal lining that can cause ulcer disease. It is now known that *H. pylori* infection accounts for 80 to 90 percent of gastric ulcers (the remainder being the con-sequence of high doses of aspirin or other nonsteroidal anti-inflammatory agents) and more than 95 percent of duodenal ulcers. Of even greater significance, it is the leading cause of stom-ach cancer.

"This Guy Is a Madman"

Marshall's first formal sting of rejection by organized medicine's or-thodoxy of belief in the causation and treatment of peptic ulcer dis-ease was felt in January 1983 when he submitted a report for the meeting of the Australian Gastroenterology Society in which he con-tended that bacteria might be responsible for ulcers. Although fifty-nine of the sixty-seven submissions for this meeting were accepted, Marshall's was not. The society had neglected the wisdom of Claude Bernard, the nineteenth-century founder of experimental medicine, who famously said, "If an idea presents itself to us, we must not reject it simply because it does not agree with the logical deductions of a reigning theory."[7] In a wry understatement Marshall noted, "It was clear I was thinking very differently from the gastroenterologists."

He was encouraged by a colleague to present his findings to a meet-ing in September in Brussels of infectious disease specialists gathered to focus on campylobacter infections. This paradigmatic shift in drawing the attention not of gastroenterologists but of infectious disease specialists at an international workshop was the stepping-stone to eventual acceptance. Marshall presented evidence that he had found a new bacterium in the stomach and that it resembled the campylobacter species. This part of his presentation was accepted by the audience as clear and appropriate.[8] However, to Martin Blaser, an American physician expert in infectious diseases, Marshall's claim regarding the bacterial causation of peptic ulcers without the presentation of any scientific evidence was "the most preposterous thing I'd ever heard. I thought, this guy is a madman."[9] In time, Blaser became a dedicated researcher in the field.

Warren and Marshall's landmark study, "Unidentified curved bacilli on gastric epithelium in active chronic gastritis," was published later that year in the *Lancet,*[10] in an unusual format: two separate letters. Warren's letter described work on the bacteria that he had conducted alone before collaboration with Marshall. Marshall's described their joint work. The two men published a joint report the following year indicating the bacterial cause of gastritis and of gastric and duodenal ulcers.[11] "I was certain," Marshall said, "that it would immediately gain universal acceptance." His confidence proved to be unduly optimistic. Indeed, most gastroenterologists viewed the hypothesis that peptic ulcers are caused by bacteria with incredulity. Nevertheless, growing interest was generated by reports published within the year by other investigators confirming the presence of gastric spiral bacteria and their association with gastritis.

PHYSICIAN, HEAL THYSELF (BUT FIRST MAKE THYSELF SICK)

Marshall and Warren faced one especially big hurdle. They had not firmly established that the microorganism *H. pylori* was the cause of the disease in question. As declared by Robert Koch, the German scientist who established the bases of bacteriology in modern medicine in the latter part of the nineteenth century, three conditions are required in order to prove causation: the organism must be shown to be constantly present in characteristic form and arrangement in the diseased tissue; pure cultures of the organism must be obtained; and the pure culture must be shown to induce the disease experimentally.

Marshall and Warren had satisfied postulates 1 and 2, correlating *H. pylori* infection with gastritis and peptic ulcer. Marshall tried to create an animal model to satisfy Koch's third postulate, but both the rats and the pigs that he tried to infect proved to be resistant to the bacteria. In desperation, in July 1984 he undertook a self-experiment that was reported in the *Medical Journal of Australia.*[12] Self-experimentation is, in fact, a noble tradition in the history of medicine by means of which many advances and discoveries have been made. This maneuver was necessary, in Marshall's words, for "closing the circle."

After gastroscopy found no bacteria or inflammation present in his stomach, Marshall swallowed a foul-tasting brew of *H. pylori*–laden broth after inhibition of gastric acidity. After a week, he suffered first vomiting and then, for about a week, headaches and putrid breath. Ten days after his symptoms began, follow-up gastroscopy with biopsy documented that he had acute gastritis and that the spiral bacteria had established themselves in his stomach. On the fourteenth day, Marshall began treating himself with an antibiotic and bismuth. His symptoms promptly cleared and another endoscopic biopsy documented resolution. This infection had induced an acute self-limited gastritis. Marshall had fulfilled Koch's third postulate, on himself!

With the tenacity of a bloodhound, Marshall made intense efforts to enhance his data. Outbreaks of a condition known as acute epidemic gastritis had been reported in the previous several years in various places. With characteristic boldness, Marshall took it upon himself to contact these sources and obtain the microscopic sections from these cases. Failure to identify the presence of spiral bacteria would be a serious setback to his hypothesis, but he was enormously relieved and encouraged to find that most indeed harbored the bacteria.

By 1985 a leading textbook, *Bockus Gastroenterology,* could dismiss the psychosomatic hypothesis with the declaration "the philosophic and metaphysical jargon that plagues the psychologic analysis of ulcer disease is of no value in our understanding the disease, nor in predicting which individuals are at risk and when, nor in the therapy of ulcers."[13] Stress and diet faded from the picture.

With the increasing recognition by microbiologists of Marshall and Warren's research, gastroenterologists and medical epidemiologists undertook further investigations of its clinical importance. The initial skepticism of clinicians was based on their bedrock beliefs regarding excess acidity. They had rejected the bacterial hypothesis of ulcers as inconsistent with what was already known.

But by the 1990s the focus was no longer on whether *H. pylori* causes ulcers but rather on how it does so and how it can be eradicated. Moreover, its mode of transmission, global prevalence, and role in causing gastric cancer and a form of gastric lymphoma were clarified. A torrent of research with the publication of thousands of articles ensued. Ulcers treated with the popular acid-inhibiting H2 blockers introduced back in the 1970s often recurred and therefore required repeated courses of the drugs, which provided a steady stream of profits to pharmaceutical firms. Marshall was convinced that this profit motive kept the drug companies, which funded much of the ulcer research at the time, from supporting the *Helicobacter* discovery.[14]

In a 1988 report in the *Lancet,* Marshall and coworkers showed that combined bismuth and antibiotic therapy healed ulcers far better than one of the most commonly prescribed acid-blockers.[15] It was major proof that eradication of *H. pylori* cures ulcers. Both Austrian and Dutch researchers reported definitive clinical trials of ulcer cure with antibiotics, improved with the addition of an acid inhibitor. One American gastroenterologist who, a few years earlier, had challenged the validity of Marshall's work was now converted: "We scientists should have looked beyond Barry's evangelical patina and not dismissed him out of hand."[16] A seven-year follow-up examination established the long-term benefits of *H. pylori* eradication.

Events rapidly unfolded. On the basis of these findings, expert panels convened by the National Institutes of Health and the American Digestive Health Foundation endorsed the conclusion that all patients with gastric or duodenal ulcers who are infected with *H. pylori* should be treated with antimicrobial and antisecretory agents. In 1995 Barry Marshall was awarded the prestigious Albert Lasker Clinical Medical Research Award. Marshall was finally vindicated after years of being scorned as, in the words of one eminent gastroenterologist, a "crazy guy saying crazy things."

"THE UNIMAGINABLE HAS HAPPENED"

Intensive epidemiologic studies showed in the early 1990s that *H. pylori* has a worldwide distribution and appears to be the most common chronic bacterial infection of humans, present in almost half of the world population, though usually without causing disease. Most carriers are without symptoms. A bacterium discovered only a few years earlier and greeted with disbelief was now proved to be globally endemic. Marshall was stunned: "I did not expect it to be the world's most common bacterial infection [in man]." *H. pylori* has been found in mummies thousands of years old. Today in the United States, it is found in 30 percent of adults and more than half of people over age sixty-five. Most people carrying it do not get ulcers, of course, so *H. pylori* is clearly not the sole cause of ulcers by itself. But there is no doubt that ulcers can be cured by attacking *H. pylori* with antibiotics and, more important, this cure is permanent!

H. pylori is believed to be transmitted orally, by means of fecal matter, through the ingestion of waste-tainted food or water. Thus the bacterium is most widespread in places where clean water is lacking. (About 80 percent of the population in most developing countries have the bacteria.) But other modes of transmission are interesting. Researchers have found *H. pylori* in the dental plaque of Indians and in the prechewed food some African mothers feed their infants. It is also possible that it is transmitted from the stomach to the mouth through reflux (in which a small amount of the stomach's contents is involuntarily forced up the esophagus) or belching, common symptoms of gastritis. Although unlikely, the bacteria could then be transmitted through oral contact, as in kissing.

The serendipitous laboratory error in Perth less than a decade earlier next led to another major discovery. Through worldwide epidemiologic studies, a stunning revelation came to light: *H. pylori* was clearly shown to also be a major cause of cancer of the stomach. With persistent *H. pylori* infection over the course of decades or years, the glands of the stomach lining are destroyed and the mucosal cells undergo a particular reactive transformation, changes that predispose them to cancer. Gastric cancer is second only to lung cancer as a cause of cancer death worldwide, killing hundreds of thousands of people each year. It is particularly prevalent in Asia and many developing countries. In the United States it is only the ninth most common.

People with *H. pylori* gastritis over decades have a risk of developing gastric cancer that is up to six times greater than among people without this infection, but the malignancy actually develops in only a tiny percentage. David Forman of the Imperial Cancer Research Fund in England had originally thought that Marshall's claim for various stomach diseases, including cancer, was a "totally crazy hypothesis." Now, along with others, he is a convert to the fact that *H. pylori* infection is a major factor in gastric cancer and in ulcers. By 1993 a status report in *Gastroenterology* could proclaim: "It would have been unimaginable that such seemingly diverse diseases as gastritis, gastric ulcer, duodenal ulcer, and . . . gastric carcinoma would all be different manifestations of an infection with a bacterium. The unimaginable has happened."[17]

MARSHALL'S ODYSSEY

Marshall's odyssey illustrates several characteristics that are typical of revolutions in medical science. Earlier observers had seen bacteria in the stomach but their importance was simply not recognized. The inability of these observers to connect the dots meant that opportunities for a major breakthrough were lost. After examination of random samples from limited segments of the stomach derived in a blind fashion, the bacteria were declared a mere contaminant in 1954 by a respected gastroenterologist and therefore of no particular interest. His premature conclusion slammed the door on further investigation. The development of the new technology known as fiberoptic endoscopy was necessary to allow Warren to inspect targeted samples for pathologic changes and thereby serendipitously observe the plentiful colonies of spiral bacteria deep in the mucus layer. Yet gastroenterologists remained uninterested.

Marshall was a youthful maverick, not bound by traditional theory and not professionally invested in a widely held set of beliefs. There is such a thing as being too much of an insider. Marshall viewed the problem with fresh eyes and was not constrained by the requirement to obtain approval or funding for his pursuits. It is also noteworthy that his work was accomplished not at a high-powered academic ivory tower with teams of investigators but instead far from the prestigious research centers in the Western Hemisphere.

The delay in acceptance of Marshall's revolutionary hypothesis reflects the tenacity with which long-held concepts are maintained. Vested interests — intellectual, financial, commercial, status — keep these entrenched. Dogmatic believers find themselves under siege by a new set of explanations. The stage is set, in the scholarly phrase of the historian of science Thomas Kuhn, for "incommensurable competing paradigms where proponents live in different worlds."[18]

Given the confluence of multiple factors, are certain discoveries inevitable? Some think so. One investigator in the field was of this opinion: "Somebody was going to find the bacterium sooner or later. But the fact is he (Marshall) did find it. And he wrote it up and he pushed it."[19]

In 2005 Marshall and Warren were awarded the Nobel Prize for Physiology or Medicine. No further proof was needed to disclaim the tongue-in-cheek charge by some that Western Australia is "the wrong end of the wrong place."

Part II

The Smell of Garlic
Launches the War on Cancer

It leaves the impression that all shots can be called
from a national headquarters; that all, or nearly all,
of the really important ideas are already in hand. . . .
It fails to allow for the surprises which must surely lie ahead
if we are really going to gain an understanding of cancer.

— A COMMITTEE OF THE INSTITUTE OF MEDICINE,
NATIONAL ACADEMY OF SCIENCES,
ON THE NATIONAL CANCER ACT AND THE "WAR ON CANCER"

9

Tragedy at Bari

In late 1943 the Adriatic port of Bari on the southeastern coast of Italy was crowded with Allied tankers and munition ships to support the invasion of Italy following the victories in North Africa and Sicily. To supply these operations, convoys carrying thousands of tons of ammunition and aviation gasoline were rushed into the Bari harbor. By December 2, 1943, every dock was occupied, and additional ships were anchored along the entire length of the seawall waiting to be unloaded. The SS *John Harvey* was moored at the harbor's eastern jetty. The port was so crowded that the ships were tied up touching each other, and long lines of supply trucks extended to numerous storage depots.

Eastern Italy was then under British jurisdiction. In order to expedite the unloading of cargo at Bari, no blackout was in force in the harbor. Rather, the port was fully illuminated. The British were convinced that the German air force in Italy was so depleted that it posed no threat in the area. British Air Marshal Sir Arthur Coningham went so far as to proclaim, "I would regard it as a personal affront and insult if the Luftwaffe would attempt any significant action in this area." He tragically underestimated the enemy.

Indeed, the Royal Air Force (RAF) had no fighter squadrons based in Bari. Yet the importance of the shipments at Bari to the Allied advance was obvious to German commanders. High-flying German reconnaissance planes documented the arrival of the convoys and the crowding in the harbor. A fleet of German bombers was rapidly organized from scattered airfields in northern Italy and ordered to attack Bari at dusk from the east.

On the evening of December 2, 1943, in a raid lasting twenty minutes, 105 Junker 88s skimmed the water and destroyed the Bari harbor. Only a few ships needed to be hit; with their vast cargoes of ammunition and fuel, each ship's explosion ignited others. One after another, they detonated in a deadly series. The surface of the harbor became a sheet of flame as the oil from broken petroleum lines, gasoline, and other cargo burned.

Axis attack on Bari, Italy, December 1943.

Hundreds of men thrown into the Adriatic desperately tried to reach safety. As they struggled in the sea or on lifeboats, the oily water stirred up by the series of explosions continuously doused them. It coated their bodies and went into their eyes and nostrils and down their throats. Some had trouble breathing from the fumes carried by billowing clouds of smoke.

In the midst of this maelstrom, some noticed a curious smell. Several of the survivors from the harbor would later recall what they described as a "garlicky odor." And across the harbor, as a very heavy cloud of smoke drifted over the town, a German prisoner of war turned to an American military policeman and said in perfect English, "I smell mustard gas." The MP looked startled. He took a deep breath and coughed. "Hell, that's garlic."[1]

Hastily formed rescue squads pulled more than a thousand men from the oily bitter-cold water. These survivors were wrapped in blankets and laid out in rows of stretchers on the ground in improvised dressing stations. Many suffered from burns and were in various stages of shock. Hospital personnel handled the influx of casualties as best they could. With the onrush, there was no time to remove the oil-soaked clothes from the survivors, wash their bodies clean of the slime, and provide them with fresh gowns or new uniforms. They were kept wrapped in blankets for as long as twenty-four hours.

Soon many began showing new symptoms: stinging eyes, excessive tearing, and eyelid spasms that led to difficulty in seeing. Areas of the skin showed brawny edema, painless at first; later, superficial layers of skin peeled off in sheets. Most striking was the puzzling nature of the shock in some of the patients. Even though their pulses were weak and their blood pressure extremely low, they did not present the typical picture of shock and showed no response to the usual measures. The doctors were struck by a consistent apathy in these men. In a not unusual instance, a patient would state that he felt rather well, even though his pulse was barely perceptible and his blood pressure very low; then he would promptly and quietly die.

Four survivors died the first day, nine the next, eleven on the third day. A few individuals who had been well enough to clean themselves of the oil and perhaps change their uniforms saved their lives by these simple acts. By the end of a month, eighty-three men had died. Since many of the survivors had severe eye irritations, it was strongly suspected that the German aircraft had used chemical bombs.

Lieutenant Colonel Stewart Alexander, a twenty-nine-year-old medical officer trained in chemical warfare and then on the staff of General Dwight D. Eisenhower's Allied Force Headquarters in Algeria, was dispatched to Bari on December 7 to investigate. Upon entering the hospital ward in Bari, he was immediately struck by an unexpected smell. Was it, of all things, garlic? Then he remembered he had encountered the same odor two years earlier when he had undertaken research on the effects of mustard gas.

Upon examining the patients, he first noticed unusual skin lesions. Their distribution confirmed Alexander's notion that a chemical toxin had touched the victims' skin. The burns clearly followed the pattern of exposure to the oily slime on the surface of the harbor. Men who had been immersed in the oil solution and then wrapped with a blanket suffered burns all over their bodies. If only the legs or arms had been exposed, burns were restricted to these parts. If a survivor had been splashed by the oily water from the harbor, the affected areas had first- and second-degree burns. Blister tracks could be traced on the chest or back where contaminated water had trickled down. Those engulfed by the black clouds of smoke had vapor burns on their exposed skin and in their armpits and groins.

Alexander speculated that if the chemical agent had been liquid mustard, it would have been diluted by the seawater, and the concentration in the different areas of the harbor would have varied considerably. Diluted mustard would cause different symptoms than those caused by concentrated mustard.

In Alexander's attempt to get to the bottom of what had happened, two things conspired to mislead him: British port authorities denied that any ship in the harbor had been carrying a chemical agent. And on the third or fourth day after the air strike, a bomb casing containing mustard that was recovered from the bottom of the harbor was assumed to be German.

Alexander was relentless in trying to determine exactly what had happened. He reviewed the manifests of the ships, searching for the type of cargo that might indicate the delivery system for mustard, and ordered chemical analysis of an oil sample from the harbor, autopsies with tissue samples, and a sketch of the harbor showing the anchorage of the ships. The autopsies confirmed chemical as well as thermal burns of the skin and toxic vapor injury of the lungs.

The sketch was crucial in correlating the hospital deaths with the ship positions, showing that the greatest number of mustard-induced deaths occurred among the personnel closest to the *John Harvey.* Another break resulted from the discovery that the recovered bomb casing was American, not German. Finally, the British port officials admitted that the *John Harvey,* which had been in the center of the moored vessels, had been carrying 100 tons of mustard gas! Not one member of its crew survived. The port authorities blundered terribly in failing to notify the struggling medical staffs in the overcrowded hospitals of the gas disaster. Colonel Alexander had been forced to uncover the hidden existence of the mustard gas step by step.

Gas warfare had been outlawed in 1925 by the Geneva Protocol. Nevertheless, before the actual outbreak of World War II, U.S. Secret Service information indicated that both the Germans and the Japanese were preparing to use poison gas.[2] In response to ominous reports that Hitler planned to resort to the use of poison gas against any invasion of southern Europe, President Franklin Delano Roosevelt announced Allied policy regarding this in August 1943. After decrying such inhumane weapons, Roosevelt warned that the United States would undertake "full and swift retaliation in kind. . . . Any use of poison gas by any Axis power, therefore, will immediately be followed by the fullest retaliation upon munition centers, seaports, and other military objectives throughout the whole extent of the territory of such Axis country." Consequently, mustard gas was stored at various depots in combat areas around the globe for reprisal use.

Chemical warfare had been notoriously introduced by the Germans during World War I. On April 22, 1915, they released a lethal cloud of chlorine gas on the French lines at Ypres in Flanders. The gas caused death by stimulating overproduction of fluid in the lungs, leading to drowning. By 1917 they had developed and employed, also at Ypres, a new diabolically efficient blistering agent and systemic poison, dichloroethyl sulfide. Called "yellow cross" because of its shell marking, it was fired at the enemy by artillery. The oily brown liquid evaporated slowly, remaining effective for several hours to several days. The heavier-than-air vapor sank into the trenches and lingered as a poison cloud. No mask provided protection against it. It burned through clothing, blistered the skin, destroyed vision, and choked out life.[3]

The poison agent was also given other names. To commemorate the site where it was introduced, some termed it Yperite, but the most common name given to the chemical that has come down to us is "mustard gas," based on its smell, which resembles that of garlic or mustard in high concentration.[4]

It has been estimated that more than 9 million shells filled with mustard gas were fired by both sides during World War I, causing some 400,000 casualties. Among these, the fatality rate was very low, 2.2 to 2.6 percent, with death most often due to pneumonia.

Immediately after the Bari air raid, more than 800 men were hospitalized. Later it was discovered that 617 of the injured were suffering from mustard exposure. Of these, 83 died. This fatality rate of more than 13 percent was much higher than the rate encountered in World War I. It was obvious that most of the deaths were due principally to the men's exposure for prolonged periods, which was unlike the experience in the earlier war. In this case, men were literally saturated with the lethal solution. The dilution of the mustard gas in the oil-soaked surface of the water and its slow absorption over the ensuing hours caused medical effects that had not been previously encountered.

In his report dated June 20, 1944, Alexander emphasized that the "systemic effects were severe and of greater significance than has been associated with mustard burns in the past." A conspicuous feature had been the *overwhelming destruction of white blood cells*. Beginning on the third or fourth day, there was a critical drop in one type of cell — the lymphocytes — after which the rest of the white blood cells disappeared.[5] In lymphomas and leukemias, there is an overproduction of white blood cells by the diseased bone marrow. Alexander's observations suggested at once the significance of mustard compounds for the possible treatment of cancers of the tissues that form white blood cells.

Meanwhile, important advances were being made on the home front. At Yale University, secret wartime research on the toxic effects of mustard exposure had been undertaken, and the huge set of medical data — including tissue blocks from the Bari disaster — provided crucial information. In 1942 the U.S. government's Office of Scientific

Research and Development had contracted with Yale to do research directed toward the treatment of gas casualties.

Colonel Cornelius P. Rhoads, on leave from the directorship of Memorial Hospital in New York City, was the chief of the Medical Division of the Army Chemical Warfare Service. With keen insight, he recognized from Alexander's report the opportunity to further investigate the use of chemical agents against cancer. Rhoads was a pathologist with particular experience in hematology, or blood disorders. Drs. Alfred Gilman and Louis S. Goodman,[6] pharmacologists at Yale, were assigned to the study of systemic effects of mustard gas in animals in order to develop a suitable antidote. They looked at modified mustard compounds and centered their research on a group known as nitrogen mustards. They learned how the mustard compounds change in the body through a series of chemical transformations.[7] Their studies underscored the remarkable sensitivity of normal lymphoid tissue to the lethal action of the nitrogen mustards. "The clinical observations on the casualities of the Bari disaster," Rhoads declared, "illustrate as adequately as any example can, the effects of the mustard compounds on blood formation."

Impressed by the way in which white blood cells were consistently depleted whenever a subject was exposed to mustard for a period of time, they speculated how this toxic effect could be used to advantage. To these investigators, thinking creatively was a natural reflex. It required the mental formulation "I have a solution! What is the problem?" Perhaps nitrogen mustard could be used as a new approach to battling cancer, for which surgery and radiation were then the only effective therapies. So they turned to their colleague Thomas Dougherty in the Department of Anatomy and suggested that he conduct experiments on mice. As they put it, "The problem was fundamental and simple: could one destroy a tumor with this group of cytotoxic agents before destroying the host?"[8]

THE MOUSE THAT ROARED

Dougherty used doses of nitrogen mustard in a mouse in which he had transplanted a lymphoma (tumor of lymphoid tissue) that was expected to kill the animal within three weeks of transplantation. The tumor was so advanced that its size almost dwarfed that of the mouse. After two doses, the tumor began to soften and shrink. Eventually it disappeared, and the mouse became frisky. "This was quite a surprising event," Dougherty exclaimed.

A recurrence regressed under further doses, but another recurrence did not, and the animal died eighty-four days after transplantation. Astonishingly, this unprecedented prolongation of life was never matched in later studies on a large group of mice bearing a variety of transplanted tumors, although good remissions were frequently obtained. Years later, Dougherty reflected on this singular incident: "The very first mouse treated turned out to give the best result. . . . I have often thought that if we had by accident chosen one of these leukemias in which there was absolutely no therapeutic effect, we might possibly have dropped the whole project."[9] Fortunately, this was the mouse that roared.

In December 1942 a therapeutic trial with intravenous nitrogen mustard was undertaken by the Yale group in a forty-eight-year-old silversmith in the terminal stages of a highly malignant lymphoma that had become resistant to radiation. Massive enlargement of multiple lymph nodes throughout his body was causing dire symptoms: those in his face and neck made chewing and swallowing impossible; those in his armpits made him unable to bring his arms down to the side; and those within his chest blocked the return of blood to his heart, causing his head and neck to swell. The patient's response to the treatment was dramatic. The bulky tumor masses receded. But the disease eventually recurred and the man died. Five other patients in the terminal stages of a variety of malignant diseases were subsequently treated at Yale, with poor results.

The drug had the Merck trade name Mustargen but was identified as Compound X in hospital charts. Nitrogen mustards were still classified as "top secret," and investigators were sensitive to the view at the time that "in the minds of most physicians, the administration of drugs, other than analgesic, in the treatment of malignant disease was the act of a charlatan."[10] Between 1943 and 1946, several medical institutions carried out controlled clinical studies under a mantle of military secrecy. The multicenter trial established a pattern for future research to determine the effectiveness of different treatment protocols by accumulating enough patient data. Restrictions on publication of data were finally removed, and two major reports came out in 1946. Alfred Gilman published the first review in the open literature of the classified wartime investigations,[11] and Cornelius Rhoads, as chairman of the Committee of Growth of the National Research Council, summarized the results of treatment of cancers with nitrogen mustard in a multicenter study of 160 patients, emphasizing their therapeutic value in cancers of the white-blood-cell-forming tissues.[12] Other reports of clinical investigations conducted during the war years were also released.

In 1949 Mustargen became the first cancer chemotherapy agent approved by the FDA. In that period, proof of safety, not of effectiveness, was the standard for approval. By 1958 a full clinical investigation clarified the applications, benefits, and consequences of nitrogen mustards. Regression of tumors was often temporary, and they eventually became resistant to further treatment. Such drug resistance continued to be a vexing problem as other chemotherapeutic drugs were developed over subsequent decades. Furthermore, the agents had to be administered very conservatively, since the cure could so easily be worse than the disease.

Nitrogen mustards remain useful today in Hodgkin's disease, other lymphomas, and acute and chronic lymphocytic leukemias. They belong to a class of cancer drugs called alkylating agents. (This term is based on the fact that they add an alkyl group — a type of alcohol radical — to compounds with which they react.) They stop the multiplication of cancer cells by forming permanent cross-links between the two strands of DNA in the cell's nucleus so that the cell can't divide. Currently, four other major types of alkylating agents with actions similar to nitrogen mustard are also used in the chemotherapy of cancer of several other organs.

The bombing at Bari was the worst Allied shipping disaster since Pearl Harbor. Yet, to this day, most World War II history books do not refer to it. The devastation at Bari was kept secret for years after the end of World War II, largely because of the presence of the deadly chemical agents. The American and British governments released only limited announcements. The incomplete defense of Bari was certainly a major wartime blunder, but Allied secrecy was imposed for decades, particularly because they had deadly chemical agents in the field, ready for prompt offloading if the need arose. Churchill commanded that all British records be purged of any mention of mustard. In 1961 the National Academy of Sciences attempted a follow-up survey of the survivors, but the project was officially impeded and blocked.[13]

Advances in the chemotherapy of cancers testify to the breakthrough the episode provided. The investigation of the Bari survivors' mysterious burns and skin ailments led to the observation that alkylating agents, such as nitrogen mustard, cause marrow and lymphoid depletion, which led to their successful use in treating certain kinds of cancers. According to the American Cancer Society, "the age of cancer chemotherapy was initiated. . . . From this [Bari] disaster, a chemical agent with anticancer activity was serendipitously discovered."[14] Nitrogen mustard became a model for the discovery of other classes of anticancer drugs.

It's said that every cloud has a silver lining. The cloud over Bari on that tragic day was very toxic indeed, but the silver lining that was fashioned by creative scientists led to lifesaving treatments for millions of people.

10

Antagonists to Cancer

The most common form of childhood cancer is leukemia, a cancer of the white blood cells produced in the bone marrow. Surgery is impossible and radiotherapy not very effective. In the late 1940s acute leukemia in children was an incurable and rapidly fatal disease. Patients were treated with transfusions and then sent home to await death in a few months from massive hemorrhage and infection.

Dr. Sidney Farber, a pathologist at Children's Hospital in Boston, had a particular interest in cancer in children. He based his work on an observation made halfway around the globe fifteen years earlier. In 1933 Lucy Wills, a British physician working in India, identified a type of anemia in textile workers that she attributed to their severe poverty and grossly deficient diet. She found that the anemia responded to the consumption of Marmite, a food product made from purified yeast that has been available since around 1900. It is typically slathered on toast, mainly by the British. The butt of many jokes, it has a strong odor, a saline taste, and the color and consistency of axle grease. The product derives its name from the French word for the logo pictured on the jar, *la marmite,* meaning "stewpot." Wills inferred that Marmite must contain an unidentified nutrient or vitamin. She subsequently found that the anemia also responded to crude liver extracts.[1]

The mystery nutrient in both Marmite and liver is now known to be folic acid. The term "folic acid" was coined in 1941, following its isolation from spinach (L. *folium* = leaf), but it was not until 1946 that its chemical structure was identified. We know today that folic acid — one of the B vitamins — is a dietary essential, present especially in fresh green vegetables, liver, yeast, and some fruits. Essential for blood formation, which occurs in the bone marrow, its deficiency leads to anemia. Studies during World War II had shown that certain anemias, in which large immature cells filled the bone marrow, responded to treatment with folic acid. Farber believed a similar solution could be discovered for leukemia. At the time, there was little understanding of the role of folic acid in the normal workings of the body. The history of its biochemical isolation and clinical application to chemotherapy is a saga of clinical stumbles, misdirection, and serendipity.

In the early 1940s cancer researchers at Mount Sinai Hospital in New York were involved in an extensive screening program to find agents that could cause regression of tumors. The program involved some 18,000 mice bearing both transplanted and spontaneous malignancies. One of the agents tested was a liver extract, thought to be folic acid, provided by Lederle Laboratories. The researchers came upon dramatic and unexpected findings: The liver extract given intravenously to mice strongly inhibited the growth of tumor cells.[2] This result was both surprising and confusing, as it appeared to contradict the known attributes of folic acid. What the researchers did not recognize at the time was that the extracts contained not folic acid but rather an *antagonist* of folic acid, a slightly different substance chemically, that effectively competes with and blocks folic acid's absorption into the cell. Unbeknownst to the observers at the time, the rapidly dividing tumor cells were being deprived of folic acid in the biosynthesis of DNA.

Researchers were then tragically misled by the erroneous conclusion that the development of cancerous cells occurs because the cells lack folic acid and get out of control. Thinking that an anticancer agent had at last been found, Lederle Laboratories quickly synthesized various folic acid compounds, and Farber and colleagues arranged clinical studies of patients with advanced cancers. Farber hoped the substance would provide the cure for leukemia.

Devastation rather than triumph ensued. The most striking result was the astonishing postmortem observation that eleven children with acute leukemia in this group showed the reverse effect of that intended — that is, their bone marrow was bursting with masses of newly formed white blood cells and their organs densely riddled with cancerous white blood cells. Since this was a totally unexpected result of these folic acid compounds, Farber termed this an "acceleration phenomenon" in the leukemia process. The sterile phrase undoubtedly hides the shock, disappointment, and guilt he must have felt.

Still, he imaginatively saw this tragedy as an opportunity: If the cells were *deprived* of folic acid, the growth of leukemia cells might be arrested. Perhaps folic acid absorption could be blocked through the use of antimetabolites. An antimetabolite, or antagonist, is a structurally similar compound that inhibits (antagonizes) the action of a metabolite by competing with it for the site of action in the body.

By 1948 Farber obtained several antagonists of folic acid. Aminopterin provided the first truly hopeful response. Out of the sixteen children treated, all of whom were suffering from acute leukemia, ten experienced unquestionable remissions.[3] The announcement of this great success was met with as much incredulity as celebration. Many observers, long inured to the futility of treatment of nonsolid tumors such as cancer of the blood cells, simply did not believe the results. Moreover, it was felt that such a breakthrough could not have been achieved by a forty-five-year-old pathologist who worked in a hospital basement laboratory with little in the way of funds, staff, or equipment.

On the other hand, practicing physicians and pediatricians were enthusiastic. Farber had achieved the first clinical remission with chemotherapy ever reported for childhood leukemia. A milestone in the history of chemotherapy had been established. With maintenance therapy, patients lived for an average of eight or nine months longer, and perhaps one in a hundred was cured.

In 1949 methotrexate was developed as a safer folic acid antagonist. By interfering with folic acid, methotrexate causes white cells to die of vitamin deficiency. By the mid-1950s it produced the first cure — not just remission but actual cure — of a solid tumor, choriocarcinoma, including, astonishingly, its lung metastases.[4] Choriocarcinoma is a malignancy in women derived from the fetal part of the placenta that has a great propensity to spread to the lungs.

The success of the folic acid antagonists led to the development of other antimetabolites for different cellular metabolic pathways.[5] Further insight was provided by a patient who had relapsed after treatment with a folic acid antagonist but who then responded to a course of treatment with another antimetabolite. This observation provided the basis for the introduction of combination chemotherapy. The basic idea of combination drug therapy is to mix drugs, all of which attack the tumor, but each of which has different side effects. These combinations are more effective than the individual drugs. Learning to improve the drug schedule produced longer remissions and then cure after cure. With time, courses of as many as four powerful drugs at once for periods of many months yielded progressively better results. Currently, most children with acute lymphocytic leukemia, the most common malignancy of childhood, are cured.[6]

Thus, what started as nutritional research in the 1930s led to the demonstration that antimetabolites could kill cancer cells, as well as the discovery of the effectiveness of combination chemotherapy.[7]

11

Veni, Vidi, Vinca

The Healing Power of Periwinkle

The discovery of a totally new chemotherapeutic agent derived from a natural plant came about through fortuitous circumstances that led to an outstanding example of serendipity.

In 1952 Dr. Clark Noble of Toronto received from a patient in Jamaica a small envelope full of leaves. These were from a subtropical periwinkle plant with the botanical name *Vinca rosea,* used by the locals to make a tea that appeared to help their diabetes. A physician in Jamaica, Dr. C. D. Johnston, was convinced that drinking extracts of the periwinkle leaves helped his diabetic patients to lower their blood sugar. Indeed, teas made from periwinkle were used by diabetics in places as geographically diverse as the Philippines, South Africa, India, and Australia, following the customs of traditional folk medicine.[1] A proprietary herbal preparation, Vinculin, was even marketed in England as a "treatment" for diabetes.

Intrigued and determined to find out if there was anything to the claims, Clark Noble decided to have the leaves analyzed. And he knew exactly where to send them: to biochemist James Collip's laboratory at the University of Western Ontario in London, Ontario. Dr. Collip had worked with Frederick Banting and Charles Best in Toronto and had distinguished himself for the extraction and purification of the hormone insulin so that it could be used in the treatment of diabetes.

Years earlier, Clark had won a coin flip against his brother

Robert for the opportunity to work with Best, who needed only one of them as an assistant. It fell to Robert Noble, an endocrinologist now working in Collip's laboratory, to investigate extracts of *Vinca*. When given orally to rabbits and diabetic rats, the extracts had no effect on blood sugar or on the disease. With the intention of increasing the possible effectiveness, Noble then injected concentrated water extracts within the abdominal cavities of the rats. "They survived about 5 days, but then died rather unexpectedly from diffuse multiple abscesses," severe bacterial infections, as "apparently some natural barrier to infection was being depressed." This phenomenon, observed only by chance, was shown to be due to profoundly depressed bone marrow function and gross depletion of white blood cells. Noble immediately grasped that these consequences would be beneficial to people with diseases in which bone marrow overproduced.

Dr. Johnston in Jamaica continued to send a supply of dried periwinkle leaves, collected by Boy Scouts on camping trips in the jungle. The amount, however, proved inadequate for separation and purification of active factors. But by growing the plant in greenhouses in Ontario, Noble and his coworkers were able to extract a new alkaloid in pure crystalline form, which proved highly active in experimental animals, not only causing marked depletion of bone marrow and a dramatic decrease in white blood cells but also acting to shrink transplanted tumors. Taking its origin and its effect on white blood cells into account, scientists named the alkaloid vincaleukoblastine, which was later shortened to vinblastine.

The discovery was announced at a meeting of the New York Academy of Sciences in 1958 in a report entitled "Role of chance observations in chemotherapy: *Vinca rosea*."[2] Few could recall any such candid title in presenting a discovery. The conference was organized to review new projects for the large-scale screening of chemicals for antitumor properties, but Noble seized upon the opportunity to underscore the value of serendipity:

> This paper may seem somewhat unorthodox in a monograph on screening procedures. However, we are eager to present the role of chance in obtaining new leads for

chemotherapy and to illustrate this with a specific example. The results of our research, which are presented here in detail for the first time, should not be considered in terms of a new chemotherapeutic agent, but rather in terms of a chance observation that has led to the isolation of a substance with potential chemotherapeutic possibilities.

Clearly, Noble was trumpeting the value of having *stumbled* across a discovery. He claimed that if he and his colleagues had restricted their search to plants with suspected anticancer properties, they would have missed the periwinkle. Big Science is not a sine qua non.

The cancer worker in the smaller institution or the academic department must view with awe the vast chemotherapeutic screening projects in progress in the United States; at the same time, however, he must consider what contribution he is in a position to make. Perhaps the role of chance observation is neglected in his consideration of ways of searching for new agents. Although somewhat irregular in comparison with the systematic prediction, synthesis and screening of an entirely new series of compounds, chance observations may well be worthy of greater consideration than they have received.

Perhaps reflecting the organizer's lack of interest in a paper so titled, Noble's report to the meeting was presented at midnight, but he related this fact good-naturedly: "By this time, in the early morning, the audience, besides the chairman, had been reduced to [my two co-investigators], the janitorial staff, a few scattered listeners, and a small cluster of scientists."[3] The latter was a group from the Eli Lilly Company who, noting his listing in the program, had previously arranged to meet with Noble to discuss their preliminary data. Gordon Svoboda and his colleagues had followed an almost identical approach after they too had discovered that *Vinca* extracts lacked antidiabetic activity, but in a general screening program at Eli Lilly their activity against a transplanted acute lymphocytic leukemia in the mouse was revealed.[4] A fruitful collaborative effort between Noble's group and the Eli Lilly research staff was undertaken. In 1961 Svoboda reported the isolation of vincristine, an alkaloid with almost identical chemical structure, but with different toxicity and range of clinical applications.

At about this time, the botanical name for the plant was changed to *Catharanthus roseus,* but it remains commonly referred to as *Vinca rosea.* Pure isolation of the alkaloid required gargantuan efforts. For example, fifteen tons of dried periwinkle leaves were required for the production of one ounce of vinblastine. Biological screening was enormously facilitated by the availability of a mouse leukemia, P-1534, that was exquisitely sensitive to the drugs. Other animal tumors were equally responsive, and clinical trials on humans were soon undertaken with rewarding results. At the time, active compounds could move from the laboratory with remarkable speed. Both vinblastine and vincristine gained FDA approval within three years of their discovery.

The periwinkle alkaloids function as poisons at a critical stage of cell division, preventing the cancer cell from reproducing.[5] Vincristine used together with steroids is presently the treatment of choice to induce remissions in childhood leukemias. Vinblastine, in a regimen with other agents, has resulted in an astonishing cure rate of over 90 percent in patients with testicular carcinomas. Employed in adult Hodgkin's lymphoma along with other drugs, the *Vinca* alkaloids have raised the five-year survival prospect to 98 percent.

Rather than preparing the compounds synthetically, a long and expensive process, Eli Lilly continues to process the agents from plants, using around eight tons of *C. roseus* annually.[6] What started as a false trail in medicinal folklore led to a chance discovery that resulted in triumphs for combination chemotherapy.

12

A Heavy Metal Rocks

The Value of Platinum

How a precious metal was shown to be useful against cancer is a fascinating story of a chance occurrence exploited by an astute researcher. The discovery arose from an unlikely source in an obscure biophysics laboratory and in a surprising way.

In the mid-1960s Barnet Rosenberg, a biophysicist at Michigan State University, was studying the effects of electric currents on *E. coli* bacteria growth. His goal initially was to sterilize medical supplies and to preserve food. He carefully chose the bacterium, *E. coli*, particularly common in the gastrointestinal tract and a frequent contaminant. *E. coli* is easy to grow in the lab and it reproduces very quickly. Furthermore, its chromosome structure was well understood. He devised a simple visual method to evaluate the effects of an electrical current upon bacterial growth by stirring the *E. coli* into a nutrient broth in glass chambers. And he selectively used platinum wire electrodes in the brew because of their presumed chemical inertness — all logical and discrete actions. Actively reproducing in the culture medium, the bacteria turned the brew cloudy. When Rosenberg zapped it with an electrical current, the solution cleared within two hours. The findings were easily reproducible. Zapping repeatedly turned the turbid culture clear.

Growth of *E. coli* in a chamber of culture medium ceased within two hours after the small electrical charges went through it. When

Rosenberg examined the medium under his microscope, he saw that the *E. coli* had ceased dividing, and the bacterial cells were essentially in an arrested stage of growth. Understandably, Rosenberg first attributed this effect to the electric current, but further experiments, often marked by false leads and blind alleys, finally led him to a startling realization.

In the original design of the experiment, Rosenberg had purposely chosen platinum as the material for the electrodes — because of its presumed chemical inertness — and a specific voltage to eliminate electrolysis effects and electrode polarization. In Rosenberg's refreshingly candid words: "both are mistaken ideas which led, via serendipity, to the effects described" and the realization that the inhibition of bacterial reproduction was due, not to the electrical effects, but instead to the unexpected release of platinum ions into the culture solution from the electrodes.[1] What caused the beneficial effect was the material that he was using to convey the electrical current. Had he chosen an electrode device made of silver or aluminum or chromium, his experiment would have led nowhere.

At this point, Rosenberg could have methodically continued along the original plan for his research. Fortunately, he saw the potential usefulness of his discovery to cancer therapy and reported his findings to the National Cancer Institute. Excited researchers subsequently found various platinum compounds to be very effective agents in many advanced tumors because of an amazing ability platinum has: it blocks cell division.[2]

Cisplatin, a platinum salt with the chemical *cis* configuration, proved to be of particularly great clinical value,[3] and it became the foundation for curative regimens for advanced testicular cancer, as exemplified in Lance Armstrong. By the time the then twenty-seven-year-old cyclist sought treatment for one testicle swollen to twice its normal size ("I'm an athlete, I always have little aches and pains," he told one reporter), the cancer had spread to his lungs and brain. After being cured by cisplatin, Armstrong powered to his seventh Tour de France victory. Cisplatin also works well against ovarian cancer and cancers of the head and neck, bladder, esophagus, and lung. Thus, an unlikely source of bacteriologic research led to an unexpected discovery of another major category of chemotherapeutic agents.

As Good as Gold

How another precious metal was discovered as a medical treatment is also a fascinating example of serendipity. Injections of gold salts are commonly used today in patients with rheumatoid arthritis, a frequently debilitating disease that attacks the joints, destroying cartilage and inflaming the lining between joints. It affects about 2.1 million people in the United States, usually women between the ages of twenty-five and forty.

The use of gold salts as a treatment to arrest the progress of the disease and to induce remission began with an observation in 1890 by Robert Koch, the father of modern microbiology. He noted that gold salts inhibited the growth of tuberculosis bacilli in test tubes. Early researchers believed tuberculosis initiated rheumatoid arthritis. In the 1920s, based on this mistaken belief, rheumatologists began administering gold injections to their patients with good results. It became clear later that this had nothing to do with tuberculosis, but rather that gold therapy acts by slowing disease progression and damage to joints. Exactly how gold does this is still not well understood.

13

Sex Hormones

Some people collect stamps, others coins, others fine art. Charles Huggins, a professor of surgery at the University of Chicago School of Medicine, collected prostatic fluid from dogs. His interest was in the biochemistry of seminal fluid. Secretions of the prostate gland form much of the ejaculatory fluid. The research was at times frustrated by the formation of prostate tumors in some of the dogs. Dogs are the only species besides humans known to develop cancer of the prostate — an incidental fact that turned out to be fortuitous.

After castrating a few dogs to see what effect testosterone injections might have on production of prostate fluid, he noticed something totally unexpected. Some tumors regressed, specifically those in elderly dogs whose tumors had arisen spontaneously (in other words, tumors that had come about on their own as opposed to being introduced into the dog by an experimenter). Huggins wondered, could it be that these tumors were hormone-dependent? As castration would obviously not have been a treatment option for humans, he injected estrogen into the dogs to produce, in effect, chemical castration. This, too, caused the tumors to regress. He had stumbled over a new threshold. This "Eureka!" moment led to the realization that some tumors are, in fact, hormone-dependent. Professor Huggins reported his results in the *Journal of Experimental Medicine* in 1940.[1]

He then began experimenting with injecting estrogen into men with prostate cancer that had spread to their bones. To his delight, the tumors regressed and the bone pain was relieved. A synthetic estrogen, stilbestrol, was as effective as naturally occurring estrogens. By 1950, in a cooperative study involving many surgeons, some 1,800 successful cases had been reported, leaving little doubt that a treatment had been found that at least delayed the progress of an otherwise intractable cancer. A breakthrough had been achieved. The discovery that estrogen therapy can control the growth and spread of prostate cancers provided the first clinical evidence that some human tumors respond to their hormonal environment. And for the first time, a synthetic drug was shown to work against cancer. Huggins, who had not been originally working on cancer, was awarded the Nobel Prize in 1966 for this valuable finding that saved many lives.

Prostate cancer is the second most common form of cancer. (In 2003, about 220,000 cases were diagnosed in the United States, and almost 29,000 men died of the disease, according to the American Cancer Society.) Huggins's breakthrough showed that some cancers are dependent on sex hormones. A turning point had been reached. A product of the human body itself rather than toxic or radioactive agents could be exploited as a cancer cure.

Tamoxifen for Breast Cancer

A powerful cancer drug, tamoxifen, arose unexpectedly from infertility research. In the late 1960s a synthetic compound was used in some countries to treat infertility in women because of its estrogenlike activity, which stimulated ovulation.[2] Research then came across an analog, tamoxifen, with an unanticipated value: it acted as an antiestrogen by virtue of its ability to block estrogen receptors. Tamoxifen came to be widely used in the treatment of breast cancer. It helps only women with estrogen-dependent breast cancer, which accounts for about 60 percent of breast cancers.[3]

14

Angiogenesis
The Birth of Blood Vessels

In August 1945 a report was published in the *Journal of the National Cancer Institute* that seemed to have little import at the time. Two researchers had noticed that tumor cells somehow elicit the growth of new blood vessels from their host. Furthermore, they speculated that a "specific substance" is produced by the tumor cells.[1] Their report aroused virtually no interest. The circulatory system was not viewed as a fertile field for research, and intense focus on cancer research did not develop for another two decades. The report was a striking example of prematurity in discovery.

Nearly twenty years later, in the early 1960s, a twenty-eight-year-old rabbi's son and Harvard Medical School graduate named Judah Folkman unexpectedly and indeed accidentally came up with a novel strategy to combat cancer. He was working at the National Naval Medical Center in Bethesda, Maryland, investigating potential substitutes for blood in transfusions for use in hospitals on aircraft carriers, when he made a momentous observation.

His experiments were set up so that blood substitutes flowed through the blood vessels of rabbit thyroid glands that were being kept alive in a glass chamber. At one point Dr. Folkman noticed that tumor cells implanted into the glands stopped growing when they were still quite small and then became dormant. However, when the dormant tumor was transplanted into a living mouse, the tumor grew rapidly. This observation led Folkman to develop a new theory about cancer. He hypothesized that in order to grow beyond 2 or 3 millimeters (slightly larger than the head of a pin), tumors had to form their own blood vessels. Folkman began to suspect that tumors recruited blood vessels by releasing factors that stimulate the sprouting of minute capillaries from nearby vessels.[2]

He pursued his idea, termed "angiogenesis," with investigations over a period of forty years in the hope of answering a number of questions: How are these new blood vessels recruited? What factors bring about their formation? Does angiogenesis enhance the tumor's capability to expand and infiltrate locally, and to spread to distant organs in the process called metastasis? Can the tumor growth be controlled by combating these factors?

As detailed by Robert Cooke in his 2001 book *Dr. Folkman's War*,[3] the successful answers to these basic questions took Folkman through diligent investigations punctuated by an astonishing series of chance observations and circumstances. Over decades, Folkman persisted in his genuinely original thinking. His concept was far in advance of technological and other scientific advances that would provide the methodology and basic knowledge essential to its proof, forcing him to await verification and to withstand ridicule, scorn, and vicious competition for grants. Looking back three decades later, Folkman would ruefully reflect: "I was too young to realize how much trouble was in store for a theory that could not be tested immediately."[4]

At the age of thirty-four, Folkman accepted the position of surgeon in chief at Boston's Children's Hospital and professor of surgery, the youngest in the history of Harvard University. Despite the honor, Folkman was considered "just a surgeon" and certainly not in the mainstream of bench research, and therefore had trouble getting published. His quest was out of step with the National Cancer Institute's march toward chemotherapy, viral and chemical causes of cancer, and tumor immunology. Besides, blood vessels were then generally considered "just plumbing," bringing nutrients in and wastes out, unlike today, when the blood vessel system is appreciated as a dynamic organ in itself.

Capillaries are essentially tiny tubes — thinner than a strand of hair — of one layer of endothelial cells (the lining cells of all vessels). In order to understand what stimulates the new growth of capillaries toward a tumor, scientists had to be able to culture endothelial cells in the laboratory. But conventional wisdom held that this was not possible. Studies by a Japanese team in the mid-1960s had been unsuccessful in culturing human endothelial cells. Then fortune intervened.

Researchers at the University of California at San Francisco, working on factors that initiate cell division (not specifically in the linings of vessels), had success in growing endothelial cells from cow aortas in culture dishes. "Serendipitously," a member of the team explained, "cow and pig turned out to be very easy endothelial cells to grow."[5] Encouraged by their success, Folkman's group was able to culture large numbers of endothelial cells from human umbilical cords and, by altering the nutrient fluid, found that they could control human blood vessel growth.[6] This five-year effort provided the keystone to the purification of factors that stimulate development of new blood vessels.

GRISTLE FOR THE MILL

Virtually all tissues in the human body require blood vessels to provide nutrients for their metabolism and function. Notably, one substance that has no blood vessels is cartilage. Cartilage is present between the segments of the spine, where the disks act as shock absorbers, and at the ends of long bones, where it provides a bearing surface to reduce the friction between moving parts. In the first week of an embryo's life, blood vessels are found in cartilage tissue to help it grow, but then a fascinating change occurs. Later in fetal development these blood vessels shrink and ultimately disappear. For Folkman, this phenomenon raised a crucial question: Is there an intrinsic property of cartilage that makes this happen? Is it a source, in other words, of an angiogenesis inhibitor?

A member of Folkman's team, Robert Langer, a chemical engineer, searched for two years for such a biochemical blocker. Eventually he worked with large amounts of cartilage, which is abundant in sharks. The skeletons and fins of sharks are made of pure cartilage, and traversing it are tough strands of large proteins. From these, Langer, in an exciting breakthrough, was able to extract a protein mixture that inhibited angiogenesis,[7] but it proved in time to be too weak to develop into clinical use.

Much to the dismay of Folkman and Langer, the large international chemical company W. R. Grace chose to exploit the finding relating proteins from shark cartilage to the inhibition of angiogenesis by marketing a food supplement product made from shark cartilage through health food stores, despite the fact that any active ingredients would not be absorbed by the body. The FDA prohibits food supplements (other examples include vitamins, minerals, fiber, garlic, and unsaturated fish oils) from making any therapeutic or prophylactic claims. The so-called remedy was promoted later by a widely distributed book, *Sharks Don't Get Cancer,*[8] further raising false hopes in cancer victims.

The Laetrile Hoax

The most notorious of commercially driven, unorthodox therapies was Laetrile, whose benefits as a cancer treatment were proclaimed by its advocates from the 1950s into the 1970s. This was abetted by the widely publicized quest of the movie star Steve McQueen, who was dying from cancer. Laetrile was a concentrated extract of a cyanide-containing substance, amygdalin, prepared from the kernels of apricot pits and bitter almond, after which amygdalin is named (*amygdale* is the Greek word for "almond"). Incredibly, its use in cancer therapy was based on the idea that cancer cells are much richer than normal cells in an enzyme that breaks down Laetrile to release toxic cyanide, which would then destroy the cancer cells. It was also trade-named "vitamin B-17," despite not being a vitamin and being nutritionally worthless. Laetrile became a billion-dollar-a-year industry, and the Laetrile cult was branded as "perhaps the most bizarre, ruthless, deceptive, misleading, and dangerous health cult to come along this century."[9] In November 1977 the FDA mailed a warning notice to every doctor and a total of nearly a million health professionals in the United States, placing them on notice that Laetrile was worthless and poisonous. The results of clinical trials conducted for the National Cancer Institute in 178 patients at four major medical centers were published in the *New England Journal of Medicine* in 1982 with the conclusion that Laetrile not only was ineffective as a treatment for cancer but frequently resulted in cyanide toxicity approaching a lethal range.[10]

A totally unexpected finding led remarkably to the isolation and purification of the first factor that specifically stimulated the growth of new blood vessels. Two biochemists in Folkman's laboratory had worked intensively on the chemically complex fluid derived from cartilaginous tumors grown in a number of rats. Expecting it to inhibit endothelial cell growth, they saw that the substance actually did the opposite: it truly stimulated growth of endothelial cells. This was dramatic support of Folkman's concept that tumors send out chemical signals that cause blood vessels to grow and extend new branches to nourish them. This natural and potent growth-stimulating protein was subsequently isolated by Napoleone Ferrara, a scientist working for a biotechnology company, and named VEGF (pronounced "VEJ-eff"), for vascular endothelial growth factor.

INSPIRATION FROM CONTAMINATION

Another major chance event occurred in November 1985 to a biologist in Folkman's laboratory: a fungus contaminated a culture dish of endothelial cells. Rather than discarding the dish, he paused to examine it and found that the endothelial cells at the site had stopped growing. (One can only marvel at the resemblance of these events to Alexander Fleming's accidental experience leading to penicillin.) Extracts of the fungus were then shown to arrest the growth of tiny blood vessels in chicken embryos. Investigation had failed to turn up an angiogenesis inhibitor, but accident now provided one. This finding resulted in the production of a compound called TNP-470. In animals, the drug blocked a wide range of tumors, slowing growth by as much as 70 percent. Clinical trials undertaken in the United States indicated by the mid-1990s that it worked best in humans in combination with chemotherapy or radiation.

"THIS GUY'S REALLY ON TO SOMETHING!"

For more than a century, physicians had puzzled over a particular phenomenon that had been observed repeatedly. The biologic aggressiveness of tumors varies enormously: some cancers, once they become evident, grow slowly, whereas others spread readily to distant organs. And some metastatic colonies grow to only a small size and then lie dormant. Yet sometimes, following surgical removal of a primary tumor, its metastasis loses any restraint and grows explosively. Could there be an innate mechanism? Could the influences of angiogenesis and possible inhibitory agents explain the phenomenon?

The solution came through a fortuitous chain of events upon which even a Hollywood screenplay could not improve. At a bustling annual meeting of the American Association for Cancer Research in 1987, Noël Bouck, a molecular biologist at the Northwestern School of Medicine in Chicago, who was suffering in a new pair of high heels, sought respite for her tired feet in an empty lecture hall. Before she realized it, the room quickly filled up with other attendees, there to hear a presentation on angiogenesis by none other than Judah Folkman. Bouck had not heard of him, since her cancer research work was directed along other lines, but she could not gracefully escape. As she listened to him, against her will, she began to think "this guy's really on to something!" She quickly became a convert to his theory on cancer. Returning to her laboratory, she redirected her own research to focus on angiogenesis and, within two years, she and her colleagues discovered the first angiogenesis-inhibiting substance produced by a tumor.[11]

Bouck's discovery pointed the way to solving the riddle of metastatic dormancy for Folkman. It propelled his thinking that the primary tumor's secretions might be harnessed as cancer drugs to suppress the growth of both the primary and small metastases.

Painstaking work over the next several years led to the identification of an inhibitor of endothelial cell growth. Calling it angiostatin (from the Greek, meaning "to stop blood vessels"), Folkman's team showed that it inhibited the development of metastases. Thirty-two years of perseverance at last not only validated and extended his original insight but also pointed the way to a new approach to cancer management. The findings were published in the fall of 1994 as the cover article in *Cell* and were widely cited as a landmark contribution.[12] One year later, his research team found another inhibitor they named endostatin that was shown to be a powerful agent against a variety of tumors implanted into mice. The news spread like wildfire, publicized in the national media in the spring and summer of 1998. The National Cancer Institute undertook the planning of clinical trials.[13]

Folkman's long years of persistence led to a new era in which dozens of pharmaceutical and biotechnology companies are now actively pursuing angiogenesis-related therapies. At least twenty-four inhibitors are being clinically tested at more than a hundred medical centers in the United States. One of these, Avastin, was approved in February 2004 for colorectal cancer and was shown to also be effective against breast cancer and lung cancer. It is given along with a generic chemotherapy drug. Avastin, marketed by the biotechnology company Genentech, is the first approved cancer drug that works by choking off the supply of blood that tumors need to grow. It blocks the protein in the body called vascular endothelial growth factor (VEGF) that promotes blood vessel growth. Avastin had sales in the United States of $676 million in its first twelve months on the market, the highest first-year sales of any cancer drug ever up to that time. As of early 2006 it appeared to be on its way to becoming the best-selling cancer drug ever.

The hope is that, as knowledge of tumor angiogenesis progresses, cancers can be detected through elevated levels of angiogenic molecules in the blood long before clinical symptoms appear. Folkman's perseverance in the face of determined skepticism and often outright opposition by colleagues was finally rewarded with acclaim and general acceptance. His work represents one of the major breakthroughs in cancer research in the twentieth century.

15

Aspirin Kills More than Pain

A vibrant new hope arose among cancer specialists in the early 1990s and continued throughout the beginning of the new millennium. It offered something perhaps better than a cure: the possibility of preventing cancers from even forming. Such promise came from a humble nonprescription drug found in medicine cabinets across America. For over a century, people have reached for aspirin to relieve headaches or back pain. In the 1990s a host of studies affirmed that aspirin might also lower the risk of developing colon cancer by as much as 40 percent.

Colorectal cancer is the third most common type of cancer and the second leading cause of cancer deaths in the U.S. In 2006 the American Cancer Society predicted, about 148,610 cases would be diagnosed in the United States and 55,170 people would die of it.

Colon cancer progresses through recognizable stages. It changes from a tiny polyp, or adenoma — a benign overgrowth of cells in the lining of the colon — to a larger polyp, a precancerous growth, and then to a cancer that infiltrates the wall of the colon. The final stage is metastasis, when the cancer spreads through the body.[1] About 20 percent of people over the age of fifty have polyps. Polyps generally grow about a millimeter a year, and once they get to 10 millimeters (1 centimeter) they have a higher chance of being malignant. It typically takes about ten years for a benign polyp to become cancerous. While most polyps never develop into cancer, this understanding of the potential sequence provides an important rationale for public health screening: if polyps can be detected and removed by colonoscopy, colon cancer can be prevented. (The value of screening by colonoscopy was underscored in the summer of 1985 when a colon polyp harboring cancer was removed from President Ronald Reagan, who, in his characteristically plainspoken way, explained to the public, "I don't have cancer. I had something inside me that had cancer. And they took it out.")

Over several years it had been incidentally noted that patients taking aspirin regularly for arthritis had a decreased incidence of colon cancer. But since this primary nonsteroidal anti-inflammatory drug (NSAID) was sold cheaply over the counter, drug companies had no incentive to invest in research to determine whether it truly had anticancer effects.

Then, in the late 1970s, a physician stumbled upon amazing evidence. Dr. William R. Waddell at the University of Colorado Health Sciences Center was treating a thirty-one-year-old woman with a rare inherited condition known as familial adenomatous polyposis, or FAP. In this disease, the colon is carpeted with hundreds — in some cases, thousands — of polyps. Since these polyps invariably undergo cancerous changes in time, doctors had surgically removed her colon when she was only twenty-three. As sometimes happens to people with this condition, within a few years, noncancerous scarlike tumors called desmoids grew in her abdomen. Waddell placed her on an anti-inflammatory drug for an upset stomach, and, much to his surprise, the tumors disappeared. Wondering if the startling effect could have been caused by the anti-inflammatory, he prescribed the drug to three of the woman's relatives who also had FAP. Amazingly, the colonic polyps either completely disappeared or shrank dramatically.

Waddell knew that he had stumbled upon a major finding, but his report was rejected by major medical journals. Finally he published it in the *Journal of Surgical Oncology* in 1983.[2]

Six years elapsed before Dr. Francis M. Giardiello, a colon cancer specialist at Johns Hopkins University School of Medicine in Baltimore, began a clinical trial testing an NSAID in twenty-two patients with FAP. His results showed a marked decrease in the number and size of colon polyps, which he described as "astounding." His 1993 report in the *New England Journal of Medicine* drew national attention, spurring basic research and clinical trials.[3]

Older NSAIDS, such as aspirin and ibuprofen, work by inhibiting a pair of enzymes called COX-1 and COX-2 (for cyclooxygenase) needed to produce hormonelike substances called prostaglandins, which cause pain and inflammation. Research disclosed that the blocking of COX-2 is what actually relieves pain, whereas blocking COX-1 prevents platelets from clumping and so can cause bleeding and stomach ulcers. In response to the new finding, pharmaceutical companies designed drugs to selectively inhibit COX-2. Pfizer came up with Celebrex, and Merck, notoriously, with Vioxx. A long-term study of the effects of Vioxx in inhibiting colon polyp growth surprisingly revealed an increased risk of heart attacks and strokes, and Merck withdrew Vioxx from the market in September 2004. Pfizer continues to market Celebrex under highly restrictive guidelines. Celebrex has received FDA approval for treatment of FAP. Although their polyps are greatly reduced in number, FAP patients still need surgery.

And then a finding provided the stunning explanation for aspirin's ability to shrink polyps: colon polyps make huge amounts of COX-2, and the increased prostaglandins can help cancers flourish. Inhibiting COX-2 not only decreases pain, it may slow cell growth, promote cell death, and prevent tumors from generating blood vessels for nourishment.[4] Furthermore, there is increasing evidence that drugs that block COX-2 might thwart not only cancers of the colon but also those of the breast, lung, bladder, skin, and esophagus. These cancers also make greatly increased quantities of COX-2.

As of the beginning of 2006 the National Cancer Institute has about forty studies under way that are designed to find out whether COX-2 drugs can treat or prevent various other forms of cancer. Although the Vioxx study on colon cancer was halted, no increased risks turned up in other studies, and they were continuing.

Thus, the chance observation of a single doctor thrust commonplace drugs to the forefront of cancer research.

16

Thalidomide

From Tragedy to Hope

A German pharmaceutical firm, Chemie Grünethal, stumbled across thalidomide in the 1950s while doing experiments aimed at developing new antibiotics. Searching for a simple, inexpensive method for manufacturing antibiotics from peptides — the bonds that hold amino acids together to form biologically active molecules — they produced a new molecule they called thalidomide. The company's research program was headed by an ex-Nazi officer, Dr. Heinrich Mückter, a medical scientist for the army of the Third Reich, who seized upon a surprising finding: the drug seemed to have a calming effect in animals. Based on that finding alone, in 1954 the company gave out free samples of the drug across Germany. Then, without having done any tests on either its safety or its effectiveness, the company began marketing thalidomide in 1957 as the first over-the-counter sedative. A British pharmaceutical firm, Distillers Company, signed up to distribute the drug in the UK the following year. By 1961, after a massive marketing campaign, thalidomide was the best-selling sedative in Germany and was being sold in forty-six countries throughout Europe, Asia, Africa, and the Americas.[1]

A well-known public health disaster ensued, as pregnant women who took it for morning sickness and as a sedative soon saw its horrifying effects. Thalidomide is a teratogen, an agent that causes malformation of the fetus. In 1960 a physician in Liverpool, England, created a registry after seeing five children born without arms. The drug was withdrawn from the German and UK markets in December 1961. Two weeks later, in a one-hundred-word letter published in the *Lancet,* Dr. William McBride, an Australian obstetrician, reported severe congenital abnormalities in a fifth of women taking thalidomide during pregnancy.[2] Infants born with severe cardiac and gastrointestinal malformations often did not survive their first year.

The striking defect that characterized many survivors was severely malformed or deficient extremities, a condition known as phocomelia. Belated experiments investigating thalidomide's toxicity showed that the drug could harm the embryos of mice, rabbits, chickens, and monkeys. Sadly, these experimental results were too late to alert doctors to the risks of the drug in pregnant women. Between 1957 and 1962, as many as twelve thousand children were affected, mostly in Europe, Canada, and Japan. Through the vigilance and courage of a legendary FDA official, Dr. Frances O. Kelsey, the drug was blocked in 1961 from being distributed in the United States. This incident stimulated more stringent drug regulation and strengthened the FDA through new laws. The thalidomide catastrophe was so shocking that it chilled the postwar euphoria of an imminent medical utopia.

Remarkably, within three months of McBride's alert, Gerard Rogerson, a doctor from Shropshire, England, raised the possibility, in another letter to the *Lancet,* that since thalidomide inhibits growing tissue in these circumstances, it might be investigated as an anticancer drug.[3] However, it was not until 1994, thirty-two years later, that Robert D'Amato, a member of the Judah Folkman laboratory, discovered that thalidomide acts as a mild inhibitor of the growth of new blood vessels critical to tumor formation. Thalidomide stunted fetal limb development by inhibiting the growth of blood vessels. Its potential as an anti-angiogenesis agent was overshadowed by the discovery of angiostatin.[4]

Thalidomide's most striking effect appears to be in multiple myeloma, an aggressive cancer of the plasma cells — derivatives of white blood cells in bone marrow — that affects 50,000 people a year in the United States.[5] The cancer can lead to bone pain, anemia, kidney failure, and recurrent infections. Thalidomide currently appears to be the first new drug in three decades to have a beneficial effect on multiple myeloma. Its mechanism of action is not clear. Although the drug blocks growth of new blood vessels, this action does not appear to explain its effect. Rather, it may have a potent influence on the immune system or may even kill myeloma cells directly. It is used along with stem cell transplants.

Thalidomide is marketed by Celgene, a New Jersey–based pharmaceutical company, and most of its sales are for multiple myeloma. Since thalidomide has been known for decades and the composition can't be patented, Celgene has patented its elaborate distribution system. This involves physician training and patient education, which includes mandatory contraceptive measures. Thalidomide has many possible side effects and complications; therefore, any real hope for its use lies in the new drugs that might be derived from it — so-called analog drugs. One analog, Revlimid, is more potent and has fewer side effects.[6]

17

A Sick Chicken Leads to the Discovery of Cancer-Accelerating Genes

In 1909 a Long Island poultry farmer brought a Plymouth Rock hen that had developed a conspicuous tumor in its right breast to Peyton Rous, a young medical researcher studying cancer in animals at the Rockefeller Institute in New York City. The farmer, referred by his local farm bureau, feared that this might indicate an infection, such as chicken cholera, that could threaten his entire flock.

Dr. Rous took a biopsy and determined it was a malignancy of the muscle known as a sarcoma. Seizing the opportunity to work with a cancer that had spontaneously arisen in an animal, he initiated research directed toward finding the mechanism through which the tumors in one animal might be transmitted to another, in the hope of preventing it from happening. His first step was to implant a small piece of the tumor into another hen of the same species. To his excitement, a sarcoma developed.

In a series of classic experiments, he ground the tumor tissues to an ultrafine consistency and passed the material through silica filters with pores so small that they would eliminate all possible cancer cells and bacteria. As the cell-free tissues still caused the disease, he assumed the cause to be a virus. (The concept of a virus at this time was poorly understood. A virus was seen as an invisible poison that seemed to have a life of its own.)[1] Rous was able to transmit the tumor through several generations of hens. He was the first to demonstrate that a virus could cause a malignant tumor.[2]

Rous published his findings in 1911 in the *Journal of Experimental Medicine,* the official publication of the Rockefeller Institute.[3] The scientific community scoffed at his theories, as the prevailing dogma held that most cancers were not contagious or the result of infection. Experience had long implicated environmental agents, such as chimney ash, but how this or other chemicals might cause malignant tumors remained a mystery. Some cancers were known to run in families, but there was no understanding of the hereditary factors. "These revolutionary findings were generally disbelieved," Rous recalled decades later in a 1966 interview. His peers believed that either the growths were not tumors or the agents were tumor cells the filters had let through. Discouraged by the lack of support, Rous soon abandoned these experiments, bitterly deeming cancer research "one of the last strongholds of metaphysics."

The idea of a viral cause for cancer was, for the most part, put into the mental attic of researchers for decades thereafter. It was not until the 1940s that the Rous sarcoma virus was identified by electron microscopy. As twentieth-century medicine progressed, researchers came to better understand viruses and even came to attribute viral origins to some tumors as well as to leukemia.[4] In 1955 the first issue of the new journal *Virology* appropriately carried an article on Rous's work on chicken sarcoma.

Fifty-five years after his breakthrough discovery, at age eighty-five, Peyton Rous was awarded the Nobel Prize in 1966. The insight — the "pregnant hint," as Walter B. Cannon liked to say — had to await a favorable intellectual climate and technologic advances.

To reproduce, cells and many viruses must copy their DNA into RNA. The Rous sarcoma virus, like some other tumor-causing viruses, was known to consist not of DNA but of RNA. As early as 1963, Howard Temin conjectured that this virus's lifestyle was to insert itself into a host cell's DNA, where it would reproduce viral RNA. This idea — revising the accepted concept of the flow of genetic information from DNA to RNA — was largely rejected until 1970 with the independent discovery by Temin and David Baltimore of reverse transcriptase, an enzyme that directs such an operation. Such viruses came to be termed "retroviruses," and it was speculated that the viral RNA in a host's cell not only facilitates the virus's replication but might remain dormant until expressed as a cancer-causing infectious agent, perhaps activated by external carcinogens such as radiation, various chemicals, or even other viruses — in effect, commandeering the cell's genetic machinery.

It was amid this scientific ferment that Bishop and Varmus came on the scene. In 1970 Harold Varmus, who was interested in the genetic basis of cancer, began working as a postdoctoral fellow in the laboratory of virologist and biologist J. Michael Bishop at the University of California at San Francisco (UCSF). Following up on Rous's newly appreciated work, Bishop and Varmus set out to study the cancer-causing gene within the Rous virus. "Harold's arrival changed my life and career," Bishop said. Their relationship soon became one of equals, and they would make their major discoveries as a team.

Viruses are subcellular forms of life — basically mere packets of genes, generally fewer than a dozen. In 1975 Bishop and Varmus made a startling observation. The virus they were studying had somehow appropriated a gene that sparked malignant growth in the host cells of chickens. The Rous sarcoma virus contains only four genes. Three of these are used to reproduce the virus; the fourth is the gene that induces the cancerous growth. They had come upon an elemental secret of cancer, a gene that can switch a cell from normal to cancerous growth.

What was the source of the gene? Does a cell itself harbor such genes? A piece of the mystery unfolded (like "lifting a corner of the veil," in Einstein's famous phrase) when they found that the gene was present in healthy chicken cells as well as infected ones. The gene was not a native component of the virus, but rather at some point during the virus's cellular passages — either as it moved from cell to cell in one host animal or as it passed from one chicken to another — it had picked up an RNA copy of a chicken gene.

The California scientists undertook a manic search for this gene, screening every species they could get their hands on. "I for one," Bishop acknowledged, "failed to foresee the eventual outcome."[5] They explored the DNA of ducks, turkeys, geese, even one of the world's largest and most primitive birds, the flightless Australian emu, and encountered startling results. All had the gene.[6] They looked at the cells of mice, cows, rabbits, and fish. The gene that was once thought of as a "chicken gene" was, in fact, present in every one of these species. Finally, they screened human DNA and were excited to find it there as well. (The researchers were startled by the vigorous skepticism of some whose incredulous response was "Are you trying to tell us that a chicken gene is also in humans?" Bishop was flabbergasted by this biological naiveté on the part of accomplished scientists. Had Darwin, he wondered, labored in vain?)

By now, reality was staring them in the face. Clearly, the gene is present throughout the evolutionary scale, having persisted for half a billion years, and belongs to the normal genome. The gene is not inherently cancer-causing but functions as a regular part of the cellular machinery, probably in connection with regulation of its development or growth. Retroviruses picked up these normal cellular genes and instigated changes that caused them to become cancerous. Such viruses thus carry a mutated cancer-promoting gene donated by a previously infected cell and may trigger the process of cancer when they infect other cells. The Rous virus, through an accident of nature during the course of viral propagation, served as the vector for the cancer-causing genetic hitchhiker, originating in a cell.

"From these findings," wrote Varmus, "we drew conclusions that seem even bolder in retrospect, knowing they are correct, than they did at the time."[7] Their new model points to the cellular origin of cancer-causing genes. When the ordinary gene malfunctions, it is transformed into an "oncogene" that can cause tumor formation. ("Onco-" stems from the Greek *onkos,* meaning "mass" or "bulk.") A whole new paradigm was identified. Oncogenes are normal cellular genes that control cellular growth and when mutated become cancer-causing.

Thus, in each cell there is a potential time bomb, latent cancer genes that may be activated immediately or many years later. Mutations may set off the time bomb, directing one renegade cell to overmultiply and become malignant. Cancer represents a series of events at the level of the genome. Susceptibility is sometimes genetic and heritable, but cancer can also arise through chance or, more often, when promoted by chemical and mutagenic carcinogens or retroviruses.

Bishop and Varmus shared the Nobel Prize in 1989, only thirteen years after their major discovery.[8] Their new model of the oncogene is a milestone that probes the ultimate origins of cancer, directing attention to a site deep within the cell. Their totally unexpected finding triggered a revolution in cancer research that continues today. It set in motion an avalanche of research on fundamental factors that govern the normal growth of cells and new insights into the complex group of diseases that we call cancer. It led to the discovery of scores of different oncogenes and their functioning through protein messengers that offer targets for drug treatment and, someday, it is anticipated, ways of preventing cancer altogether.

Like Riding a Bicycle

In 1993 Varmus described how strongly research gripped him: "It's an addiction. It's a drug. It's a craving. I have to have it."[9] A long-distance bicyclist, Varmus drew a compelling metaphor for scientific research from this activity: "Long flat intervals. Steep, sweaty, even competitive climbs. An occasional cresting of a mountain pass, with the triumphal downhill coast. Always work. Sometimes pain. Rare exhilaration. Delicious fatigue and well-earned rests."[10] After winning the Nobel Prize, Varmus became director of the National Institutes of Health for a number of years and then went on to head the Memorial Sloan-Kettering Cancer Center in New York.[11]

18

A Contaminated Vaccine Leads to Cancer-Braking Genes

On April 12, 1955, following nationwide clinical trials on 1.8 million American schoolchildren, Dr. Thomas Francis Jr. of the University of Michigan announced that the Salk vaccine against polio — at the time the leading crippler of children — was "safe, effective, and potent." Euphoria swept the country. As Richard Carter noted in his biography of Jonas Salk, "People observed moments of silence, rang bells, honked horns, blew factory whistles, fired salutes, kept their traffic lights red in brief periods of tribute, took the rest of the day off, closed their schools or convoked fervid assemblies, therein drank toasts, hugged children, attended church, smiled at strangers, forgave enemies."[1]

Up until 1961, the culture for preparing the lifesaving polio vaccine used kidney cells from rhesus monkeys from India. When a measles infection (from human contact) spread throughout the monkey colony, Merck switched to using African green monkeys.

Such an event might seem unfortunate, but what happened next is another example of how serendipity works its magic. Thorough testing uncovered the surprising finding that the first monkey species harbored a previously undetectable virus that was harmless to them but capable of killing and replicating in the cells of the second species. The conclusion was both startling and distressing: the Salk polio vaccine, in use since 1953, was badly contaminated with a virus. Had it not been for the chance measles infection and the subsequent switch to African green monkeys, the virus would not have been detected.

Alarm spread when both Merck and NIH researchers showed that the virus caused cancer when injected into newborn hamsters. The shocking reality could not be dismissed: The polio vaccine, which acted to reduce or eliminate poliovirus epidemics all over the world, was contaminated with a virus that initiated cancer in hamsters. Mass immunizations with injections of three doses of vaccine were so popular that about 450 million doses had been administered in the years 1955–59. The United States Bureau of Biologics acted swiftly to eliminate the virus from poliovirus seed stocks and grow poliovirus solely in the African green monkey cells.

Meanwhile, amid the intellectual ferment in molecular biology of this era in cancer research, Arnold Levine, a molecular biologist at Princeton University, was intrigued by this newly discovered virus shown to transform normal cells into cancer cells. Levine started by thinking that the virus's protein shell might induce an antibody in the blood. The antibody could be detected as a biomarker to indicate the presence and severity of the infection and even the results of treatment.

Levine's search led him in 1979 to unexpectedly discover a new protein in the blood. He named it simply p53 because the protein has the molecular weight of 53,000 hydrogen atoms.[2] For ten years after its discovery, Levine followed the p53 gene and its protein down a false trail, believing it to be an oncogene, a cancer-accelerating gene. He found that putting the gene into normal cells (using rat embryos) made them cancerous. In 1983 he cloned the gene for the first time and demonstrated what appeared to be its tendency to induce tumors. But Levine was wrong about the actual nature of p53.

Finally, in 1989, the real breakthrough occurred. While pursuing the genetic causes of colorectal cancer in humans, Bert Vogelstein at Johns Hopkins University found a mutation in the p53 gene.[3] As Levine reviewed the data, he experienced what can only be described as a "Eureka!" moment. In a conceptual leap, he realized that p53 was not a gene that stimulates cancer but a gene that *suppresses* tumors.[4] He had been led astray because the clones he had used in the early experiments were unwittingly composed of mutants. It was as though he was struck by lightning when he came to understand that the p53 gene had to be mutated in order to foster cancer transformation within a cell — that is, its normal function is to *inhibit* division and growth of tumor cells.

The normal p53 protein acts as an emergency brake to arrest any runaway growth of cells that may have acquired cancerous tendencies and as a damage-control specialist to induce cell death under specific circumstances, such as in the presence of DNA damage. It has been dubbed the "guardian of the genome" because of its vital biological functions. After ten years of dedicated research, "suddenly," Levine ruefully noted, "p53 became a hot ticket in cancer research." A single mutation in the 135th position of its 393 amino acids can eliminate the surveillance capability of the protein and allow a cancer to grow. In its mutated form, it is found in more than 50 percent of human cancers, including most major ones. Among the common tumors, about 70 percent of colorectal cancers, 50 percent of lung cancers, and 40 percent of breast cancers carry p53 mutations. The p53 gene is also linked to cancers of the blood and lymph nodes. Put another way, of the ten million people diagnosed with cancer each year worldwide, about half have p53 mutations in their tumors. The gene's inactivation through mutation enables tumor cells, in a famous phrase, "to reach for immortalization."

In 1993 *Science* magazine named p53 the Molecule of the Year. It has generated much enthusiasm for the development of strategies for the diagnosis, prevention, and cure of cancers.

Arnold Levine is a round-faced, high-energy, fast-thinking, fast-talking individual who commands deep affection from his laboratory team. He went on to become the president of Rockefeller University in New York. Would he consider the discovery of p53 serendipitous? "Absolutely," he frankly responds, noting the frequency of chance discoveries and citing the observations of Richard Feynman, the Nobel laureate in physics, on how scientists commonly write their articles in such a way as to "cover their tracks" when it comes to how they stumbled upon the truth.[5]

PROBING THE GENOME FOR THE RENEGADE PATHWAY

A major focus resulting from such advances in recognizing the influences of genetic mutations upon cancer is the field of targeted drug therapy. In the human cell's intricate inner circuitry are dozens of molecular chains of communication, or "signaling pathways," among various proteins. It is now understood that there are roughly ten pathways that cells use to become cancerous and that these involve a variety of crucial genetic alterations.

Research is actively directed toward developing drugs able to interfere with the molecular mechanisms that drive the growth in a tumor and that are attributable to these mutated genes. A handful have been approved for use. The best known is Gleevec, which has a dramatic effect on an uncommon kind of leukemia called chronic myelogenous leukemia (CML) and an even more rare stomach cancer, gastrointestinal stromal tumor (GIST). CML, however, is exceptional in that it is due to a single gene mutation, and the drug is able to block its specific tumor-signaling mechanism. Unfortunately, over time, further mutations circumvent the molecular signal that Gleevec blocks, building drug resistance. Most cancers are more complex, with multiple mutations that require a multipronged attack.

Picking the right targets is critical, and the most promising new diagnostic technology is the DNA microarrary, or "gene chip." Gene chips can identify mutations in particular genes, as well as monitor the activity of many genes at the same time. It is hoped that this precise molecular biology will lead to the design of a range of drugs that target highly specific signaling pathways of cancer.

Cancer was long thought of as one disease expressed in different parts of the body — breast, lung, brain — but researchers are now teasing apart the myriad genes and proteins that differentiate cancer cells, not just from healthy cells but also from each other. The aim is to classify cancers mainly by their genetic characteristics, by which pathway is deranged, not by where in the body they arise or how they look under a microscope. The ideal is to fine-tune the diagnosis and treatment of cancer for more "personalized medicine."

But knowledge of how to do that is still lacking in most cases. The magic bullet has yet to be found. Many of the new cancer drugs are targeting things in cancer cells that may or may not be driving that cancer. The drugs in use by many patients in clinical trials work for only a minority. In truth, scientists cannot really identify, in most cases, the Achilles' heel of a cancer cell or understand very well the targets at which they are shooting. And as the tumor grows, the target is ever changing. The successes claimed seem, at times, to be merely instances of drawing a bull's-eye around the spot where the arrow landed.

Even here, serendipity may be grafted upon imperfect knowledge. The story of the drug Iressa illustrates the point. Iressa blocks the activity of an enzyme that stimulates cell division. As the enzyme is overabundant in 80 percent of lung cancers, researchers hypothesized that the drug would be an effective treatment for most lung-cancer patients. They were both disappointed and perplexed when only 10 percent had a strong response. It then became evident that the patients who did respond well not only had lots of the enzyme, they also had a mutant form of it. By contrast, those patients who had either no response or a partial response had only the normal version of the enzyme.

In the recent case of a promising experimental drug targeted against melanoma, serendipity played a major role. The drug, developed by the German company Bayer and by Onyx Pharmaceuticals, a California biotechnology company, goes by the awkward code name BAY 43-9006. The drug is the first to block a protein called RAF, one of a family of enzymes called kinases, which relay a cascade of signals leading to cell growth. The drug is most effective in kidney cancer, but *not* mainly by blocking RAF. Thanks to "dumb luck," in the words of Dr. Frank McCormick, founder of Onyx and director of the cancer center at the University of California at San Francisco, it turns out that the drug also blocks a protein involved in the flow of blood to the tumor. In another discovery unforeseen by Bayer and Onyx, scientists in Britain found that RAF was mutated in about 70 percent of melanomas. When combined with chemotherapy, BAY 43-9006 shrunk melanomas in seven of the first fourteen patients.[6]

In his 1998 book *One Renegade Cell,* Robert Weinberg, a pioneering scientist at MIT, put these discoveries and potentials into perspective:

> Until recently, the strategies used to find the genes and proteins that control the life of the cell have depended on ad hoc solutions to formidable experimental problems, cobbled together by biologists who lacked better alternatives. Time and again, serendipitous discoveries have allowed new pieces to be placed in the large puzzle. . . .
>
> Discoveries of critically important genes often have depended on little more than dumb luck. . . .
>
> Those who engineer these successes [technologies, gene mapping of cancers, targeted drug therapies] will view the discoveries of the last quarter of the twentieth century as little more than historical curiosities. . . . We have moved from substantial ignorance to deep insight.[7]

Nevertheless, can one doubt that serendipity will continue to play a role?

19

From Where It All Stems

In the late 1950s treatment of leukemias and other cancers of the blood involved radiation to kill cancerous blood cells and subsequent bone marrow transplantation to replace them. It was known that bone marrow transplants (a procedure then in its infancy) replenished the essential cells of the blood system, but there was no understanding of the source of these cells.

The three different types of blood cells had been identified in the late nineteenth century: red blood cells, which transport oxygen throughout the body; white blood cells, which protect against germs; and platelets, which keep us from bleeding. It was known that they came from inside the bones — in the bone marrow — where trillions are made every day. Immature forms could even be identified. But how they were produced was a great mystery.

Meanwhile, outside the laboratories, the Cold War was raging, and many people feared a nuclear war. Researchers were looking for ways to treat people, most likely military personnel, who might be exposed to whole-body irradiation from nuclear weapons. Ernest McCulloch, a physician, and James Till, a biophysicist, at the Ontario Cancer Institute in Toronto, were part of this effort. They set out to measure the radiation sensitivity of bone marrow cells and to determine how many bone marrow cells were needed to restore blood cell production in irradiated mice.

The design of their study called for a group of the irradiated mice to be killed ten days after transplantation of various numbers of marrow cells. Normally, on a weekday, a lab technician would be the one to "dispatch" the animals, cut them open, collect the specimens, and hand them over to the scientists. As luck would have it, in 1959 day 10 fell on a Sunday. McCulloch himself, not a laboratory technician, came in that afternoon to do the dirty work. When he opened up his irradiated mice that Sunday, McCulloch was intrigued to find nodules on the surface of their spleens.[1]

This observation, which easily could have been overlooked, revealed changes that would have remained undetected within the bone marrow except under the most meticulous examination. The nodules were foci of new blood cell formation. Blood cell formation in the bone marrow, sequestered deep within the cavities of the bone, occurs among its interweaving lattice of supporting connective tissue only at specific scattered sites. In the mouse, however, blood cell formation occurs also in the spleen. The observation itself was thus not very surprising. What was surprising was that individual nodules contained dividing cells, some of which were differentiating into the three main types of blood cells: red cells, white cells, and platelets. This finding meant that something capable of making all blood cell types, a blood-forming stem cell, was trapped in the spleen. Moreover, it was as rare as Waldo among the transplanted marrow cells, since as many as 10,000 of these had to be injected for each nodule observed.

Establishing an accurate ratio between the number of marrow cells transplanted and the number of nodules observed suggested the possibility of a single formative source.[2] But how to prove it? The nodules had newly formed blood cells not only of the various types but also of varying stages of maturity. Where did these colonies stem from? McCulloch and Till would have to work backwards, like tracing a family tree back to the primary ancestor. And then fortune smiled. With the assistance of Andrew Becker, a graduate student, they devised an ingenious and elegant experiment using mutated cells that arose from irradiated bone marrow. By documenting that the blood cells in each spleen nodule bore a particular genetic signature, the investigators proved that diverse blood cells come from individual stem cells.[3]

Since there is no specific feature that allows stem cells to be identified by microscopy, this discovery was a huge conceptual achievement. McCulloch proudly called this "hematology without a microscope."[4]

The concept of stem cells had been around since at least the early 1900s, but efforts to find them using the light microscope and histological stains had failed because the stem cells were so few in number and indistinguishable from other blood cells by appearance alone. Said Till, "Our work changed the emphasis from . . . form to . . . function. . . . We stumbled across a way of looking at the developmental potential of stem cells by looking at the descendants they could give rise to."[5]

McCulloch and Till established the two main properties of stem cells: self-renewal and differentiation into specialized cells that have limited life spans. Genes in the stem cells and factors in the tissue environment are important in promoting normal stem cell duplication and specialization.

When a stem cell divides, each new cell has the potential to either renew itself for long periods through continued cell division or, under certain physiologic or experimental conditions, become another type of cell with a more specialized function. Unlike embryonic stem cells, which are defined by their origin in a three-to-five-day-old embryo (called a blastocyst), adult stem cells in mature tissues are of unknown origin. The adult tissues in humans that are now believed to contain stem cells include bone marrow, blood, brain, blood vessels, skeletal muscle, skin, and liver. There is a very small number of stem cells in each tissue, where they may remain dormant (nondividing) for many years until they are activated by disease or tissue injury. The discovery of stem cells has opened up the exciting possibility of cell-based therapies to treat disease. Some examples of potential treatments include replacing the dopamine-producing cells in the brains of Parkinson's patients, developing insulin-producing cells for Type 1 diabetes, and repairing damaged heart muscle following a heart attack with cardiac muscle cells.

But there is a dark side to stem cells, one that offers a fascinating line of study for researchers and a ray of hope for the human race.

Growing evidence suggests that stem cells are also the wellspring of cancer. A research group in Madrid reports that stem cells taken from adults can turn cancerous if they are allowed to multiply for too long outside the body.[6]

A better understanding of the mechanism whereby undifferentiated stem cells become differentiated may yield information on how cancer arises through abnormal cell division and differentiation. Cancer stem cells might continually replenish tumors. Indeed, this might explain why treatments that can shrink tumors don't always cure the disease. Perhaps the treatments are less effective at killing the cancer stem cells. Cancer stem cells use a variety of biochemical pathways to survive. Identification of the biochemical switches they use to reproduce could provide the target for a silver-bullet drug.

In 2005 McCulloch and Till received the Albert Lasker Award for Basic Medical Research. In an article published a month later, they paid tribute to the vastly undercredited elements of both luck and astute observation: "We weren't deliberately seeking such cells, but thanks to a felicitous observation, we did stumble upon them. Our experience provides yet another case study of both the value of fundamental research and the importance of serendipity in scientific research."[7] Reflecting on his work in an interview in 2006, McCulloch said he had always thought that the idea of "the scientific process" was overblown.[8] He knows that "typically a successful scientist may start with an experimental design but then makes an unexpected observation that leads a prepared mind to follow a chance event."[9]

20

The Industrialization of Research and the War on Cancer

The word "cancer" derives from the Greek *karkinoma* and the Latin *cancer*, for "crab," suggested by the long, distended veins coursing from lumps in the breast. During the first half of the twentieth century, "cancer" was an almost unmentionable word. Public figures were never described as dying from cancer, and obituaries obliquely referred to a "prolonged illness." An overpowering mood of powerlessness and fear, mixed with a sense of shame attached to the suffering caused by the disease, fueled a widespread cancerphobia.

The ramifications of this pervasive fear are graphically described by James Patterson in his book *The Dread Disease,* a cultural history of cancer in the United States.[1] Patterson includes an experience recounted by George Crile Jr., a surgeon at the Cleveland Clinic. A seventy-five-year-old woman who suddenly could not speak was referred to Dr. Crile by her family physician, who suspected that thyroid cancer had spread to her brain. After a series of tests, Crile informed the family that the disease was far worse than cancer, which he might have been able to treat. "There is nothing to be done," he told her adult children. "Your mother has suffered a stroke from a broken blood vessel. The brain is irreparably damaged. There is no operation or treatment that can help." After making this difficult speech, Crile was taken aback by what came next.

The oldest daughter leaned forward, tense, and with a quaver in her voice, asked,

"Did you find cancer?"

"There was no cancer," Crile replied.

"Thank God!" the family exclaimed.[2]

By midcentury the stage was set for a divisive conflict over the nature of cancer research that would persist for decades. It involved a fundamental distinction between a *targeted categorical disease approach* and a *basic research approach*. The former requires a coordinated attack on a particular disease, whereas the second relies upon independent scientific studies of the processes of the human body. On one side of the conflict were those favoring centralized management and specifically targeted research. Even though such an approach typically involved a needle-in-the-haystack search for a useful agent, supporters were confident that triumphs were within reach. Others — particularly those in academic medicine — were highly skeptical of the role of "regimented direction" of biological research, fearing that it would lead to a certain loss of serendipity and would favor a technological approach over independent, curiosity-driven basic research.[3] As early as 1945 the medical advisory committee reporting to the federal government on a postwar program for scientific research emphasized the frequently *unexpected* nature of discoveries:

> Discoveries in medicine have often come from the most remote and unexpected fields of science in the past; and it is probable that this will be equally true in the future. It is not unlikely that significant progress in the treatment of cardiovascular disease, kidney disease, cancer, and other refractory conditions will be made, perhaps unexpectedly, as the result of fundamental discoveries in fields unrelated to these diseases. . . . Discovery cannot be achieved by directive. Further progress requires that the entire field of medicine and the underlying sciences of biochemistry, physiology, pharmacology, bacteriology, pathology, parasitology, etc., be developed impartially.[4]

Their statement "discovery cannot be achieved by directive" would prove to be sadly prophetic.

At this time there arose two figures who would play a commanding role in biomedical research in the United States, Albert Lasker, a wealthy advertising magnate, and his wife, Mary. As president of Lord & Thomas Company, the world's largest advertising firm, Albert had ironically pioneered for the Lucky Strike account the promotion of smoking by women in the 1920s and 1930s with the slogan "Reach for a Lucky instead of a sweet." In 1942 the couple established the Albert and Mary Lasker Foundation to support biomedical research. It was based on organizing a small lobby of key professional and legislative people that would persuade Congress to allocate funds for a national effort. In 1946 the Lasker Foundation began giving out the highly prestigious Albert Lasker Awards in Medical Research. The most prominent spokesman in cancer was Dr. Sidney Farber, the scientific director of the Children's Cancer Research Foundation in Boston and the discoverer in the late 1940s of the success of the antifolic acids in combating acute childhood leukemia.

Albert Lasker's death from colon cancer in 1952 reinforced his wife's determination that medical research provide answers, and Mary became an increasingly influential figure. She had energized a small, rather inert American Society for the Control of Cancer founded in 1913 by an elite group of surgeons and gynecologists into becoming the massive publicity and fund-raising machinery of the American Cancer Society and added businessmen to its board.

The U.S. government had undertaken support of cancer research in the late 1930s with the establishment of the National Cancer Institute (NCI). Even at this early stage, skepticism about the role of centralized management with "regimented direction" arose. Several doubted the wisdom of large-scale federal support for biomedical research. During the war years, it was a somnolent enterprise until funds from the defunct Office of Scientific Research and Development were transferred to the National Institutes of Health (NIH).

Shortly after the war, Alfred P. Sloan Jr., chairman of the board of General Motors, and Charles Kettering, GM's vice president and research director, gave $4 million to establish at Memorial Hospital in New York the Sloan-Kettering Institute of Cancer Research. It became the largest private cancer research facility in the world. Cornelius Rhoads, its director, was known for his energy, initiative, and tenacity in the face of great challenges. He was driven by two compelling beliefs: chemical agents could be found that would stop cancer cells from dividing, and research could be organized on the principles of the leading industrial labs of the day, like Bell Labs, which linked scientific inquiry with efficiency and high output. (The industrialists who sat on his board encouraged him a great deal in the latter belief.)

Scientists at Memorial Sloan-Kettering tested more than 1,500 forms of nitrogen mustard and other alkylating agents on animals and occasionally on patients between 1946 and 1950. Basic and clinical research was so industrialized that by 1955 about 20,000 chemical agents had been tested.[5] Rhoads was featured in a *Time* cover story in 1949 as a confident investigator-administrator and pictured with a crew cut and a white lab coat against a background illustration of a sword with serpents wrapped about its hilt (the symbol of the American Cancer Society) plunging through a crab (cancer).

Rhoads zealously stated, "Some authorities think that we cannot solve the cancer problem until we have made a great basic, unexpected discovery, perhaps in some unrelated field. I disagree. I think we know enough to go ahead now and make a frontal attack with all our forces."[6] Some may have found his disdain for "unexpected discovery, perhaps in some unrelated field" surprising in a man who had been involved with the serendipitous medical benefits of the military disaster at Bari just a few years earlier.[7]

By 1955, under lobbying pressures spearheaded by Rhoads and Sidney Farber, the center of drug testing had shifted from Sloan-Kettering Institute to the NCI. The NCI was the jewel in the crown of the federally supported NIH in Bethesda, Maryland, which funds most biomedical research. The hopes raised by the enormous potential breakthrough that the two drugs, nitrogen mustard and methotrexate, could bring about cannot be exaggerated: it was believed that what sulfa drugs and penicillin had done for the scourge of infectious disease, chemotherapy would do for cancer.

In 1953 an article in *Look* magazine predicted that cancer would be conquered within a decade. *Newsweek* anticipated that a vaccine against cancer would shortly be developed. In 1958 *Reader's Digest* wrote approvingly of experiments being conducted on Ohio prisoners who had volunteered for injection of cancerous cells to test their immunity. The article considered this experiment a "history-making research project that could lead to a breakthrough in the struggles to understand this dread killer."[8] In the same year, *Life* quoted John "Rod" Heller, then the head of NCI, as saying, "I've spent many years in cancer research. Now I believe that I will see the end of it. We are on the verge of breakthroughs."

By 1957 the chemotherapy program was immense, taking up nearly half of the NCI's budget and testing thousands of chemicals each year.[9] Yet even the director of the NCI's drug development program, Kenneth Endicott, was highly skeptical of the approach. "I thought it was inopportune, that we didn't have the necessary information to engineer a program, that it was premature, and well, it just had no intellectual appeal to me whatever," he said a decade later.[10]

In 1971 the U.S. government finally launched an all-out "war on cancer." In his State of the Union address in January 1971, President Richard Nixon declared: "The time has come in America when the same kind of concerted effort that split the atom and took man to the moon should be turned toward conquering this dread disease. Let us make a total national commitment to achieve this goal."

As the country debated a bill known as the National Cancer Act, the air was filled with feverish excitement and heady optimism. Popular magazines again trumpeted the imminent conquest of cancer. However, some members of the committee of the Institute of Medicine, a part of the National Academy of Sciences, which was asked by the NCI to review the cancer plan envisioned by the act, expressed concern regarding the centralization of planning of research and that "the lines of research . . . could turn out to be the wrong leads." The plan fails, the reviewers said in their confidential report, because

> It leaves the impression that all shots can be called from
> a national headquarters; that all, or nearly all, of the

really important ideas are already in hand, and that given the right kind of administration and organization, the hard problems can be solved. It fails to allow for the surprises which must surely lie ahead if we are really going to gain an understanding of cancer.[11]

The vigorous lobbying efforts, testimonies, and debates in congressional hearings in the eleven months before the act's final passage remain highly instructive today and provide a template for the differing perspectives and dilemmas regarding the nature of creative discoveries. During the hearings, long-simmering opposing viewpoints were sharpened.[12] A war within the war flared up over how much centralized control the proposed national cancer authority would exert. What is the best way to direct the path of meaningful research? Is it accomplished best under central direction, or should it be left more independently to curiosity-driven scientists? What role does serendipity play in medical discoveries? Can discoveries be anticipated and therefore targeted for funding and clinical trials, or do they often arise unexpectedly from basic research in a variety of disciplines? How important is the revelation of a fundamental finding in one field — chemistry, physics, pharmacology, physiology — to another? Is the term "war" in this context even appropriate, given that cancer is not a single entity and has multiple causes and manifestations?

In numerous testimonies before congressional committees, Sidney Farber of Boston Children's Hospital had long served as a forceful spokesman for clinically targeted research reflecting, it was believed, the immediate needs of patients. Impressive and dignified, he spoke deliberately and had a flair for the dramatic anecdote. Farber argued that it was not always necessary to know the cause of a disease in order to cure it or even to prevent it. His stand was vigorous: "The history of medicine is replete with examples of cures obtained years, decades, and even centuries before the mechanism of action was understood for these cures — from vaccination, to digitalis, to aspirin."[13]

Sol Spiegelman, director of Columbia University's Institute of Cancer Research, stated that "an all-out effort [to cure cancer] at this time would be like trying to land a man on the moon without knowing Newton's law of gravity." The emphasis, many declared, should clearly be on basic biological research rather than on centralized direction, which inhibits creative minds. A particularly compelling case was made by Francis Moore, M.D., professor of surgery at Harvard Medical School and surgeon in chief at Peter Bent Brigham Hospital in Boston. Moore's opinion was diametrically opposed to Farber's. Moore's logical review of medical history at the hearings pointed out that if there had been a diabetes institute in the late nineteenth century, it would not have supported the work of Paul Langerhans on the pancreas, work that ultimately led to the discovery of insulin.[14] At the time, the relation between the pancreas, insulin, and diabetes was simply not known. Similarly, he noted, a government institute on polio would likely have not supported the work of Dr. John Enders in the late 1940s in attempting to grow the mumps virus, work that found the method that ultimately proved critical to producing the polio vaccine. Moore argued that many medical advances come from creative research by "often young people, often unheard of people" in universities.[15]

Nevertheless, the act was passed on December 23, 1971, signed into law by President Nixon, and hailed as a bountiful Christmas gift for the American people.[16] When the dust settled, the National Cancer Institute was given substantial autonomy and markedly increased funding.

The scope of the NCI's enterprise was breathtaking. By 1960 the program had begun screening 30,000 compounds annually for antitumor activity. These required almost 300,000 tests using three different types of mouse tumors. There were two major sources of chemicals: random "off the shelf" selections and active programs of synthesis. By 1970 the potential of some 400,000 drugs had been explored.

Over the years, the American Cancer Society tried to sustain the public's hope for continuous progress, often using phrases such as "on the threshold of a golden age of medicine" and "a cure is just around the corner." At times the evangelism even became reckless. In 1977 a former president of the American Cancer Society defended the society's aggressive promotion of X-ray mammography by declaring that, even though there might be a potential of increasing a woman's risk of breast cancer in the future, "there's also an excellent chance that by that time science will have learned to control the disease."[17]

In time, the odds against finding the needle in the haystack proved overwhelming. Only about 1 in 2,000 of those compounds screened for activity would be selected for clinical trials; the odds of a compound making it as a commercially available anticancer drug was at best 1 in 10,000.[18] The NCI conducted extensive national clinical trials. Phase I, generally involving a few dozen volunteers, determined the drug's toxicity and maximum tolerated dose. Phase II, involving perhaps a few hundred patients, assessed any benefit and established the schedule of treatment. Phase III studies, typically involving thousands of patients, compared the treatment's safety and benefits with standard therapies. If the results were favorable, the drug would be submitted to the FDA for approval.[19]

Despite this Herculean effort and enormous expense, only a few drugs for the treatment of cancer were found through NCI's centrally directed, targeted program. These attacked the fast-growing, disseminated forms of cancer, such as leukemias and lymphomas, found primarily in young individuals.

Over a twenty-year period of screening more than 114,000 plant extracts, representing about 15,000 species, not a single plant-based anticancer drug reached approved status. This failure stands in stark contrast to the discovery in the late 1950s of a major group of plant-derived cancer drugs, the *Vinca* alkaloids — a discovery that came about by chance, not through directed research.

By 1980 C. Gordon Zubrod, director of the medical cancer program at the NIH Clinical Center, was forced to pay homage to the contribution of serendipity. He woefully concluded: "For the most part, the original discoverer of biological activity [of cancer drugs] was not aiming directly at discovering drugs for treatment of cancer. The original rationale for the biological study of a drug that later is proved to cure cancer may come from almost anywhere — basic research, targeted research, university, industry — but mostly from areas other than cancer. So one cannot program the initial discovery of drugs."[20]

Michael Shimkin, a longtime NCI physician-investigator, in 1983 astutely likened the national cancer activities to the Strategic Air Command: "One third of the force is always in the air, coming and going to project site visits, reviews, conferences, and meetings in which the same people talk on the same subjects to the same audience. . . . It is time to try longer investments in smaller laboratories, and leave them alone to reach for the brass ring."[21]

In 1986 a report in the *New England Journal of Medicine* shattered the myth of progress and boldly declared that "we are losing the war against cancer."[22] This assessment was based on hard statistics. Combination chemotherapy had yielded amazing success against childhood leukemia, but the overall number of these cases was too small to have had any real impact on national cancer statistics. By this time, the government had spent upward of $8 billion through a vast national superstructure. One critic estimated that a *trillion* dollars had been spent on cancer treatment and research since the beginning of Nixon's war on cancer.[23]

In 1997 biostatisticians John C. Bailar III and Heather Gornik reported an updated and more sophisticated analysis entitled "Cancer Undefeated" in the same journal.[24] This showed a 6 percent *increase* in age-adjusted mortality due to cancer from 1970 through 1994. The best strategy against cancer, they urged, was prevention, which could be promoted by convincing people to quit smoking and by urging doctors to use screening methods, such as mammography, Pap smears, and colonoscopy, to detect cancers while they were still curable.

Only Taxol, originally isolated from the bark of a yew found in the Pacific Northwest, finally achieved FDA approval in 1992 for use in the treatment of breast, lung, and ovarian cancers.[25] Indeed, Taxol was the greatest treasure find in the government's thirty-year screening program. Marketed by Bristol-Myers, it proved to be the first billion-dollar blockbuster drug in the history of cancer chemotherapy.

Overall, the few dramatic increases in cure rates and patient longevity occurred in a handful of less common malignancies — including Hodgkin's lymphoma, some leukemias, cancers of the testes and thyroid gland, and most childhood cancers. Sadly, these successes have not had a major impact upon the inexorable prevalence of cancer in our society. And the fact cannot be escaped that many of these successes came about in the early days of the war on cancer, or before war was even declared.

The Pap Smear

In 1928 the Greek-American pathologist George Papanicolaou, while investigating the reproductive cycle by using vaginal smears obtained from human and animal subjects, observed the presence of cancer cells in people who were not known to be sick. That a cancer could be found this early, through the examination of cells (cytology), was a totally revolutionary idea. Its elaboration is a classic example of medical serendipity by which an important discovery about cancer came from research in an unrelated area. (Papanicolaou said the first observation of cancer cells in a smear of the uterine cervix was "one of the most thrilling experiences" of his entire scientific career.)

Papanicolaou went on to develop the famous Pap smear test — so called from the first three letters of his name — which was not routinely adopted until the early 1940s. Cervical cancer, once the leading cancer killer among American women, is now, thanks to Pap smears, a distant seventh.

In June 2006 the FDA approved a new vaccine — Gardasil, made by Merck — which had proved highly effective at preventing cervical cancer in women. It works by providing immunity to two types of a sexually transmitted virus (human papillomaviruses) that cause 70 percent of the cases of the disease.

By the mid-1990s the National Cancer Institute redirected money to fundamental research into molecular processes, the relationship between diet and cancer, and prevention. The statistics, however, remain alarming. In 2004 almost 564,000 Americans died of cancer, and nearly 1.4 million new cases are expected to be diagnosed in 2006. In recent years, cancer has surpassed heart disease as the top killer of Americans younger than 85. In people 45 to 64 years old, cancer causes more deaths than the next three causes (heart disease, accidents, and stroke) put together. Such dismal statistics are all the more upsetting when compared with those for heart disease and stroke. Age-adjusted death rates for these diseases, which strike mostly older people, have been slashed in the United States by an extraordinary 59 percent and 69 percent, respectively, during the last five decades.

Following the testing of nearly half a million drugs, the number of useful anticancer agents remains disappointingly small. Expressions of discontent with the methodology of research and the appalling paucity of results were, over the years, largely restricted to the professional literature. However, in 2001 they broke through to the popular media. In an impassioned article in the *New Yorker* magazine entitled "The Thirty Years' War: Have We Been Fighting Cancer the Wrong Way?" Jerome Groopman, a respected clinical oncologist and cancer researcher at Harvard Medical School in Boston, fired a devastating broadside. "The war on cancer," he wrote, "turned out to be profoundly misconceived — both in its rhetoric and in its execution. The high expectations of the early seventies seem almost willfully naïve." Regarding many of the three-phased clinical trials, with their toxic effects, he marveled at "how little scientific basis there was and how much sensationalism surrounded them." Groopman concluded that hope for progress resided in the "uncertainty inherent in scientific discovery."[26]

Groopman's pull-no-punches article ripped off the curtain to expose the failings of those claiming to be wizards. The sad reality is that chemotherapy drugs are primitive, blunt-edged weapons of mass destruction. They interfere with the cellular process involved in the rapid divisions of cancer cells, but also act on other body cells with high metabolic rates and high cell turnover, including the bone marrow, intestinal wall, and hair follicles. Chemotherapy's ill effects on the bone marrow — the very foundation of the body's immune system — often result in blood-deficiency diseases (diminished white blood cells, decreased platelets, aplastic anemia) that lead to overwhelming infections. Furthermore, chemotherapy is of little or no use against many common cancers, including those that invade the gastrointestinal and respiratory tracts. It was never a very good answer to the problem of cancer.

Three years later, in 2004, *Fortune* magazine published a cover story lamenting the sad fact that "we're losing the war on cancer" due to a misguided approach. The emphasis, it asserted, should be on prevention and early detection.[27]

Interferon

Ironically, the NIH's effort in the 1960s and 1970s to find a cancer virus led to a discovery that was of great value years later in another field. By the 1980s this effort had run its course. Only limited success had been achieved, most notably the discovery of virus particles in cells of lymphoma patients in Africa — the Epstein-Barr virus, the causative agent of Burkitt's lymphoma — and of an antiviral protein given off by cells when they were attacked by viruses. This protein, dubbed interferon, became an overhyped biotechnology product. But the viral work brought scientists a wealth of knowledge about retroviruses, which can copy their own RNA information into the DNA of the host cell. When HIV arrived in the United States, scientists were able to promptly identify the retrovirus, work out its biomechanisms of growth and replication, and target its cellular weak spots.

The truth remains that over the course of the twentieth century, the greatest gains in the battle against cancer came from independent research that was not under any sort of centralized direction and that did not have vast resources at its disposal. As we have seen, such research led to momentous chance discoveries in cancer chemotherapy and a greater understanding of the mechanisms of the disease that have resulted in exciting new therapeutic approaches.

21

Lessons Learned

What has the saga of the discovery of anticancer drugs taught us?

The reigning image for all diseases in Western culture is that of war. It is in the area of cancer that military metaphors are most often applied. We speak of how the disease "strikes" or "attacks," setting in motion the "body's defenses." We speak of "struggle" and "resistance" in the "battle against cancer." Modern medicine hunts for "magic bullets" in the "crusade" or "war" against this malignant foe. In exploring the cultural mythologizing of disease in her book *Illness as Metaphor*, writer Susan Sontag condemns the militaristic language around cancer as simultaneously marginalizing the sick and holding them responsible for their condition. Yet one overwhelming fact persists: cancer will kill more Americans in the next fourteen months than perished in all the wars the nation has ever fought — *combined*.

Biomilitarism, however, oversimplifies the problem by clouding the fact that cancer is a catchall term for a class of more than 110 separate malignancies. One military metaphor labeling America's cancer campaign as a "medical Vietnam" proved to be woefully apt. Francis Moore's congressional testimony in 1971 and the Institute of Medicine's analysis on "the war against cancer" were sadly prophetic. No clinically important scientific breakthrough, Moore argued, had ever resulted from such directed funding.

It is clear that programmatic research dictates unhealthy rigidity that inhibits creative minds and constrains a critical feature of the discovery process: serendipity. The more narrowly a scientific goal is defined, the more the discovery process is obstructed. Scientific discoveries so often happen when they are least expected.

Cancer has for too long been treated with some combination of surgery, radiation, and chemotherapy in a program derided by some as "slash, burn, poison." In this war on cancer, in which all cancers were considered the same, this was a scorched-earth policy. A new understanding of this enormously complicated disease was desperately needed. Fortunately, by the 1970s, molecular biologists made revolutionary insights into the characteristics within the cell that lead to the initiation of cancer. But these resulted from determined efforts in the fields of cell biochemistry, tumor virology, and medical genetics.

There are more than ten thousand billion cells in the human body. In most tissues, cells are continually dying and being replaced, and control of the equilibrium is critical in maintaining the normal architecture. The complex biological blueprints for the behavior of cells are controlled by their genes. Humans have 20,000 to 25,000 distinct individual genes, and genetic information is carried in DNA molecules. The human genome is estimated to include 3.08 billion units of DNA. The complexity of what the cell's machinery can generate is indicated by the fact that a few thousand genes can make trillions of combinations of proteins. Cells multiply ten million billion times during the course of each human lifetime. Mutations, random changes in the design of a gene's DNA structure, occur constantly, because there is an intrinsic error rate when DNA is copied. In extreme errors, the cell will self-destruct; otherwise, the aberrant cell survives.

Starting in the last decades of the twentieth century, sophisticated genetics and molecular biology have been aimed toward a more precise understanding of the cell's mechanisms. Yet, even here, chance has continued to be a big factor. Surprising discoveries led to uncovering cancer-inducing genes (oncogenes) and tumor-suppressing genes, both of which are normal cellular genes that, when mutated, can induce a biological effect that predisposes the cell to cancer development. A search for blood substitutes led to anti-angiogenesis drugs. Veterinary medicine led to oncogenes and vaccine preparations to tumor-suppressor genes. In one of the greatest serendipitous discoveries of modern medicine, stem cells were stumbled upon during research on radiation effects on the blood.

Experience has clearly shown that major cancer drugs have been discovered by independent, thoughtful, and self-motivated researchers — the cancer war's "guerrillas," to use the reigning metaphor — from unexpected sources: from chemical warfare (nitrogen mustard), nutritional research (methotrexate), medicinal folklore (the vinca alkaloids), bacteriologic research (cisplatin), biochemistry research (sex hormones), blood storage research (angiogenic inhibitors), clinical observations (COX-2 inhibitors), and embryology (thalidomide).

Part III

A Quivering Quartz String Penetrates the Mystery of the Heart

Experimental ideas are often born by chance,
with the help of some casual observation.
Nothing is more common, and this is really
the simplest way of beginning a piece of scientific work.

— CLAUDE BERNARD, 1865

22

An Unexpected Phenomenon

It's Electric!

As in the great age of exploration, when the coastal contours of a continent were first outlined and thereafter its rivers, mountains, and valleys penetrated, so was the terra incognita of the heart and the circulatory system explored. Long considered a "sacred organ," the heart was conventionally viewed as the body's spiritual center. Simply to consider the emotions attributed to the organ reveals its profound importance in man's mind: heartache, heartbroken, heartfelt, heartless, heartsick, heartwarming. Researchers were traditionally fearful of the potential dire consequences of undertaking investigations directed to unveil the heart's mysteries.

The first to ponder them was Galen, the second-century Greek physician. Forbidden to dissect human cadavers by Church authorities, he could only guess at how the blood and the heart work in tandem. He theorized that the blood passes through invisible pores in the heart's septum (the tissue that divides the heart's ventricles), with the heart acting as a massive pump. Central to his beliefs was that the heart impressed the blood with vital spirits. It was not until 1628 that William Harvey in England discovered that the heart repeatedly pumps the blood through a closed system of arteries and veins. Because blood coursed in a circle in the body, it was referred to as circulation or the circulatory system. But for hundreds of years thereafter, that was all doctors knew. Even three centuries after Harvey's discovery, most doctors could not accurately diagnose a heart condition.

In 1856 two German scientists, Albert von Kölliker and Heinrich Müller, accidentally discovered electrical activity in cardiac muscle.[1] Working with frog preparations, they had excised a leg nerve with its attached muscle and had just opened the chest wall of a second frog when they were called out of the laboratory. Upon returning, they encountered an astonishing and wholly unexpected phenomenon. The muscle of the excised preparation from the first frog was contracting along with the heartbeat of the second frog. The cause was evident: they had inadvertently dropped the first frog's excised nerve end on the exposed surface of the heart of the second frog. They accidentally discovered that the heart produces an electrical current with each beat.[2]

It was Augustus D. Waller, a British physiologist working at London's University College and at St. Mary's Hospital Medical School, who first recorded the heart's electrical deflections. Waller was an unconventional and somewhat flamboyant figure, "endowed with outspoken friends and enemies." In contrast to other physicians of the time, who always dressed somberly in morning coats and silk hats, he sported a double-breasted blue jacket and a gray beard that made him look exactly like a skipper in the merchant navy. Photographs show him habitually smoking a cigar and invariably accompanied by his bulldog, Jimmy, who not only was his constant companion but also served in Waller's initial animal experiments.

Waller showed that the currents of the heart could be studied without opening the thorax (the upper part of the torso) simply by putting electrodes, to which a capillary electrometer was attached, on the animal's body. In 1887 he was able to record a human heart: "I dipped my right hand and left foot into a couple of basins of salt solution which were connected with the two poles of the electrometer and at once had the pleasure of seeing the mercury column pulsate with the pulsation of the heart."[3]

The instrument Waller used, the Lippman capillary electrometer devised fifteen years earlier in Berlin, consisted of a column of mercury within a very thin glass tube (capillary), one end of which was immersed in a solution of dilute sulfuric acid. The electrical discharges from the heart caused a fluctuation of the mercury level due to the change in surface tension, and its changing level was recorded on sensitized paper. Inexplicably, however, Waller never recognized the enormous potential of his observations for clinical use. Several years later he would admit, with more than a touch of remorse, "I certainly had no idea that the electrical signals of the heart's actions could ever be utilized for clinical investigation."[4]

"The Secrets of the Heart"

This lost opportunity was seized by Willem Einthoven, a recent medical school graduate of the University of Utrecht in the Netherlands. Einthoven (pronounced Aynt-ho-fen) was born in 1860 in Java, one of the major Indonesian islands, then part of the Dutch East Indies. The name was Dutch, but the family traced its ancestry to Jews who had migrated to the Netherlands from Spain during the Inquisition. After the death of his father, a military physician, Willem's mother brought the family back to Utrecht.

Appointed chairman of the physiology department at the University of Leiden in 1886 at the age of twenty-six, Einthoven noted the disadvantages of the recording system that made measurements erratic and inaccurate: the friction and inertia of the mercury column, its insensitivity, and the arduous task of making long and detailed mathematical analysis and reconstruction into a curve that he called an electrocardiogram. Most notably, the unit was extremely sensitive to vibration. Large horse-drawn carts clattering over the cobblestones in the street outside his laboratory would shake the wood-framed building and jostle the mercury level in the capillary electrometer, resulting in multiple inaccuracies.

Einthoven removed floorboards and had a hole dug ten to fifteen feet deep and filled with large rocks in an attempt to build a stable base for the unit. These efforts, however, were to no avail. He would have to invent his own reliable instrument that could accurately record the tiny electrical charges that accompany each heartbeat. His challenge was to pick up the faintest electrical signals from the body and use them to produce a permanent recording of the beating heart. After six years of work, he successfully developed the string galvanometer, a delicate thread conducting the heart's minuscule electrical current through a magnetic field.[5] In our electronic age, we can only look back with admiration at the investigation of bioelectricity with self-assembled crude equipment.

Critical to its success was the development of an extremely thin string sensitive enough to detect the subtle changes in the magnetic field brought about by a heartbeat — and these changes had to be recorded accurately. Einthoven devised the right string in an ingenious way, using a bow and arrow and some quartz. First he attached a piece of quartz to the tail end of an arrow. He heated the quartz, and, as it melted, he shot the arrow across the room, creating a fine string of quartz.[6]

A mere 0.0001 inches wide, the microscopically thin wire could not be seen with the naked eye. Einthoven solved this problem with yet another clever design. When he illuminated the quartz string against a light background, the difference in light intensity occasioned by its faint shadow could be perceived by the eye. Coated with nebulized silver for purposes of conduction, the string was suspended between the two poles of a large electromagnet. Oscillations of the wire, responding to cardiac electrical deflections, were recorded by an optical system that focused the shadow of the string on a uniformly moving photographic glass plate. Time lines were projected onto the photographic plate by means of large spokes of a bicycle wheel that rotated and interrupted the beam of light at regular time intervals. Linear deflections of the string were then converted to voltage.

Einthoven's galvanometer was a huge construction. Its complexity may have looked like a Rube Goldberg cartoon as invented by his wacky character Professor Lucifer Gorgonzola Butts. It weighed about 600 pounds, occupied two rooms, and required five people to operate it. Nevertheless, its most delicate part, the string galvanometer, was the most sensitive electrical measuring instrument yet devised.[7] A subject would immerse his hands or feet in large bowls or buckets of saline used as electrodes. This forerunner of the minuscule portable

The Bow and Arrow behind Glass

The seminal importance of Einthoven's quartz string is memorialized in the Boerhaave Museum in Leiden, the Netherlands' National Museum of the History of Science and Medicine. Wandering through it, passing microscopes and globes, surgical instruments and pistons, one is struck by a four-foot-high glass exhibit displaying Einthoven's original bow and arrow, whose inscription emphasizes that "the recording of the first electrocardiogram was a typical do-it-yourself job."

ECG instruments of today yielded three waveforms, called P, QRS, and T. The first human electrocardiogram was recorded by the instrument on November 18, 1902. The quality of the records was superb, and it became the standard method until the 1950s, when direct-writing instruments became available.

To demonstrate the clinical usefulness of his electrocardiogram, Einthoven arranged for it to be linked to the university hospital one mile away by underground telephone lines to transmit the signals from the bedside so that he could undertake a systematic study of cardiac abnormalities.[8] Professional jealousy, however, abrogated the arrangement between Einthoven and the hospital after a few years. The physician in charge of the hospital's department of medicine was embarrassed by Einthoven's ability to diagnose a heart irregularity before doctors could discern it at bedside and felt further that Einthoven was unfairly getting all the credit for this innovative technique.

Eventually Einthoven was able to put together a deal with the Cambridge Scientific Instrument Company of England, headed by Horace Darwin, the youngest son of Charles Darwin. Einthoven's model was redesigned by the company and reduced considerably in size to a table model in 1911. An early prototype was loaned to London physican Thomas Lewis in 1909 and installed under a staircase in the basement of the University College Hospital in London.[9] (In 1913, at the age of only thirty-two, Lewis published the first definitive textbook on clinical electrocardiography.) By 1925, a total of 450 of these machines had been installed in the United States.

Knowledge of coronary artery disease and myocardial injury had evolved very slowly. The chest pain that occurs when the heart muscle is deprived of oxygen is called angina pectoris, or simply angina. This is brought about by occlusion or spasm of a coronary artery. If the heart muscle actually dies — "infarcts" — the condition is sometimes fatal. Once known in medical terminology as coronary thrombosis, such "heart attacks" are now simply referred to as "MIs," short for myocardial infarctions.

The classic signs and symptoms of angina were described as early as 1772 by the English physician William Heberden, who stressed the temporary "sense of strangling and anxiety with a painful and most disagreeable sensation in the breast, which seems as if it would extinguish life, if it were to increase or continue." Nitroglycerin for its relief was initiated more than one hundred years later.[10]

It was not until 1912, however, that someone made the first bedside diagnosis of a heart attack. James Herrick, a Chicago internist, was called to the bedside of a banker in severe distress with extreme chest pain, cold skin, nausea, and a racing pulse. After consulting with other physicians, he was not even sure what organ was causing the problem. He and his colleagues pondered many possibilities, including acute pancreatitis and pneumothorax (punctured lung). Herrick finally guessed that there was a clot in the coronary artery. The patient subsequently died, and, much to the pathologist's surprise, Herrick was proved correct at autopsy. Herrick reported the case in the *Journal of the American Medical Association*. His paper "Certain Clinical Features of Sudden Obstruction of the Coronary Arteries" was a giant step in the medical understanding of heart disease, but it stirred little interest. Writing a quarter of a century later, Herrick explained:

> The fate of the early paper was a surprise to me and a keen disappointment. I did not realize, as I do now, that in medical history, as in the history of the growth of ideas in general, while some new facts are accepted as soon as

announced, or at least attract enough attention to be subjects for discussion, very often others are passed unnoticed or unapproved until again brought forward in a more striking form or with more convincing proof and at a time when the medical world is more ready to listen.[11]

Doctors began diagnosing heart attacks at their patients' bedside, but they couldn't always be sure that the symptoms they saw were not due to causes other than myocardial infarction. It was several more years before cardiologists began using Einthoven's ECG to more specifically diagnose heart attacks.

Even as late as 1924, Einthoven himself was not fully aware of the diagnostic importance of his instrument. Samuel Levine, a Harvard cardiologist, related an incident that occurred during a visit from Einthoven:

He [Einthoven] and I were seated in the electrocardiographic room chatting about one thing or another, when Miss Bertha Barker, our devoted and efficient technician, brought in a wet electrocardiographic tracing that she had just taken and developed. She interrupted our conversation and asked me whether she should telephone the medical house officer on the wards and tell him that the patient had an acute coronary thrombosis. The understanding in the laboratory was that if a tracing were taken and showed certain changes with which Miss Barker was quite familiar, she was to telephone the intern directly and not wait until the following morning at 9 o'clock, when I usually read all the tracings of the previous day. When I looked at the electrocardiogram she had just taken, I confirmed her diagnosis of acute coronary thrombosis and she left. On overhearing this conversation, Professor Einthoven was amazed. He remarked, "Do I understand correctly that this lady who is not a physician can make a diagnosis of acute coronary thrombosis from the electrocardiogram, without seeing the patient?"[12]

Einthoven's astonishment must have been due, as Levine supposed, to the fact that he was unaware, like many others in Europe, of the developments in the United States concerning the clinical and electro-cardiographic diagnosis of myocardial infarction.

Sharing the Prize

One event that occurred after Einthoven received the Nobel Prize in Physiology or Medicine in 1924 speaks volumes about his integrity. In the construction of his string galvanometer and laboratory experiments over many years, Einthoven was rather clumsy with his hands and relied very much on the collaboration of his chief assistant K. F. L. van der Woerdt. Years later, when he received the $40,000 in Nobel Prize money, Einthoven wished to share it with his assistant but soon learned that the man had died. He sought out the man's two surviving sisters, who were living in genteel poverty in a kind of almshouse. He journeyed there by train and gave them half of the award money.

Einthoven spoke in a jocular manner of his string galvanometer: "Do I not read all the secrets of the heart?"

In 1926 the Cambridge Scientific Instrument Company introduced the first truly portable ECG, an eighty-pound machine requiring wheeled trolleys. Today, small electronic ECG monitors are ubiquitous in physicians' offices and hospital coronary care and intensive care units and surgical suites.

In time, however, it became increasingly clear that Einthoven's electrocardiograph did not truly reveal "all the secrets of the heart," and medical conquistadors arose with a more aggressive approach to directly probe the heart itself.

23

What a Catheter Can Do

A twenty-five-year-old surgical intern in Germany, Werner Forssmann, famously performed the first human cardiac catheterization in 1929 — on himself! — in an astounding example of self-experimentation in medicine.

In medical school, Forssmann had been inspired by a study undertaken by two French physiologists decades earlier, in 1861. Étienne Marey and Auguste Chauveau had inserted a thin tube into the jugular vein of a conscious, standing horse and guided it into its heart, from which they recorded pressure readings of the heart's chambers. This pioneering procedure was accomplished without disturbing the horse's heart. The print from an old engraving depicting the event had captured Forssmann's imagination. "I hoped that it should be possible to transfer this procedure to man."

Forssmann received his medical degree from the University of Berlin in 1929. That year, he interned at a small hospital northwest of Berlin, the Auguste-Viktoria-Heim in Eberswalde. He pleaded with his superiors for approval to try a new procedure — to inject drugs directly into the heart — but was unable to persuade them of his new concept's validity. Undaunted, Forssmann proceeded on his own. His goal was to improve upon the administration of drugs into the central circulation during emergency operations.

The circumstances of the incident on November 5, 1929, revealed by Forssmann in his autobiography, could hardly have been more dramatic. The account reflects Forssmann's dogged determination, willpower, and extraordinary courage. He gained the trust of the surgical nurse who provided access to the necessary instruments. So carried away by Forssmann's vision, she volunteered herself to undergo the experiment. Pretending to go along with her, Forssmann strapped her down to the table in a small operating room while his colleagues took their afternoon naps. When she wasn't looking, he anesthetized his own left elbow crease. Once the local anesthetic took effect, Forssmann quickly performed a surgical cutdown to expose his vein and boldly manipulated a flexible ureteral catheter 30 cm toward his heart. This thin sterile rubber tubing used by urologists to drain urine from the kidney was 65 cm long (about 26 inches). He then released the angry nurse.

They walked down two flights of stairs to the X-ray department, where he fearlessly advanced the catheter into the upper chamber (atrium) on the right side of his heart, following its course on a fluoroscopic screen with the aid of a mirror held by the nurse. (Fluoroscopy is an X-ray technique whereby movement of a body organ, an introduced dye, or a catheter within the body can be followed in real time.) He documented his experiment with an X-ray film. Forssmann was oblivious to the danger of abnormal, potentially fatal heart rhythms that can be provoked when anything touches the sensitive endocardium, the inside lining of the heart chambers.

There are two conflicting accounts of the responses of his fellow staff members. Forssmann's version was: "News spread like wildfire in a hospital. Suddenly [a fellow intern] burst in [to the X-ray room], half asleep and his hair all tousled: 'You idiot, what the hell are you doing?' He was so desperate he almost tried to pull the catheter out of my arm. I had to give him a few kicks in the shin to calm him down."[1]

Quite another version was offered by another colleague intern, who claims that he heard a rumor that Forssmann had just committed suicide by pushing a catheter through a vein and into his heart. Bursting into Forssmann's room, he found him silent and pale, staring at the ceiling, with the bedsheets soaked in blood and the catheter still in his arm. The catheter was pulled out, but Forssmann refused to divulge if his intentions were suicidal or investigational. In his colleague's experience, Forssmann was always "a rather queer, peculiar person, lone and desolate, hardly ever mingling with his co-workers socially. One never knew whether he was thinking or mentally deficient."[2]

Forssmann's report in the leading German medical journal garnered him not hosannas but instead fierce professional criticism and scorn. In response to a senior physician who claimed undocumented priority for the procedure, the twenty-five-year-old Forssmann was forced to provide an addendum to his publication one month later. Rigid dogmatism and an authoritarian hierarchy characterized the German medicine of that day. The human heart, as the center of life, was considered inviolable by instrumentation and surgery.

Forssmann's surgical career was subsequently severely curtailed. Nevertheless, he continued his experimental pursuits with injection of X-ray dyes through a catheter into dogs and his own body. Since hospitals at that time did not have special quarters for experimental animals, Forssmann's mother cared for the dogs in her apartment. Forssmann would inject a dog with morphine, put the sleepy animal in a potato sack, and take it by motorcycle to the hospital. As in the experiments by the French physiologists, Forssmann would insert a catheter through a vein in the dog's neck and into its heart. He would then inject dyes and attempt X-ray documentation.

Forssmann had assured himself of the safety of the iodine-containing contrast solution in humans by pressing it against the lining of the inside of his mouth for several hours, without reaction. On subsequent experiments on himself, the catheter course did not always oblige his intentions. On the first attempt, the catheter tip deflected into a neck vein rather than toward the heart. When the contrast dye was injected after proper positioning of the catheter, he felt only a mild irritation of the nasal membranes, an unpleasant taste in his mouth, and a transient dizziness. When he presented the tentative results of his studies — on repeated successful passages of a catheter into the heart chamber in both dogs and himself — at the annual meeting of the German surgical society in Munich in April 1931,[3] he received an icy response. There was no applause and no discussion. It was said that Forssmann stopped his self-experimentation only when he had used all of his veins with seventeen cutdowns.

It was not Forssmann but two New York physicians, André Cournand and Dickinson Richards, a few years later, who advanced cardiac catheterization in humans in 1940 at Bellevue Hospital. Cournand readily acknowledged Forssmann's explorations and, years later, contributed the preface to Forssmann's autobiography. He described Forssmann as "not lacking in pride of self, a man at once disputatious, full of resources and will power, and endowed with physical courage, if not with great political perspicacity."

In 1956 the Nobel committee announced it would award that year's prize for Medicine to Forssmann, Cournand, and Richards. Plucked out of obscurity a quarter century after his exploits and told of the news, Forssmann, now a country doctor in the Black Forest, told a reporter, "I feel like a village parson who has just learned that he had been made bishop." Cournand stated in his Nobel lecture that "the cardiac catheter was . . . the key in the lock."[4] With the presentation to Forssmann, it was stated that he had been "subjected to criticism of such exaggerated severity . . . based on an unsubstantiated belief in the danger of intervention, thus affording proof that — even in our enlightened times — a valuable suggestion may remain unexploited on the grounds of a preconceived opinion."[5]

A Milestone in Visualization

To fulfill the demands put forward by the new surgeries for congenital heart defects, the next imperative associated with cardiac catheterization in the 1950s was the need to develop a procedure known as "selective angiocardiography." This procedure involves the injection through the catheter of a radiopaque substance, or "contrast medium" (loosely termed "dye" by some), into the heart or aorta to enable visualization of the heart's interior and abnormalities of the aorta. A safe, reliable technique for entrance into the arterial system was essential. At the time, a surgical cutdown was the most widely used method of introducing a catheter into an artery or vein before guiding it to the desired location and injecting contrast medium.

This problem was ingeniously solved by a thirty-two-year-old radiologist, Sven-Ivar Seldinger, of the Karolinska Institutet of Stockholm. He devised a method to introduce a catheter into the vascular system (usually within the groin) through the skin without the need for a surgical incision to expose a vessel. The procedure is simpler, safer, and has little risk of complication.

Using Seldinger's technique and the intravascular injection of a contrast medium, doctors can see not only the position and course of a vessel but especially the presence, degree, and localization of atherosclerotic obstructions. Seldinger's concise nine-page report in the prestigious Swedish journal *Acta Radiologica* in 1953 established a medical milestone.[6]

Seldinger's technique can be likened to Harvey's discovery of the circulation of blood — an explosive illumination that enabled vast advances. It became widely adopted for catheterizations of the vascular system and led to modern cardiology and cardiovascular surgery.

The Tip Flips by Mistake

As the great vessel, the aorta, carries the richly oxygenated blood from the contracting left ventricle of the heart, its immediate first branches nourish the heart muscle itself. These branches, likened to a crown (*corona* in Latin) as they arise from just above the heart, are named the coronary arteries. Doctors very much wanted to be able to see inside the aorta to look for leaking valves, blockages, and other abnormalities.

Investigators never dared to directly enter the openings of the coronary arteries with a catheter and introduce contrast material to enable their visualization by X-rays. This fear was based on two perceived calamitous effects that would deprive the heart muscle of oxygen: spasm of the coronary artery induced by the catheter and replacement of its blood by an unoxygenated iodine-containing contrast fluid. A host of indirect methods were attempted at numerous research centers over several years, but no investigator would attempt to selectively catheterize a coronary artery in a human. Instead, the tip of the catheter was often placed within the thoracic aorta as close as possible to the openings of the coronary arteries to deliver the contrast fluid.

Then, in 1959, a brief abstract appeared in the journal *Circulation*.[7] It was based on an oral presentation at the annual meeting of the American College of Cardiology by F. Mason Sones Jr. and associates from the Cleveland Clinic. Its opening sentence declared that "a safe and dependable method has been devised for contrast visualization of the coronary arteries to objectively demonstrate atherosclerotic vessels." What stirred only a ripple of interest at the time was shortly recognized as a landmark achievement. Now the location, number, and severity of blockages of the coronary arteries could be identified. This would advance cardiology as much as the electrocardiograph had done half a century earlier.

Today, coronary artery disease is the number one killer in the Western world. In the United States, about 17 percent of the adult population has some cardiovascular condition. Each year, 650,000 Americans suffer a first heart attack. About 450,000 have a recurrent attack. Heart disease occurs because cholesterol-containing plaque accumulates in the walls of the coronary arteries and eventually becomes a blockage that prevents blood from reaching cells. "Atherosclerosis" is the term used for the deposits of fatty plaques on the walls of arteries.

How Sones's method had been "devised," however, was not openly acknowledged until many years later, after the technique had been universally adopted in the revolutionary advance of reconstructive coronary surgery.

Mason Sones, a pediatric cardiologist, had been recruited to the Cleveland Clinic nine years earlier to start a cardiac catheterization laboratory, having learned the technique at the Henry Ford Hospital in Detroit. Complementing cardiac catheterization with angiocardiography, his work initially involved congenital heart disorders in children and rheumatic heart disease in adults. Sones was a perfectionist who worked prodigiously. His frequent comment about his own work was "it's not good enough."

Sones's catheterization laboratory in the basement of the clinic would have looked downright medieval to twenty-first-century observers. Sones had excavated a deep pit in the floor to accommodate a six-foot-long unit designed to enhance fluoroscopic images. Looking like a submarine commander peering into his periscope, he would sit within the pit at a level below the patient, who would be laid out on the X-ray table. It was only in this manner that he could view and photograph images of the heart with the image-intensifier. Another operator would inject dye into the patient's heart chamber, carefully avoiding the opening to the coronary artery.

On October 30, 1958, Sones was working with a twenty-six-year-old man with rheumatic heart disease. His assistant inserted the catheter into the patient's artery and directed it up through the aorta into the heart with the tip of the catheter just above the aortic valve. From his place in the pit, Sones watched in horror as the tip of the catheter flipped around and dye was injected directly into the coronary artery. This was a frighteningly large amount, never imagined for use in the human heart. Fearing a cardiac arrest, he rushed out of the pit to the patient's side to be able to open his chest and massage the heart. The patient had no heartbeat for several frightening seconds. He was still conscious, so Sones asked him to cough. To everyone's relief, his heart started again, without any further complications.[8]

The frightening misadventure taught Sones that, contrary to views held at the time, non-oxygen-carrying fluid could be injected into a major coronary artery safely. He realized he had discovered a technique for obtaining clear and detailed pictures of the entire coronary circulation. "During the ensuing days I began to think that this accident might point the way for the development of a technique which was exactly what we had been seeking. If a human could tolerate such a massive injection of contrast directly into a coronary artery it might be possible to accomplish this kind of opacification with small doses of more dilute contrast agent. With considerable fear and trepidation we embarked on a program to accomplish this objective."[9] He devised a catheter with a flexible tapered tip that permitted easy direct entry.

By 1962 Sones had successfully performed selective coronary

arteriography with small doses of contrast medium — 4 to 6 ml — in more than a thousand patients. Characteristically, he waited until he felt confident enough to report his clinical results with a large series. A brief paper on his technique and experience was published by the American Heart Association in its widely circulated monthly leaflet *Modern Concepts of Cardiovascular Disease.*[10] Despite the enormous importance of the work, the report was written in modest style and was only four pages long. Its impact was explosive, leading to the rapid growth of the technique during the 1960s.

In 1967 Sones reported that he and his colleagues had performed coronary arteriograms on 8,200 patients, representing all types of atherosclerosis. In more than 99 percent of the cases, both coronary arteries could be seen. Branches as small as 100 to 200 microns were readily visualized.

All who knew Mason Sones were alert to his forceful and frequently bumptious personality. A small, chubby man who "swore with gusto," his dominant characteristic was a zealous striving for perfection, accuracy, and truth. This was often manifested at medical meetings as open disagreement with a lecturer making a scientific presentation. He quickly acquired a reputation for being aggressive and even disrespectful.

One particularly memorable episode occurred at an annual scientific meeting of the American Heart Association. A slide was introduced by a guest speaker with the words, "In this normal coronary arteriogram . . ." He was abruptly cut off by Sones, who stood up in the back of the room and declared, "That's not normal! There's a stenosis [narrowing] in one of the coronary arteries!" Sones was correct, and similar intemperate exhibitions in his relentless pursuit of truth caused more than one public speaker to suffer some indignity. Some of this was undoubtedly because, as a pioneer, he had to fight to impose his ideas to dispel myths still defended by eminent cardiologists.

To further support his passionate struggle, Sones underwent coronary artery catheterization himself and would occasionally show his normal films during meetings to emphasize the low risk of the procedure. Eventually, Sones developed cataracts as a result of too much radiation exposure.[11] (Interventional cardiologists and radiologists now use protective eyeglasses.)

As expressed by Donald Effler, Sones's surgical colleague at the Cleveland Clinic, "figuratively speaking, Sones's catheter pried open the lid of a veritable treasure chest and brought forth the present era of [coronary] revascularization surgery."[12]

Once sites of blockage in the coronary arteries could be determined precisely, surgical techniques to bypass the obstruction could be devised. Blood flow could then be diverted around and beyond the obstruction to carry critical oxygen to the heart muscle.

In May 1967 Mason Sones's surgical colleague, René Favaloro, performed the first coronary artery bypass graft (which, because of its initials CABG, is referred to in medical jargon as a "cabbage") as a treatment for angina.[13] The bypass is accomplished by connecting a piece of the saphenous vein, a long vein of a leg, before and after the point where a diseased coronary artery is narrowed or blocked.

Favaloro, originally from a small town in the pampas of Argentina, became intrigued by Sones's X-ray studies of the coronary arteries and spent hours looking at them. Astonishingly, he performed the first bypass without having tried it first on animals, but Favaloro explained that animal experiments are "not always obligatory."[14] The Cleveland Clinic became a mecca for angina patients who could benefit from surgical revascularization. A number of operating suites in a newly established Department of Cardiovascular Surgery were opened.

By the time Favaloro returned to Argentina in June 1970, he and his colleagues had performed 1,086 operations with an acceptably low mortality. About 250,000 were done in the United States in 2005. These involved single and multiple bypasses with either vein or internal thoracic arteries, sometimes utilizing microsuture techniques.

Small-Scale Sewing

The development of microsuture techniques is yet another striking example of serendipity in surgery. At the University of Vermont a young surgeon, Julius Jacobson II, was asked to sever the nerves to vessels in dogs. Researchers wanted to observe the effects different drugs would have on the animals in the absence of nerves. Jacobson concluded that the simplest approach was to cut the actual blood vessels along with their accompanying nerves. Having fulfilled the goal of severing the nerves, he then faced the challenge of reuniting the minute ends of the tiny vessels or the dogs would die. But the common methods of suturing after surgery did not apply on such a tiny scale. It must have seemed like an almost impossible undertaking.

Jacobson realized that "the problem was not in the ability of the hand to do, but rather the eye to see." With the help of a powerful microscope, he was able to suture the small vessel ends together. He described the experience as "like looking at the moon for the first time through a powerful telescope."[15] Jacobson and a colleague went on to miniaturize conventional surgical instruments and in 1960 reported the success of their techniques in animal blood vessels as small as 1.4 mm.[16] Because Jacobson found a solution he was "not in quest of," microvascular surgery was born.

24

"Dottering"

Once atherosclerosis could be visualized in the body and the obstructions surgically bypassed, the stage was set for working inside the artery. In the first internal approach to relieve an arterial obstruction, serendipity played a major role. In 1963 Charles Dotter, a radiologist at the University of Oregon in Portland, was examining a patient with an obstruction in a major artery in the pelvis. In performing a diagnostic catheter study by the Seldinger technique, Dotter inadvertently jammed the catheter through the site of obstruction, only to discover that he had unblocked the vessel! The potential therapeutic importance of the unexpected discovery was not lost on him.

Dotter, an outdoorsman and mountain climber, was a short, muscular, bald man whose bright darting eyes and hyperkinetic manner indicated his active mind. He was an imaginative innovator. After studies on the leg arteries of cadavers, Dotter conceived of removing obstructions gradually with a series of dilators of increasing size.

He got his chance to use his new technique for the first time on January 16, 1964. An eighty-three-year-old woman who had been bedridden for six months with early signs of gangrene in her left foot and toes due to blockage of the arteries in her leg was advised by her vascular surgeon to have her foot amputated. When she refused the operation, Dotter got his chance to prove the value of his approach. He progressively dilated her blocked artery. After the procedure, blood supply to the foot was restored and the patient was able to walk on her own leg without difficulty until her death three years later.[1] This technique, which came before its time, elicited skepticism in the United States as its limitations became obvious. Vascular surgeons, feeling threatened, openly ridiculed Dotter. However, European radiologists enthusiastically embraced the technique and referred to it as "dottering."

The "Crowning" Achievement: Coronary Angioplasty

Dotter's method of reopening obstructed arteries without surgery was the inspiration for a groundbreaking advance by Andreas Gruentzig, a charismatic young German, in coronary angioplasty. After receiving his medical degree from Heidelberg Medical School in 1964, Gruentzig pursued a postgraduate education that ranged over a variety of disciplines, including epidemiology and medical statistics, internal medicine, radiology, and cardiology. At University Hospital in Zurich, Gruentzig developed a technique that involved advancing a catheter with an inflatable balloon tip into a narrowed coronary artery (under fluoroscopic control) and inflating the balloon to compress the atherosclerotic plaque, reduce the obstruction, and enhance blood flow. Initially, he made the coronary balloon catheters in his kitchen. After successful animal studies followed by cadaver studies, Gruentzig presented his experimental results at the American Heart Association meeting in November 1976, where he was met with skepticism and derision.

Gruentzig performed the first human coronary angioplasty in May 1977 during bypass surgery through an incision made in the diseased coronary artery. After accomplishing this procedure in an additional fifteen patients, Gruentzig performed the first coronary angioplasty on a conscious thirty-seven-year-old patient on September 16, 1977, using the Seldinger approach. He reported on his success to a meeting of the American Heart Association later that year. (Mason Sones, who was in the audience, rushed toward the podium immediately after the presentation and, with tears streaming down his cheeks, embraced Gruentzig with congratulations.)[2]

In a brief letter to the editor, published in the February 4, 1978, issue of the *Lancet,* Gruentzig announced his breakthrough.[3] Word for word, the impact of this brief communication was as great as any article that had ever appeared in a medical publication. By 1979 his article in the *New England Journal of Medicine* on fifty coronary angioplasties in patients was immediately greeted as a triumph.[4] The medical community was by then ready to embrace the procedure — perhaps because the benefits of surgical coronary bypass had become apparent. Physicians streamed to Zurich to learn Gruentzig's procedure.

The following year, at the age of forty, Gruentzig moved to Emory University in Atlanta, where he performed approximately 2,500 angioplasties in the next five years. During this period, he was able to modify the procedure from one requiring hospitalization to one that could be performed on an outpatient basis. To demonstrate his faith in the safety and feasibility of outpatient coronary angioplasty, he asked one of his fellows to perform cardiac catheterization on him. Gruentzig jumped onto the table at 5:00 P.M., underwent the procedure, jumped off the table, picked up his wife, and arrived for the department Christmas party by 7:00 P.M.[5]

Gruentzig was an exceptionally prolific contributor to the medical literature on the subject — in 1984 alone he published thirty-eight scientific papers — and helped to popularize the procedure in the United States. His stature took on legendary proportions.

Today, more than 300,000 coronary angioplasties are performed in the United States each year, and a slew of newfangled high-tech devices have rendered the once ultrarisky surgery almost ho-hum. Its further growth is virtually guaranteed by the inevitable coronary problems of the generation of Baby Boomers heading into their sixties in the first decade of the twenty-first century.[6]

25

A Stitch in Time

Every scientist must occasionally turn around and ask not merely,
"How can I solve this problem?" but, "Now that I have come to a result, what problem have I solved?" This use of reverse questions is of
tremendous value precisely at the deepest parts of science.

— NORBERT WIENER, *INVENTION: THE CARE AND FEEDING OF IDEAS*

On a winter day in 1953, an elderly man in acute distress appeared in
the emergency room of the Columbia-Presbyterian Medical Center in
upper Manhattan. He had severe back pain and low blood pressure,
and doctors could feel a pulsating mass in his abdomen. The diagnosis was clear: the patient was suffering from a ruptured abdominal
aortic aneurysm — in medical shorthand, a "triple A." The aorta, the
large vessel coursing from the heart through the abdomen to supply
blood to the abdominal organs, had widened and ballooned due to
atherosclerotic changes. The danger from such an aneurysm was rupture of its weakened wall, with the possibility of life-threatening
bleeding into abdominal tissues.

To surgically remove the diseased segment was itself a formidable
undertaking, but the problem remained of how to close the gap to restore circulation. Historically, the surgical repair of large arterial defects had represented a frustrating challenge. Even in small localized
injuries the successful anastomosis (surgical connection) of two segments of a blood vessel had not been accomplished until the early

1900s. The main problem in suturing vessels was injury to their interior lining, called the intima or endothelium, which could give rise to fatal blood clots.

Alexis Carrel, a French surgeon, experimented for eight years to devise a method of rejoining severed blood vessels. In 1902 he introduced an ingenious technique that involved a trio of stitches. With the use of three sutures as stays, he transformed the blood vessel's round severed end into a triangle, providing the surgeon with a flat surface. This converted the circumference into a straight line, allowing the use of a continuous suture of very fine silk. This method of "triangulation" laid the foundation for procedures that are still in use today.[1]

Once it was demonstrated that the rejoining of severed blood vessels was possible, the next step was to bridge large defects caused by injury or disease. Occasional brave surgeons attempted a variety of artificial grafting materials, including ivory, glass, aluminum, and silver, but such nonpermeable, nonflexible tubes invariably clotted. By the 1940s a conventional treatment for aortic aneurysms, some the size of volleyballs, involved a tedious eight- to ten-hour surgical procedure of stuffing them with hundreds of feet of fine steel wire through a needle a few inches at a time, in the hope that clotting would occur to prevent rupture. (In this condition, clotting reduces the pressure inside the weakened walls of an aneurysm to prevent rupture.) This technique undoubtedly appeared bizarre to an observer at the time, but yet another factor added a science fiction element to the scene. In an attempt to reduce the high incidence of operative infection, the procedure was performed under ultraviolet light, requiring space lab clothing to protect the operating team from sunburn.

After World War II, natural grafts from humans stored in a few medical center artery banks were used, but these were always in short supply and did not maintain circulation for long.

The patient in the ER was seen first by the senior surgical resident, Dr. Arthur Voorhees, who called his attending physician, Dr. Arthur Blakemore. The ensuing scene was vividly recalled by a participant.[2] In the operating room, while awaiting a response to the urgent request for a graft phoned to the aorta bank at a major medical center across town, Blakemore opened the patient's abdomen and placed a clamp across the abdominal aorta above the aneurysm. It was then that a nurse entered the operating room and grimly announced, "The aorta bank is empty. There isn't an aortic graft in New York City!" Arthur Blakemore was highly respected as an intrepid surgeon with an unflappable attitude that was reinforced by a deliberate manner and a soft, slow way of speaking. He was warmly referred to by his colleagues and staff as "Blake." After a shocked silence, he quietly drawled, "Well, I guess we'll just have to make one."

This circumstance set the scene for a dramatic confluence of dedicated research initiated by chance a few years earlier by Arthur Voorhees and the first urgent clinical application of its results. After graduating from Columbia's medical school in 1946, Voorhees was attracted to the "manual engineering" aspects of surgery and was offered a research fellowship after his surgical internship by Blakemore, who as mentor and coinvestigator encouraged and supported his self-described "flights of medical and surgical fantasy."[3] This was an era of exciting developments in cardiovascular surgery, and following the devastating battlefield wounds seen in World War II, there was a fevered drive to find a suitable blood-vessel substitute. A host of favorable conditions set the scene for an explosive development in vascular surgery. These included the availability of antibiotics, blood banks, anticoagulant therapy, and the return from the war of a large number of well-trained surgeons eager for research projects and advancement in their specialties. It was during the fellowship year that Voorhees made a simple observation regarding an inadvertent finding that would revolutionize the field of vascular surgery.

In the spring of 1947, Voorhees was experimenting on dogs, working blindly — only by touch — within the chamber of a beating heart, to develop a procedure for replacing one of the heart's valves. While sewing in the valve's replacement, he unwittingly misplaced one of the silk sutures, which ended up traversing the entire cavity of the left ventricle, the main pumping chamber of the heart. Voorhees was aware that this had occurred and made a mental note to follow up by looking for it in the autopsy on the dog.

Months later, he opened the lifeless dog's heart and was startled at what he saw: a cord stretched across the chamber, its surface glistening. Within, he could discern the three-inch-long stitch. What had caused the glistening filled Voorhees with excitement. The stitch was coated with endocardium — meaning, the lining of the heart chamber had grown over the suture!

Voorhees immediately grasped the potential significance of this serendipitous observation. In a similar way, he creatively imagined the lining of a vessel extending to line the inside of a tubelike graft. Here lay the solution to bridging a large arterial defect. In a flash he could visualize the coated single silk stitch as a component of a cylindrical structure of fabric acting as a latticework of threads that might serve as an arterial prosthesis. Lined by new intima, it would allow blood flow to be restored without inducing blood clots.

To test his idea, he fashioned a silk handkerchief on a sewing machine into a tube and used it to replace a segment of the abdominal aorta in a dog. The graft remained unobstructed for one hour before bleeding occurred.

In 1948 Voorhees was assigned to Brooke Army Medical Center in San Antonio, Texas, where he continued his work on arterial substitutes on his own time. He constructed tubes from scraps of nylon parachute cloth on sewing machines borrowed from one neighbor or another and implanted them in the aortas of six dogs. Results in one dog that survived for a month were promising.

In 1950 Voorhees returned to Columbia-Presbyterian Hospital as a surgical resident. Requirements for the cloth to be used as an arterial prosthesis were few but essential. It had to be strong, inert, stable, of the right porosity to allow endothelial growth through its lattice, supple, and yet easily traversed by a fine suture needle. The Union Carbide Company supplied a bolt of Vinyon-N material, which turned out to be the cloth used in sails. This was too inert to take a dye evenly, so it had little commercial value for the company, but it had all the physical characteristics Voorhees needed. He became adept at turning, flipping, and cutting the cloth into different sizes, shapes, and configurations. Trained as a surgeon, Voorhees needed no lessons in stitching and suturing. His proficiency on a sewing machine would have made him a master tailor, but he put his talents instead to fashioning prosthetic tubes and implanting them in dogs' aortas.

He reported his successes in the *Annals of Surgery* in March 1952. Typical of an initial announcement is his dry passive voice: "It was observed in this laboratory that a simple strand of silk suture . . . became coated in a few months throughout its length by a glistening film."[4]

Voorhees planned further experiments, but did not consider using the prosthesis in humans until several months later when faced with the desperately ill man with the ruptured aneurysm and the unavailability of human aorta grafts.

Blakemore's quiet conclusion that "Well, I guess we'll just have to make one" was the signal for immediate action. Voorhees broke scrub, ran down a flight of stairs to the dog lab, hurriedly sewed a sheet of Vinyon into a tubular prosthesis, trimmed off the excess, ran back to the operating room, gave it to the nurse to sterilize in the autoclave, and scrubbed in again. Blakemore sewed in the graft and took off the clamps. The prosthesis functioned well, but the patient died from a heart attack caused by hemorrhagic shock and a diffuse clotting disorder.

Encouraged by the serviceability of the prosthesis, Blakemore placed Vinyon grafts in a succession of aneurysm patients and was thrilled to find that many were successful. A year later, he and Voorhees reported their extensive experimental and clinical work with the Vinyon graft to the National Research Council.[5] The surgical world greeted these reports with tremendous excitement. The threshold of a new era in vascular surgery had been crossed.

Throughout the United States, laboratories were established to explore the use of different textiles and methods of fabrication. Steven Friedman, a vascular surgeon, described the avidity of the reception: "Surgical meetings assumed the air of textile conventions, as surgeons readily adopted a new lexicon. Terms such as crimping, needle-per-inch ratio, and tuftal rhexis were glibly bandied about by these pioneer practitioners of arterial replacement with prosthetics."[6] Surgeons quickly adopted Dacron, a polyester fiber well established in the garment industry, over Vinyon because it was easier to sew. This fabric served as the mainstay in the surgical treatment of aneurysms for decades.

Abdominal aortic aneurysm (AAA) is an abnormal widening, or ballooning, of a portion of the aortic artery, related to weakness in its wall. It is a life-threatening condition diagnosed in about 250,000 Americans every year. If detected in advance, with a CT scan or sonogram, the condition is treatable in 95 percent of cases. Undetected, such aneurysms grow and eventually rupture, killing 75 percent of the people who experience them. AAAs are very unusual in people under the age of fifty-five, but as age advances, risk increases. More than 10 percent of men eighty to eighty-five years old experience AAAs. The condition kills at least 9,000 people a year in the United States. Abdominal aortic aneurysm was the cause of death for Albert Einstein, Lucille Ball, and George C. Scott.

Einstein's Fatalism

In December 1948 Albert Einstein, then age sixty-nine, was hospitalized suffering from agonizing abdominal pains. Physicians could feel a grapefruit-sized mass in his abdomen, and surgery confirmed an abdominal aortic aneurysm. His doctors decided not to remove it, but they monitored him and prescribed pain-relieving drugs. Two years later, doctors discovered that the aneurysm had grown.

In terms of his own health, Einstein was a determined fatalist. Told that his aorta might burst unless he took care, he brusquely replied: "Let it burst." Shortly thereafter, Einstein completed his last will and testament. In April 1955, after collapsing at home, he was taken to Princeton Hospital. The aneurysm had ruptured. He refused emergency surgery, saying, "I do not believe in artificially prolonging life." His physician and surgical consults could not dissuade him. He died on April 18 and was cremated the same day, except for his brain, which he bequeathed for research.

In the second half of the twentieth century, thousands of patients with AAAs were successfully treated with Dacron grafts. In the 1990s metallic stents replaced Dacron prostheses. These stent-grafts are introduced through a catheter in an artery in the groin and act as scaf-folds within the aortic aneurysms no longer requiring surgical resection. They serve as a sort of artery within an artery: the aneurysm is by-passed, and blood flows through the graft instead. Open abdominal incisions are thus avoided, and some medical centers are now per-forming this as a one-day outpatient procedure. Former U.S. Senator Bob Dole was treated with a graft when his AAA was discovered in June 2001. (His father had died of an aortic aneurysm).

In 1985, long after being appointed professor of surgery and chief of the vascular service at Columbia-Presbyterian, Voorhees re-vealed, "I stumbled on the notion. . . . The whole development was a serendipitous outgrowth of an observation made in a related but dif-ferent experiment."[7]

26

The Nobel Committee Says Yes to NO

The saga of the surprising discovery of how the natural production of a gas, nitric oxide (NO), in the body can cause a wide variety of actions is replete with dramatic instances of stumbling blocks transformed into stepping-stones.

The principal investigator, Robert Furchgott, an academic researcher in pharmacology, is a gentle, soft-spoken, unassuming scientist. Small and thin, he uses a hearing aid and views the world through a thick set of lenses. In a personal interview, he was unusually candid in admitting the illuminating role of numerous "accidental discoveries" along a research trail he followed for more than half a century that revolutionized vascular biology.[1]

At the beginning of Furchgott's career, the innermost layer of cells lining blood vessels, the endothelium or intima, had classically been thought to serve only as a passive protective lining. The muscle in the wall of the blood vessel surrounding the endothelium, which can contract or dilate (relax), is termed smooth muscle and is not under voluntary control. The pharmacology of smooth muscle became Furchgott's major research interest.

Furchgott's mind, trained by early experiences with serendipity, was always alert to the benefits that could accrue from accidental discoveries.[2]

THE CLUE WAS AT HIS FINGERTIPS

In 1951, while at Washington University, Furchgott started using a strip of rabbit aorta to study the effects of drugs. His goal was to characterize receptor sites in smooth muscle cells responsive to the actions of drugs. In the current era of Big Science, Furchgott's preparation now seems absurdly simple. He would cut by hand, with a small pair of scissors, strips of aorta and measure their changes in length when exposed to a variety of drugs.

A couple of decades later, in 1978, after Furchgott had moved to the State University of New York (SUNY) Health Science Center in Brooklyn, his method played a role in a serendipitous breakthrough that led to other discoveries. Furchgott himself admitted that "an accidental finding as a result of a technician's error completely changed the course of research in my laboratory."[3] The technician failed to wash out a drug from an earlier experiment, which brought about the opposite effect. Furchgott, not realizing the mistake, expected that the arteries would contract, and was surprised to see that they relaxed.

It was the first time he had seen the drugs under study produce such a reaction. He used this serendipitous observation to plan additional experiments, and it forced him to review his techniques of preparation — ones he had used for more than twenty-five years.

> In that method, as a continuous strip was being cut by scissors from an aorta held in one hand, each newly cut portion of the strip was pulled over the tip of the middle finger of that hand so that the lengthening strip could hang free from that finger and not get in the way of the part of the aorta yet to be cut. The surface of the strip being pulled over the fingertip was the intimal surface! Moreover, the full strip was positioned with intimal surface down on a piece of moistened filter paper before being cut into smaller strips. . . . Now we found that if we cut . . . strips in a different way so that the intimal surface did not make contact with fingertips or other surfaces during the preparation, then such strips exhibited [classic responses].[4]

Further study led to an astonishing realization: Furchgott's unintentional gentle rubbing of the lining of strips in the course of preparing them had scraped off the lining's endothelial cells!

Furchgott had to face another embarrassing realization. Twenty-five years earlier, he had noted the same paradoxical response to this same drug, acetylcholine, a very potent vasodilator that elicited no relaxation but only contraction of the aortic strip. At the time, it was peripheral to his main research pursuits and he did not recognize its significance. Looking back, he ruefully noted: "Little did I suspect then what I was able to show many years later — namely, that relaxation of arteries by acetylcholine is strictly endothelium-dependent, and that my method of preparing the strips inadvertently resulted in the mechanical removal of all the endothelial cells."

Furchgott quickly reported these early results and magnanimously included the name of his technician who had committed the oversight among the coauthors of the abstract.[5]

The experience led Furchgott to the quite unexpected and exciting realization that contraction and relaxation of blood vessels in response to many agents are dependent on the endothelium. The vascular endothelium eventually received the status of an organ, albeit a widespread one, responsible for widening blood vessels, helping to regulate blood pressure, initiating erections, battling infections, preventing formation of blood clots, and acting as a signal molecule in the nervous system.

It's a Gas

It was immediately apparent that stimulation of certain receptors of endothelial cells led to relaxation of the adjacent smooth muscle. The question was how this occurred. Furchgott hypothesized that the endothelial cells, in response to stimulation, release a substance that diffuses to and acts on the smooth muscle cells to cause relaxation. Pondering how to prove this, Furchgott said, "the idea came to me suddenly and unexpectedly one morning just as I was waking up."

In a simple but ingenious experiment, he made a key discovery that set the stage for future breakthroughs. A strip of artery with endothelium removed was mounted with another strip of the same length and width but with endothelial cells present — intimal surface against intimal surface. In this so-called "sandwich" preparation, the endothelium-free strip, which alone did not relax in response to acetylcholine, now exhibited good relaxation in response to that agent. In a report in *Nature* in 1980, Furchgott concluded that an unknown diffusible substance was produced in the endothelium and caused relaxation.[6]

This was a great discovery, and the hunt began in several laboratories for the identity of the endothelial factor. One of the areas of research involved nitro-compounds. Nitroglycerin had been used successfully for more than a hundred years for the treatment of angina, but its mechanism of action had not been clearly understood. Angina pectoris is a prominent symptom of a cardiac condition known as myocardial ischemia. This condition occurs when the heart muscle is deprived of the blood it requires.[7]

Precise investigative work by the pharmacologist Ferid Murad, first at the University of Virginia and then at Stanford University, showed that nitroglycerin acts by releasing nitric oxide gas. It remained only to close the loop by identifying the endothelial factor to be nitric oxide. At a scientific meeting at the Mayo Clinic in the summer of 1986, Furchgott and Louis Ignarro of Tulane University each presented this conclusion independently. (Years later Furchgott said in an interview, "To tell you the truth, it took an embarrassingly long time to understand what we had discovered.")[8]

NO is a gas most commonly known as an air pollutant formed when nitrogen burns, for example, in automobile exhaust fumes. It is a chemical relative of the anesthetic gas nitrous oxide (laughing gas). It is a simple molecule, with two atoms — nitrogen and oxygen. As astonishing as the realization was that certain cells in the human body generate a gas, it was a truly revolutionary breakthrough to finally understand that blood vessels have the ability to control their own diameter and thereby their rate of blood flow and blood pressure. A new phenomenon had been discovered: an internally produced short-lived gas with the capacity to act as a signaling molecule between cells in the body. In 1994 NO was named "molecule of the year" by the prestigious American journal *Science*.

Scientists were astounded to learn that this simple molecule could also be an important neurotransmitter, and as a result of the new insight, biological research in the field of nitric oxide exploded. As research progressed, nitric oxide was shown to regulate an ever-growing list of biological processes, including blood pressure, blood clotting, bacterial endotoxic shock, the immune system, intestinal motility, behavior, and some aspects of learning.

In 1998, twenty years after discovering the presence of the vascular relaxing factor derived from the endothelium, the eighty-two-year-old Robert Furchgott shared the Nobel Prize with Louis Ignarro and Ferid Murad. In the personal introduction to his Nobel lecture, Furchgott gave serendipity its due: "I was very lucky to stumble on unexpected results in 1978 that led to the finding of endothelium-dependent relaxation and eventually to NO."[9] His discovery arose from what he simply termed "old-fashioned pharmacological research."

Remarkably, Furgott had accomplished all his work throughout the years by using simple and easily reproducible techniques, without sophisticated instruments and technology. His experience is a reminder that an individual does not need a heavily funded laboratory equipped with state-of-the-art technology to uncover fundamental truths.

27

"It's Not You, Honey, It's NO"

Throughout history, aging in men has been associated with depleted sexual vigor. In June 1889 a dramatic announcement was made. Charles-Édouard Brown-Séquard, a renowned seventy-two-year-old physiologist at the Collège de France in Paris, reported to spellbound doctors at a meeting of the Société de Biologie that he had injected himself with a "masculinizing factor" extracted from the testicles of guinea pigs that he claimed to have dissolved in water.[1] Brown-Séquard said that animal testes harbored an invigorating principle that made him "tremendously rejuvenated," sexually and in muscular strength. Later he wrote of how his impotence had been cured.

To many, it seemed the Elixir of Life had been found. However, it is now known that the male hormone is insoluble in water, so this aqueous extract could not have contained any testosterone; indeed, many years later, Ernst Laqueur required more than a ton of bulls' testicles when he became the first scientist to isolate enough testosterone for a single course of injections.[2] Clearly, the invigoration experienced by Brown-Séquard was due entirely to autosuggestion.

Therapies for impotence throughout most of the twentieth century were awkward, uncomfortable, and at times surgical in nature. They involved injecting drugs directly into the penis, inserting suppositories into the urethra, using a pump to draw blood into the penis, constrictive bands, and surgical procedures such as implantation of penile prostheses. As recently as a few decades ago, men in need of erections were told to rub nitroglycerin paste on their genitals to increase blood flow to the organs.

A Memorable Science Exhibition

The annual scientific meeting of the American Urological Association is usually a pretty staid affair. But one meeting has entered the annals of folklore. In the early 1980s, during the course of his lecture on the effectiveness of injecting substances directly into the penis to increase blood flow, one urologist announced that he had performed such injections on himself only an hour earlier. Stepping from behind the podium, he dropped his trousers and proudly demonstrated to the audience his own erect manhood. Urologists who attended this meeting still shake their heads at the memory.

By 1918 another possibility was opened up: testicle transplantation. Dr. Leo L. Stanley, the prison doctor at San Quentin in California, actually performed such transplants in five hundred inmates. He took testicles from executed prisoners and transplanted them into live inmates. He also surgically implanted "animal glands" from rams, goats, and boars into the scrotums of some inmates. Stanley's human experimentation was no secret, and his report on it was published in the journal *Endocrinology*.[3]

For a quarter of a century, from 1918 to 1942, a dubiously certified physician, John R. Brinkley, performed "goat gland" transplants on thousands of men afflicted with impotence or flagging sexual capacities. Modifications of the procedure were offered at exclusive Swiss spas catering to the rich and famous.

Desperate men who heard about the "cure," which was heavily promoted on Brinkley's own radio station, flocked to his office in Kansas (and later, as growing criticism compelled Brinkley to relocate, in Texas and then Arkansas). Eventually his practice was branded as quackery by organized medicine.[4]

THE LITTLE BLUE PILL

Around 1986 laboratory scientists working at Pfizer's research facility in Sandwich, in the United Kingdom, developed a new chemical compound, sildenafil citrate, and considered it a candidate for the treatment of angina. In early clinical trials in 1990 and 1992, its effect on circulation and heart function was disappointing. The researchers decided to alter the dose to see if it would make any difference.

It did indeed — but not in the way the scientists expected. Rather than boosting blood flow to the heart, it affected blood flow to the penis, with startling effect: many impotent patients experienced restored penile erectile function. Pfizer, of course, was astute enough to explore what its scientists had inadvertently stumbled upon, something entirely new in the annals of medicine.

Pfizer's large-scale clinical trials, initiated in 1993 at first throughout Europe and then in the United States, proved highly successful for the treatment of impotence and validated that the drug sildenafil citrate works only in response to sexual stimulation.

Meanwhile, a National Institutes of Health panel comprised of leading urologists, psychiatrists, psychologists, gerontologists, and surgeons, recognizing that the term "impotence" was confusing and had pejorative implications, created a more accurate definition of what came to be called "erectile dysfunction" (ED). Their report was published in the *Journal of the American Medical Association* in July 1993.[5] A survey of almost 1,300 men at around the same time found that the combined prevalence of minimal, moderate, and complete erectile dysfunction in the United States was more common than previously thought — affecting about 52 percent of men — and now was considered a "major health concern."[6]

In March 1998 the U.S. Food and Drug Administration granted market clearance for Viagra, a breakthrough oral treatment for ED. This diamond-shaped blue pill is popularly thought of as the first pharmaceutical aphrodisiac. After its approval by the FDA, Viagra received an enormous amount of publicity.[7] Even Pfizer did not anticipate the impact the drug would have on the culture. Within the first month, 40,000 Viagra prescriptions were filled each day.[8] For a time, it was the fastest-selling drug in history. Worldwide sales were $788 million in its first nine months on the market in 1998. In 2004, some 250,000 to 300,000 men were taking Viagra weekly.

Anatomy of an Erection

The human penis is made up of three cylindrical bodies: the corpus spongiosum (spongy body), which contains the urethra and includes the glans penis, or head of the penis, and the paired corpora cavernosa, or erectile bodies. These two bodies, functioning as one unit, are actually responsible for the erection, as their sinusoidal tissue, fine-walled empty spaces, expands as it fills with blood. The sinusoidal tissue can be thought of as a myriad of small, interlacing vascular spaces lined by endothelial cells and smooth muscle.

The key element necessary for the production of an erection is smooth muscle relaxation within the walls of the sinusoidal tissue. This engorgement with blood from dilated arteries is dependent upon release of nitric oxide by nerves in the penis following sexual stimulation. Viagra blocks an enzyme known as phosphodiesterase-5 (PDE-5), allowing smooth muscle cells in the penis to relax, so it magnifies the effect of nitric oxide.

With studies suggesting that more than 150 million men worldwide and at least 30 million in the United States suffered from erectile dysfunction, drug companies knew the market would expand rapidly. Pfizer soon had competitors. Bayer, GlaxoSmithKline, and Schering-Plough, the makers of Levitra, which was approved for sale in August 2003, claimed their drug acted somewhat more quickly than Viagra.[9] Another competitor, Cialis, made by a partnership of Eli Lilly and the

Icos Corporation, was approved shortly thereafter and is said to be effective for thirty-six hours, as opposed to four to five hours for Viagra and Levitra.

In 2004 combined global sales of the three ED remedies reached about $2.5 billion, with Viagra claiming the lion's share of the market. It accounted for 61 percent of prescriptions, compared with 25 percent for Cialis and 14 percent for Levitra.

28

What's Your Number?

We are all used to certain numbers in the determination of our physical health. Temperature: 98.6. Blood pressure: 120 over 80. Vision: 20/20. Other familiar indices may involve caloric intake or body fat ratio. In the last several years, another number has entered the national consciousness. Even in casual conversation between friends, the question "What's your number?" is taken to mean your level of blood cholesterol. This is in widespread recognition of a factor critical to our health.

Heart disease from clogged arteries is the number one killer in the West. (It is the number two killer in the United States, second only to cancer.) Cholesterol is the primary culprit. It causes plaque to build up on artery walls. Several consequences may ensue from this buildup. As the plaques enlarge, they may induce clots to form; together they may block the coronary artery, or a piece may break off to float downstream and obstruct blood circulation. Even while the plaque remains small, it may induce the coronary artery to go into spasm, shutting off the heart's blood supply.

The underlying causes of atherosclerosis long remained elusive. The first experiments on its nature were undertaken in the early twentieth century. In 1909 a clinical investigator at the Russian Imperial Military Medical Academy in St. Petersburg found that rabbits fed a mixture of eggs and milk developed atherosclerotic plaques. He purposely chose dietary staples of rich Russians in contrast to the peasant class, but he wrongly concluded that the atherosclerosis resulted from the *protein* in these foods. Within a few years, at the same institution, the pathologist Nikolai Anichkov established that cholesterol, a soft waxy substance, was to blame.[1] Cholesterol, a pale yellow substance in dairy products and beef, is produced naturally by the liver and is essential to the creation of cell walls and certain steroid hormones.

This Russian breakthrough in understanding remained unrecognized by European and American scientists for decades for several reasons. The St. Petersburg medical school was simply unknown as a research center to Western medicine, and most of its investigators' papers were published in Russian. Moreover, the cause or development of aortic or coronary atherosclerosis was not a subject interesting mainstream Western scientists at the time. The electrocardiogram was not in general clinical use for the diagnosis of heart attack until the mid-1920s.

Then, in 1950, Western scientists learned of an important finding in a report in *Science*.[2] Stimulated by previous observations that atherosclerotic plaques contain cholesterol, phospholipids, and fatty acids, a team of researchers led by John Gofman had worked to develop atherosclerosis in rabbits by feeding them a high-cholesterol diet. (Gofman was not dissuaded by the simplistic criticism that the rabbit is herbivorous and thus ordinarily ingests essentially no cholesterol.) Cholesterol moves throughout the bloodstream by attaching itself to water-soluble proteins called lipoproteins. When the Gofman team spun the rabbits' fatty serum — that is, all the fats (cholesterol and triglycerides) in the serum, the noncell liquid part of blood — in tubes in a high-powered ultracentrifuge, they found that it separated into two distinct compartments. They designated the lighter portion of the fatty serum, composed of much more fat than protein, low-density lipoprotein cholesterol, now referred to as LDL. They designated the other portion, richer in protein and lower in fat content, high-density lipoprotein, now referred to as HDL.

In not only these atherosclerotic rabbits but also in a large group of patients who had recovered from a heart attack, they found that most of the cholesterol in the blood was carried in the LDL complexes. Since many individuals with atherosclerosis show blood cholesterol in the accepted normal range, it became apparent at this time that LDL concentration is more significant than total blood cholesterol content. For this reason, LDL is referred to as "bad cholesterol."

Awareness of the contribution of diet subsequently arose from several sources. In 1952 Louis Dublin, senior statistician at the Metropolitan Life Insurance Company, declared that Americans "are literally eating themselves to death." His exposure of the "epidemic" of heart disease was based on decades of actuarial research.[3] Further convincing proof that diet does cause human coronary disease was then demonstrated in a report on young American soldiers killed in the Korean War. Their average age was only twenty-two years, yet autopsies showed that three-fourths of them already had fat streaks in their coronary arteries. In contrast, the Korean and Chinese soldiers, having spent their lives eating no milk products and little egg or meat fat, had no cholesterol in their coronary arteries. Following President Eisenhower's widely publicized heart attack in 1955, the dangers of high-fat diets were openly discussed in the general press as well.

In 1958 a strongly worded editorial in *Circulation,* the official publication of the American Heart Association, served as a call to arms. It rebuked cardiac researchers for their inattention over the years to the key role of diet in the primary pathology of atherosclerosis.

Twenty-five years later, two dedicated young researchers, Michael Brown and Joseph Goldstein, brought about a conceptual revolution when they told the world that the fundamental problem lay not in the blood but in the body cell.

Goldstein, from South Carolina, and Brown, from Brooklyn, became close friends when both were medical residents at Boston's Massachusetts General Hospital from 1966 to 1968. Their mutual interest in pursuing an academic career in molecular biology led to one of medicine's great creative research partnerships. Both physicians joined the National Institutes of Health in Bethesda, Maryland, and worked in different labs concerned with how genes regulate enzymes — Goldstein in biochemical genetics, and Brown in digestive and hereditary diseases. Their diverse talents and training were particularly complementary and would lead to the illumination of a whole new area of biologic medicine.

By 1972 the two scientists were on the faculty of the University of Texas Southwestern Medical School, in Dallas and began collaborative research on cholesterol and cholesterol-related disease.[4] Brown and Goldstein began their studies to isolate the crucial mechanism of cholesterol metabolism by focusing on a striking genetic disorder, known as familial hypercholesterolemia, in which patients develop sky-high cholesterol. People with this disorder have blood cholesterol that is up to eight times greater than normal, yellow accumulations of cholesterol in the skin, and a greatly increased risk of recurrent heart attacks. The two researchers concentrated their efforts particularly on the homozygous form, in which a child receives a defective gene from both parents. The condition is very rare, occurring in one in one million people.

The key problem was why cholesterol builds up to dangerous levels in some people and not in others. Initially, Brown and Goldstein pursued the obvious hypothesis: that excessive cholesterol was being overproduced in the liver, perhaps by a deranged enzyme.

Brown and Goldstein decided to use tissue cultures of skin cells from patients suffering from inherited high cholesterol and from normal individuals. The NIH turned down their application for a grant, but fortunately they were able to tap other funds. They began their work by tagging the LDL molecules with radioactive iodine, which enabled them to trace their course after their introduction into the cell culture.

As conventional belief held that LDL flowed freely into cells to be assimilated and utilized, these investigators were stunned to find that the fat-laden molecules attached to specific receptor sites on the skin cell membranes. Only at these receptor sites on the cell membrane could the cholesterol-laden LDL gain entrance to the cell to be utilized. Furthermore, they noted an important distinction in what was going on in the cells of normal people versus the cells of people with inherited high cholesterol. The cells from the latter either completely lacked or were deficient in LDL receptors. In effect, these patients' blood was awash in cholesterol that couldn't be absorbed by cells.[5]

Brown and Goldstein, barely in their thirties, had discovered a new phenomenon, the mechanism by which a cell extracts cholesterol from the blood: cholesterol receptors.[6]

Like a space shuttle mooring at one of several docking stations on a satellite to deliver its contents, the LDL particle attaches to certain receptor protein sites clustered in coated pits on the surface of the cell membrane. And then a most remarkable series of events follows. The pit pinches off, and all are drawn into the cell. Inside the cell contents, the cholesterol is extracted, the protein of the lipoprotein complex is digested into its constituent amino acids, and the receptor itself is recycled back to the cell membrane. The cholesterol can be used in manufacturing cell membrane or other products. In the liver, excess cholesterol is excreted in the bile into the intestinal tract. In the other cells, it passes back into the bloodstream, where it is prevented from building up in the arteries by being transported by HDL to the liver, which leads to its removal from the body. For this reason, HDL is referred to as "good cholesterol."

The significance of the process was evident: if the regulatory mechanism of the receptors could be explained, drugs might be designed to influence that mechanism and rid the body of excess cholesterol before it clogged arteries. Of this insight, Goldstein recalled, "It was the most exciting moment of my life."

They promptly uncovered that the LDL receptors in the cultured cells are controlled by a discrete feedback mechanism as a foreman controls inventory in a busy factory. The more LDL is brought to a cell, the fewer receptors are made. If a cell is given less cholesterol, more receptor sites appear.[7]

But certain critical questions remained unsolved. How are blood cholesterol concentrations controlled? Where in the body are the functions of LDL receptors most important? How can blood cholesterol levels be lowered in patients at high risk of heart disease? The researchers were stymied at this point. All this time, they had been conducting their studies on tissue cultures of skin cells. What they needed for further progress was an animal model. Their meticulously planned basic scientific studies, a model of disease-specific research, needed a stroke of luck.

A serendipitous discovery by a veterinarian in Japan could not have been a more propitious deus ex machina in enhancing the prospect for LDL receptor research. In 1973 Yoshio Watanabe of Kobe University noticed that a rabbit in his colony had ten times the normal concentration of cholesterol in its blood. By appropriate breeding, Watanabe obtained a strain of rabbits all of which had this very high cholesterol level, and all of these animals developed a coronary heart disease that closely resembled human heart disease. The rabbits were also found to lack functional LDL receptors in their cells, as do humans with inherited high cholesterol.[8]

The serendipitous mutation in the Watanabe rabbit "proved invaluable," in Brown and Goldstein's words, in putting together the puzzle. They now were able to establish the mechanism by which LDL receptors in the liver control blood cholesterol levels in humans. Normally, the liver accounts for about 70 percent of the removal of LDLs from the bloodstream. Within the cell, LDL is broken down to release cholesterol, which then reduces the level of the enzyme responsible for making more cholesterol. This rate-controlling enzyme would prove to be the therapeutic target. Increased amounts of cholesterol within the cell inhibit production of more LDL receptors and thereby the influx of more LDL. Thus, this process normally balances the cell's own synthesis of cholesterol and the cholesterol it obtains from the circulating blood influenced by diet.

By the mid-1980s the stage was set for a dramatic coalescence of events. In 1984, after completing a ten-year study conducted in twelve centers in North America involving more than 3,800 men and costing $150 million, the NIH established that lowering blood cholesterol clearly lowers the risk of heart attack. The nation's premier health institute officially recommended that Americans lower the amounts of fat and cholesterol in their diet.[9]

It was now clear that atherosclerosis in the general population was caused by a dangerously high blood level of low-density lipoproteins, resulting from failure to produce enough LDL receptors, and that cholesterol not absorbed into the cells as it courses through the circulatory system sticks instead to the walls of blood vessels, disrupting the flow of blood to the heart and brain. It was now understood why some people can eat more cholesterol-rich foods and not have high blood cholesterol concentrations, while others on low-fat diets continue to have high levels: the difference is in the number and efficiency of the LDL receptors. Some people are born with a greater or lesser capacity to develop adequate receptors. Individuals with "low capacity" LDL receptors have higher cholesterol levels in the blood-stream and an increased risk of coronary heart disease because they remove cholesterol more sluggishly from their circulation. Once this mystery was solved, the therapeutic approach was clear. It lay in developing drugs that block cholesterol synthesis in the liver in order to stimulate production of the receptors and reduce blood cholesterol.

Brown and Goldstein promptly identified the underlying genetic mutation in the Watanabe rabbits, leading to their sequencing the human LDL-receptor gene. They found that a variety of mutations occur naturally and can be located anywhere on the gene, resulting in failure to synthesize the LDL receptor proteins or in production of deficient proteins. "The multiple mutations may be the reason for the high frequency of defective genes in the population," Brown suggests. Approximately one person in every five hundred carries a single deficient copy of the LDL receptor gene, incurring a higher risk of heart attack. One person in a million inherits defective genes from both parents, and these individuals with familial hypercholesterolemia have extremely high blood cholesterol concentrations and often die of heart attacks before their twentieth birthday.

Two Defective Genes

A Texas girl named Stormie Jones was literally "one in a million," as one in every million people in the population inherits two defective LDL receptor genes. She had stratospheric cholesterol levels and did not make functional LDL receptors — hence drugs were ineffective. She suffered her first heart attack at age six, followed within the next year by another heart attack and two coronary bypass operations. In 1984, at age eight, she underwent the first heart-liver transplant ever performed and, as Brown and

> Goldstein's theories had predicted, the presence of normal LDL receptors on the transplanted liver produced an 80 percent reduction in her blood cholesterol levels. The double transplant enabled her to live for six more years.

In 1985, thirteen years after they began their work, Goldstein, age forty-five, and Brown, age forty-four, shared the Nobel Prize in Medicine. In an interview with a local newspaper a few days after the Nobel Prize announcement, Brown openly disclosed that their brilliant basic research over thirteen years had benefited from serendipity:

> In the beginning, we started with a hypothesis that was incorrect. . . . Initially we thought that an enzyme had gone wild and was producing the excessive cholesterol. As it turned out, we found the enzyme was not the problem, it was the fact that the [body] cells had trouble getting cholesterol from a lipoprotein [and thus removing it from the blood]. . . . We did not dream there was any such thing as a receptor. It was not within the worldview of any scientist that such a thing existed.[10]

In 1987 Merck introduced a cholesterol-lowering drug, lovastatin, that works by slowing down the production of cholesterol by the liver and increasing its ability to remove LDL cholesterol in the blood.[11] Of the six statins on the American market in 2006, Pfizer's Lipitor was the most popular, taken daily by as many as 16 million Americans. It was the world's best-selling prescription medicine, with $11 billion in sales in 2004. As a group, these drugs were generating revenues of more than $25 billion a year for their manufacturers, which include, along with Pfizer and other American drug companies, Germany's Bayer and the British-Swedish company AstraZeneca.

In July 2004 aggressive new guidelines for LDL levels were proposed. These range from a rock-bottom level of 70 or lower for those with the very highest risk factors (heart disease, high blood pressure, smoking) to 160 for healthy people with little risk.[12] Large clinical trials found that lowering LDL to such levels sharply decreases the risk of heart attacks. Under the old guidelines, about 36 million people in the United States should have been taking statins, but less than half that number had been.

Brown and Goldstein are still at the University of Texas Health Center, where they are often referred to as "Brownstein" or "the Gemini twins." They plan research jointly, publish together, and share the podium for lectures. Out of the lab, they are partners at bridge. The two men brought the concept of cholesterol receptors and their specific genetic coding into the light of day. Their turn down an unexpected path is emblematic of the process of basic research: diligent pursuit of fundamental processes of nature, often without a hint of what one may unexpectedly find.

29

Thinning the Blood

Blood, to maintain its crucial roles in the body, must remain fluid within the vascular system, and yet clot quickly at sites of vascular injury. Examples of such conditions include a heart attack due to a clot formation within a coronary artery, the abnormal cardiac rhythm known as atrial fibrillation, which may be complicated by a clot within that chamber of the heart, and deep venous thrombosis, in which clots form in the leg veins (often from prolonged inactivity, such as a long flight), a piece of which may become dislodged and travel to the lungs, a frequently life-threatening event.

Drugs used to "thin the blood" to prevent clotting or to dissolve a formed clot have different mechanisms of action. The drugs were chanced upon in unexpected and sometimes very dramatic circumstances.

This Is Exactly What I Wasn't Looking For!

In 1916 a medical student at Johns Hopkins, Jay McLean, was assigned a research project to find a natural bodily substance to clot blood. Under the direction of William Howell, the world's expert in how blood clots, McLean set out to characterize factors in the human body that promote clotting. Instead, he discovered the exact opposite: a powerful anticoagulant (blood thinner). When his mentor was skeptical, McLean placed a beaker of blood before him and added the substance. The blood never clotted, but Howell remained unconvinced. McLean rushed the report of his discovery into print to establish his priority and to indicate that Howell had played a subordinate role.[1] This would be the only paper he would write on the substance.

Howell shortly labeled the substance heparin (from the Greek *hepar*, the liver) to indicate its abundant occurrence in the liver, although heparin can be extracted from many other body organs.[2] After its physiological and chemical behavior was detailed and the factor purified, heparin became clinically available in 1936, and in the 1940s its use became a standard in treating a variety of venous and arterial clotting disorders. Its major limitation was its inability to be taken orally.

The real nature of the serendipitous discovery was published by McLean forty-three years later. He wrote: "I had in mind, of course, no thought of an anticoagulant, but the experimental fact was before me."[3]

STREPTOKINASE: CLOT-BUSTING

In the early 1930s William Tillet, a bacteriologist at the New York University School of Medicine, was studying streptococci and the body's defensive ability to clump these bacteria together. At the time, the process by which blood clots are formed and dissolved was not his focus of interest. In one phase of his experiments, a particular specimen of blood initially clotted, but then, just before discarding his test tubes, Tillet happened to look at them again. To his surprise, the contents of the tubes had liquefied.

Tillet pursued this unexpected observation — which was outside his area of interest — and in 1933 reported that this species of streptococci produces a protein that inactivates a critical factor in the formation of a clot. Ultimately named "streptokinase," this protein has become useful as a clot-buster if given to a person in the initial stages of a heart attack.[4]

NOT-SO-SWEET CLOVER

One Saturday afternoon in February 1933, in the middle of a howling blizzard, a Wisconsin farmer appeared in the office of chemist Karl

Paul Link, where he presented him with a dead cow and a milk can containing blood that would not clot. The man had driven almost two hundred miles from his farm to seek help from the Agricultural Experiment Station at the University of Wisconsin in Madison. But it was a weekend, and the veterinarian's office was closed, so chance led him to the first building he found where the door was not locked: Link's biochemistry building. Many of the farmer's cows had recently hemorrhaged to death. He had been feeding his herd with the only hay he had: spoiled sweet clover.

This hemorrhagic disease had been recognized first in the 1920s after farmers in the northern plains states began planting sweet clover imported from Europe, which survives in the harsh climate and poor soil, as feed for cattle. Fatal spontaneous bleeding in the cows, along with prolonged clotting time, was traced to spoiled sweet clover.

Sweet clover is so named because the plant is sweet-smelling when freshly cut, but some strains are bitter-tasting and avoided by cattle. Link was hired to investigate this problem and found that one particular chemical in the clover, coumarin, was responsible for both characteristics. Coumarin was even used commercially to scent inferior tobacco and some perfumes. There was no suspicion that the sweet-smelling, bitter-tasting chemical was related to spoiled sweet clover disease.

Link's serendipitous encounter with the farmer forever changed the direction of his research, and it took seven years before he solved the mystery of why only *spoiled* hay caused the bleeding disorder in cattle. When clover spoils, its natural coumarin is chemically transformed into dicumarol, which would be shown to interfere with the role of vitamin K in the clotting process.[5] After engaging in some creative thinking along the lines of "I've found the solution, what's the problem?" Link and his colleagues reasoned that if too much caused a hemorrhage, a minuscule amount might prove to be a useful anticoagulant. Physicians avidly welcomed the new drug.

Link synthesized scores of coumarin variants. Noticing that one of them appeared to induce particularly severe bleeding in rodents, he developed it as a rat poison and gave it the name warfarin. It is still sold today for this purpose under the same generic name.[6] In early

1951 an army inductee tried to kill himself by eating the rat poison, but his surprising full recovery led to the testing of warfarin on human volunteers, and by 1954 it was proven more potent and longer-lasting than dicumarol. Warfarin was promptly marketed, sometimes under the trade name Coumadin, and it is now the standard treatment for all venous or arterial blood clots requiring long-term treatment. It requires careful monitoring. An estimated 2 million Americans take it each day to help prevent blood clots that could result from heart attack, abnormal heart rhythm, stroke, or major surgery.

AN ASPIRIN A DAY KEEPS THE DOCTOR AWAY

Aspirin (acetylsalicylic acid), derived from the bark of the willow tree, has for over a hundred years been marketed for its painkilling and fever-quenching properties. It is the single most commonly used medicine. In the United States alone, more than 30 billion pills are purchased annually.

In the late 1940s a California physican stumbled upon another of aspirin's powers. Tonsillectomy in children was all the rage in the 1940s when Lawrence Craven was working as an ear, nose, and throat (ENT) physician in Glendale, California. After surgery, Craven would tell his patients to chew Aspergum (containing aspirin) to relieve their sore throats. He was surprised to find that those who used more gum bled excessively. Craven speculated that Aspergum reduced the tendency of the blood to clot.

At a time in history when health-care officials were recognizing heart disease as a national epidemic, he took it upon himself to treat all his adult male patients with aspirin to prevent blockage of the coronary arteries and heart attacks. Within a few years, Craven was convinced that heart attack and stroke were occurring far less frequently among his patients than among the general population. His reports on his astonishing series of cases were largely ignored, as they were published in rather obscure regional medical journals and his trials had had no controls.[7]

The mechanism of aspirin's action was not understood until a historic 1971 report by the British biochemist John Vane, who was awarded the Nobel Prize in 1982. Prostaglandins are chemicals released by the body in response to injury. They help blood to clot. Aspirin counteracts prostaglandin's effect of making platelets stick together, thereby "thinning" the blood.[8]

Today one baby aspirin (81 mg) a day is widely recommended for the prevention of heart attack and stroke.

Just as the three princes of Serendip classically came upon things they were "not in quest of," researchers came upon unexpected discoveries enabling a series of cardiovascular breakthroughs. Concentration on one problem may direct thinking, but this may obscure the big picture. Being too focused may exclude a peripheral observation that could reveal the solution to a problem not even originally conceived.

The British geneticist J. B. S. Haldane once commented, "The world is not only stranger than we imagine, it is stranger that we can imagine." This famous quote is often used to support the notion that the mysteries of the universe are beyond our understanding. Here is another way to interpret his insight: Because so much is out there that is beyond our imagination, it is likely that we will discover new truths only when we accidentally stumble upon them. Development can then proceed apace.

Part IV

The Flaw Lies in the Chemistry, Not the Character: Mood-Stabilizing Drugs, Antidepressants, and Other Psychotropics

If we were to eliminate from science all the great discoveries that had come about as the result of mistaken hypotheses or fluky experimental data, we would be lacking half of what we now know (or think we know).

— NATHAN KLINE, AMERICAN PSYCHIATRIST

30

It Began with a Dream

In the United States today, doctors write a total of 3 billion prescriptions each year. An astonishing 10 to 15 percent are for medications intended to affect mental processes: to sedate, stimulate, or otherwise change mood, thinking, or behavior. The term "psychotropic" is used for a chemical substance with such effects.

Psychotropics are generally in four categories: antipsychotic, antianxiety-sedative, antidepressant (mood-elevating agents), and antimanic (mood-stabilizing drugs). Their introduction and subsequent widespread use created a medical and cultural revolution. Thorazine, the first major tranquilizer, appeared in 1954 and immeasurably changed the institutionalization of psychotics. The sedative Miltown was marketed a year later, to be followed within a few years by the antianxiety agents Librium and Valium, and two new classes of antidepressants, iproniazid and imipramine. Lithium, the mainstay in the treatment of manic-depressive disorder, was unexpectedly discovered in 1949 but not marketed in the United States until 1970. LSD, the powerful mind-altering chemical, was discovered in 1943, but only in the first decade of the twenty-first century is it being put to legitimate therapeutic use.

These psychotropic drugs have brought about the era of psychopharmacology, not only changing the clinical practice of psychiatry but also stimulating advances in biochemistry and the neurosciences. Particularly since the 1970s, the psychoanalytically oriented psychiatrists have increasingly yielded to the biological group. The pervasive use of psychotropics has generated revolutionary changes in the understanding and treatments of mental illness. A profound shift has occurred, away from psychiatry centered on providing custody and care of "lunatics" (asylums, locked doors, straitjackets, and solitary confinement) to a much broader conception of mental illnesses and individualized treatment options.

In 2005 a survey by the government's National Institute of Mental Health revealed the high prevalence of mental illness.[1] More than half of Americans (55 percent) will develop a mental illness at some point in their lives, often beginning in childhood or adolescence. The study found that the most common problems experienced are depression, affecting about 17 percent of the population with varying degrees of severity, and alcohol abuse, affecting 13 percent. Phobias are also common.

Through the end of 2004, there were fourteen potential new drugs for depression, eighteen for anxiety disorders, and fifteen for schizophrenia. Overall, there were 109 medications being developed for all mental illnesses.[2]

It was not until the latter decades of the nineteenth century that the complex anatomy of the central nervous system began to be understood. The Spanish neuroanatomist Santiago Ramón y Cajal showed that the direction of the nervous impulse in the brain and spinal cord is from the small end branches (dendrites) of a nerve (neuron) to the cell body of another. However, the mechanism by which nerves transmit their impulses from one to another evoked considerable controversy. How is the stimulus transmitted from one nerve cell (axon) to another across the space of their microscopic junction (synapse) or, in the case of the peripheral or autonomic nerve system, to the end organ? Is it by a chemical messenger or an electric discharge? Arguments raged in the "soup vs. sparks" debates.

The breakthrough in understanding that transmission occurs via a chemical produced by the nerve endings themselves occurred through an almost mystical experience.

Otto Loewi, chairman of pharmacology at the University of Graz in Austria, was a man of many interests. In his early school years he had studied the classics, training that he believed "widened horizons and encouraged independent thought." Throughout his life, he was an ardent music lover. He spent a postgraduate year studying art history but was persuaded by his parents to pursue medicine instead. He found research more challenging than clinical practice, and in 1909 he began his long tenure of the pharmacology chair in Graz.

Loewi had long been interested in the problem of neurotransmission and believed that the agent was likely a chemical substance and not an electrical impulse, as previously thought, but he was unable to find a way to test the idea. It lay dormant in his mind for seventeen years. In a dream in 1921, on the night before Easter Sunday, he envisioned an experiment to prove this. Loewi awoke from the dream and, by his own account, "jotted down a few notes on a tiny slip of thin paper." Upon awakening in the morning, he was terribly distressed: "I was unable to decipher the scrawl."

The next night, at three o'clock, the idea returned. This time he got up, dressed, and started a laboratory experiment. He set up in separate baths of nutrient fluid two frog hearts connected to each other by tiny glass tubes through the blood vessels that supply and drain the hearts. He left the nerve supply of one heart intact and removed the nerves from the other. Stimulation of the vagus nerves — which regulate the body's internal organs — to the first heart produced the well-known effect of slowing the heartbeat. Loewi then drained the fluid from the first heart's bath and transferred it to the second, nerveless heart. Loewi was euphoric to see that it produced the same effect of slowing the heartbeat. Clearly, a chemical substance had been secreted from the nerve ending of the vagus. Loewi named this neurotransmitter *Vagusstoff* (vagus stuff). It was identified chemically by others some years later as acetylcholine.[3]

Loewi won the Nobel Prize in 1936, but events soon overtook him. In March 1938 the Nazis marched into Austria and assumed power. Stormtroopers arrested the rotund, avuncular-looking scientist at gunpoint and jailed him along with hundreds of other Jewish males. Fearful that he would be killed, Loewi was desperate to record the results of his current research. He had just completed the very last of a series of experiments begun almost twenty years earlier to show that nerves taking messages to the brain secrete a different chemical than do those coming from the brain. He persuaded a prison guard to mail a hastily written postcard summarizing his findings to the journal *Die Naturwissenschaft,* and felt "indescribable relief" that his results would not be lost.

After transferring his Nobel Prize monies to a Nazi-controlled bank, the sixty-five-year-old Loewi was allowed to leave Austria. He was given refuge by friends and colleagues in England. In a short time he was offered a research professorship at New York University School of Medicine. In applying for an American visa, Loewi encountered a farcical situation. The U.S. consul in London required formal proof of his experience as a teacher. Loewi suggested that the consul contact Sir Henry Dale, but the consul had no idea of Dale's eminence. Perhaps the consul should consult Dale's entry in *Who's Who,* Loewi suggested. The volume was duly obtained and approval given. With visa finally in hand, Loewi politely inquired of the consul if he knew who had written the entry on Dale in *Who's Who.* Loewi then informed him that it was he himself.

To add insult to injury, as he was passing through immigration in New York, Loewi glimpsed the medical certificate from London. It read: "Senile, not able to earn his living." A wise immigration officer ignored this observation, and Loewi went on to an active career at NYU for the next fifteen years.

It was not until twenty-four years after receiving the Nobel Prize that Loewi, at age eighty-seven, revealed that the solution had come to him in a dream. He wrote that the process of his discovery "shows that an idea may sleep for decades in the unconscious mind and then suddenly return. Further, it indicates that we should sometimes trust a sudden intuition without too much skepticism."[4]

Within a few decades, some mental states of mood, thinking, or behavior came to be recognized as arising through an error in synthesis, storage, metabolism, or release of neurotransmitters in the brain. Chief among these chemicals identified were serotonin, dopamine, and noradrenaline. Released by a nerve cell at its junction with another, such a chemical transmitter is picked up by receptors, or binding sites, on the latter, initiating a series of chemical changes inside the cell with specific effects on mood, behavior, or thought. The entire system of release of a neurotransmitter into the synaptic junction and reabsorption by the neuron is mediated by a series of enzymes. It is these neurotransmitters that may be modified by the action of drugs. These brain neuroreceptors and transmitters came to be seen as factors constituting and influencing the basis of mind.

Thorazine, for example, blocks the dopamine-binding receptors in the forebrain. The antidepressant drug imipramine blocks reabsorption of noradrenaline and serotonin, prolonging the time they spend in the synapse. The longer they act on the neuron's receptors, the less depressed a patient is.

One dominant fact courses through this revolutionary development, recognized by psychiatrists, pharmacologists, and archivists alike. Serendipity, not deductive reasoning, has been the force behind the discovery of virtually all the psychotropics: Chlorpromazine (Thorazine) was unearthed in a hunt for antihistamines. The first antianxiety drug meprobamate (Miltown) was chanced upon in a search for an antibiotic. Valium was found by a scientist tidying up a lab bench. Antidepressants were uncovered as a side effect in the treatment of tuberculosis or stumbled upon in the quest for a tranquilizer. LSD was a surprising result in experiments for an ergot-derived stimulant. Lithium was happened upon through a misguided theory regarding a metabolic defect in manic states.

It's as if the muse of serendipity set the stage, inspired the main actors, and found an accepting audience.

31

Mental Straitjackets

Shocking Approaches

To fully appreciate the explosive advances in psychiatric treatment brought about by pharmacotherapy (psychotropics) since its advent in 1949, one must put them in the context of how little was known about diseases of the mind in the first half of the twentieth century. The radical nature of the treatments used for mental illness in earlier times is reflected in the title of Elliot Valenstein's chronicle of this era, *Great and Desperate Cures.*[1] These cures were often based on serendipitous clinical observations made in mental hospitals in Europe.

FEVER THERAPY

Psychotic symptoms of advanced neurosyphilis accounted for a large proportion of mental hospital admissions in the nineteenth and early twentieth centuries. The first successful treatment for any mental illness originated in the observation that certain psychotic patients improved dramatically when suffering from acute fevers caused by various diseases.[2] In 1917, on this association, Julius Wagner von Jauregg, the director of the psychiatric division of the Vienna General Hospital, introduced the malaria treatment. One of his patients, a soldier sent back from the Macedonian front with shell shock, had malaria. His blood, which contained malaria parasites, was injected intramuscularly into nine patients suffering from insanity caused by advanced syphilitic infection. Multiple episodes of high fever, intended to destroy the spirochetes (the syphilis microbes) that had invaded the brain, were controlled with quinine.

Several of the patients demonstrated impressive remissions, some of which lasted ten years. Wagner von Jauregg approached psychiatry as an experimental biologist.[3] This therapy was directed against an underlying brain disease expressing itself in psychiatric manifestations. This approach for neurosyphilitic psychosis was path-breaking in that it ended the nihilism of earlier generations who had not experimented with treatments at all. If the neurosyphilitic psychoses could be halted, perhaps psychotic illness from other causes was treatable as well.[4]

INSULIN COMA

In the 1920s, doctors in England, France, and particularly the Soviet Union dealt with the acute excitement states of catatonic and manic patients by trying to keep them in a continuous state of sleep, accomplished through the use of hypnotics. It was a short stumble to go from induced continuous sleep to induced temporary coma.

In 1933 in Vienna, insulin shock treatment — by means of hypoglycemic comas — was introduced by Manfred Sakel, who claimed to be a direct descendant of Moses Maimonides, the twelfth-century rabbi, physician, and philosopher. Insulin shock was the first treatment method ever to be directed at schizophrenia, the most common and most hopeless illness among mental patients in institutions.[5]

Shock treatment came about through a chain of serendipitous events when Sakel was working as an assistant physician at the Lichterfelde Sanitorium in Berlin. Sakel accidentally administered an overdose of insulin to one of his morphine-addicted patients who also happened to be diabetic. The level of glucose — the main source of energy for the brain — in the patient's blood dropped so severely that a mild coma ensued. Upon recovery, "the patient's mind, until then completely befogged, was absolutely clear," and his craving for morphine had subsided. Emboldened, Sakel gave insulin overdoses to other drug addicts and, in 1930, published a report on his successes.

Again by accident, he gave a significant overdose of insulin to a drug addict who also happened to be psychotic, putting him into a deep coma. When the patient came out of it, Sakel thought he detected signs of mental improvement. Had he stumbled on a new approach? Confirming by experiments in animals — conducted, according to Sakel, in his kitchen — that such comas could be controlled by prompt injections of glucose, Sakel began inducing deep comas in schizophrenics. He relocated to Vienna and, after five years of work, announced his insulin shock cure before the Vienna Medical Society in November 1933. The psychiatric world initially found the results incredible, but soon specialists came from all over the world to observe the treatment and results directly. Sakel's reputation grew with his shock treatment of Vaslav Nijinsky, the famous ballet dancer who was stricken with schizophrenia. For the first time, Nijinsky was able to leave the sanitarium where he had been confined since 1919. Nevertheless, he spent the last thirty-two years of his life in an insane asylum.

Ardently espoused and publicized by Sakel, particularly after he fled to the United States in 1936, two years before the Nazis came to power in Austria, insulin shock treatment was widely adopted. The reports of its success were exaggerated, received enthusiastically by the popular press, and not critically analyzed by the psychiatry profession.

Foster Kennedy, professor of neurology at the Cornell Medical School and Bellevue Hospital, celebrated the victory over traditional psychoanalysis, in the preface to the 1938 American edition of Sakel's work: "In Vienna, at least one man had revolted from the obsession that only psychological remedies could benefit psychological ills. . . . We shall not again be content to minister to a mind diseased merely by philosophy and words." This tribute must have offered some balm to the wound incurred when Sakel's application for membership in the psychoanalytically dominated American Psychiatric Association was initially refused. A report by an American adherent of Sakel detailing early clinical experience with insulin-coma therapy at Bellevue Hospital was published in a neurology journal after being rejected by psychiatry journals.

Doctors induced a state of unconsciousness in their patients by administering overdoses of insulin. The patient would go into convulsions and then, anywhere from thirty minutes to several hours later, be brought out with glucose administered through a stomach tube or intravenously. There were impressive results in patients in the early stages of schizophrenia who were treated with a series of as many as fifty of these hypoglycemic comas over six to ten weeks. Some patients were fully cured of their psychotic symptoms, while others saw only temporary improvement. Complications of the procedure were not widely publicized. The death rate for the treatment averaged about 6 per 1,000. Death occurred when a patient could not be revived from the coma. Nonfatal brain damage occurred in about 8.5 cases out of 1,000.

Insulin shock treatment was a turning point in psychiatry. For the first time in medical history, there was real hope for the mentally ill. It transformed mental institutions from exclusively custodial facilities to centers of treatment and rehabilitation. But it certainly was not without its drawbacks.

Extensive studies in 1939–40 showed that most patients did not derive long-term benefit from insulin shock treatments, although some did sustain a slight improvement. Interest in the procedure declined in the 1940s and disappeared with the introduction of modern drugs in the 1950s.

CHEMICALLY INDUCED CONVULSIONS

Based on observations made during autopsies on the brains of epileptics and schizophrenics, neuropathologist Ladislaus von Meduna in Budapest came to believe that there were differences in the nerve cells between the two and formulated the theory that epileptics would not suffer from schizophrenia. In other words, he believed schizophrenia and epilepsy were biologically exclusive diseases. In 1934 he began treating schizophrenia patients with Metrazol, a modified version of camphor, which is derived from laurel and has a long history of causing seizures.[6] It caused considerable pain at the injection site, but this was not the worst of its effects. The induced epileptic fit was a particularly distressing procedure because it evoked terrible fright in a patient as he felt himself seized by a powerful, all-encompassing force.

While somewhat useful in treating schizophrenia, Metrazol was found a few years later in the United States to be most effective against depression. By 1940 almost all major mental institutions had it among their treatments, but within a few years it was replaced by electroconvulsive therapy.

A TRULY SHOCKING APPROACH: ECT

The third major convulsive treatment, electroshock, was introduced in Rome by physicians Ugo Cerletti and Lucio Bini in 1938. Long interested in the possibility of inducing convulsions with electricity, they had been concerned about the safety of the procedure in humans. Cerletti wrote, "The idea of submitting man to convulsive electric discharges was considered utopian, barbaric and dangerous: in everyone's mind was the spectre of the 'electric chair.'" This fear was not unfounded, since their initial animal experiments often resulted in death. When Bini, working as the electrotechnician, realized that the electrodes originally placed in the mouth and anus resulted in the electric current passing through the heart and stopping it, he changed their position to the sides of the head and none of the animals died.

Later, on a visit to one of Rome's slaughterhouses, Cerletti and Bini observed that pigs were stunned into unconsciousness by electric shocks applied to both sides of the head before having their throats cut. After this visit, they began conducting numerous experiments on dogs by applying an electric current through the skull and inducing convulsions and then modified the technique for its use in man for depression and schizophrenia.

Their first hapless human subject was a thirty-nine-year-old schizophrenic found wandering around the Rome train station by the police. During a phase of convulsions, an individual typically suspends breathing temporarily and may turn blue — a condition called cyanosis — from the lack of oxygen. Following this first experience in a human, Cerletti described their alarm:

We watched the cadaverous cyanosis of the patient's face . . . it seemed to all of us painfully interminable. Finally, with the first stertorous breathing and the first clonic spasm, the blood flowed better not only in the patient's vessels but also in our own. Thereupon, we observed with the most intensely gratifying sensation the characteristic gradual awakening of the patient "by steps." He rose to sitting position and looked at us calm and smiling, as though to inquire what we wanted of him. We asked: "What happened to you?" He answered: "I don't know. Maybe I was asleep." Thus occurred the first electrically produced convulsion in man, which I at once named electroshock."[7]

After eleven applications of electroconvulsive therapy (ECT), the patient lost his notions of persecution and hallucinations, and was discharged from the clinic a month later and returned to his work as an engineer.

ECT was widely adopted and replaced chemically induced epileptic fits. It was recognized as an almost specific treatment for not only recent but also long-standing severe depressions, and is still used today in drug-resistant depression. Sylvia Plath's fictionalized account of an ECT session in *The Bell Jar* (1963) uses analogy to portray the feeling of loss of self: "A great jolt drubbed me till I thought my bones would break and the sap fly out of me like a split plant."[8] Better anesthetics and chemical relaxants have made ECT less traumatic.

32

Ice-Pick Psychiatry

Surgical approaches were adopted in the 1920s at a few medical centers on the rationale that certain disordered organs underlay mental illnesses. Some endocrine glands were cut away, but the most bizarre rationale in the United States was surgery to eliminate the presumed sources of infection in the body that released toxins affecting the brain: besides removal of teeth and tonsils, resections of the stomach, colon, cervix, and uterus were undertaken.[1] This was germ theory gone mad.

The era of psychosurgery was introduced by António Egas Moniz in 1938.[2] A Portuguese aristocrat, Egas Moniz held the chair of professor of neurology at the Faculty of Medicine in Lisbon from 1911 and, for many years, pursued a second career as a politician. He was briefly ambassador to Spain in 1917, before becoming minister of foreign affairs and leading the Portuguese delegation at the Paris Peace Conference. He was a celebrated neurologist who pioneered cerebral angiography, the process of X-ray examination of the pattern of arteries in the brain by injecting radiopaque solutions via its feeding vessels. He used this knowledge to develop a method of diagnosing and localizing brain tumors by visualization of abnormalities in the arterial pattern. In 1927 he announced these results.

"Gage Was No Longer Gage"

The frontal lobes of the brain, regions just behind the forehead and in the front of the motor areas, had long been considered the "silent lobes," since they lack the easily identifiable sensory and motor areas of other parts of the brain. This medical terra incognita began to be mapped with much help from the famous case of Phineas Gage.

In 1848 Gage was a twenty-five-year-old railway construction foreman who was highly respected by his men for his judgment and competence. One of his major tasks was overseeing the blasting of rock to lay new tracks. The detonation was prepared in a series of steps: drill the hole in the rock, fill it halfway with explosive powder, insert a fuse, tamp in a layer of sand to direct the explosion inward, and finally light the fuse. But on this fateful day, Gage was distracted for a critical second. Thinking that the man working with him had already inserted the sand, he began tamping the powder directly with his specially designed iron bar. A spark ignited the powder, and the explosion hurtled the three-and-a-half-foot-long tamping iron right through his head, specifically through his frontal lobe, carrying off about half a cupful of brains with it.

Poor Phineas survived for thirteen years but was never the same. He became reckless, profane, stubbornly willful, and lost any concept of the future, yet was unaware of the changes to himself. To his friends, "Gage was no longer Gage."[3] Gage's case was the first reported instance of loss of intellectual and behavioral function from damage to the frontal lobes.

In 1861 Paul Broca, the French surgeon and anthropologist, attributed the higher functions of the brain to the frontal lobes, the part of Gage's brain that had been largely destroyed: "The majesty of the human is owing to . . . judgment, comparison, reflection, invention, and above all the faculty of abstraction. . . . The whole of these higher faculties constitute the intellect."[4] Further understanding was advanced by observation of soldiers in World War I who had suffered frontal-lobe damage from gunshot wounds. These were characterized by personality changes without damage to any vital body process.

The frontal lobes are the most recently evolved part of the

nervous system. They execute all higher-order purposeful behavior. They are crucial in focusing on a goal, developing plans to reach it, and monitoring the extent to which it is accomplished. The frontal lobes free man from fixed routines of behavior and cognition, and allow for imagination, judgment, and identity. With the loss of such functions, an individual loses himself, loses his "soul," and is indifferent to or ignorant of this loss.

Egas Moniz was sixty-one when he attended the 1937 London conference at which two American physiologists from Yale University reported the results of surgically removing the frontal lobes of chimps. Before the operation, the chimps had been adept at solving certain types of puzzles, but they were also excitable. Afterwards, they were unable to solve any problem involving extensive use of short-term memory or integration of data over time. Yet they were placid and imperturbable. Egas Moniz was particularly impressed with the elimination of "frustrational behavior" — angry, impatient actions in situations of frustration — in the animals and decided to try the method on humans with anxiety states or "disturbing" social behavior. Within the next year, without conducting experiments on animals to test the safety of the procedure, he performed lobotomy — or leucotomy, as he called it — on twenty patients. (The term "leucotomy" is composed of two segments derived from the Greek: *leuco-*, referring to the white matter of the brain, and *-tome,* meaning knife.) Egas Moniz claimed that he was scientifically basing his work on an anatomical concept derived from Ramón y Cajal. In truth, however, his speculations were poorly constructed and served merely as an ill-conceived rationalization for his surgery. His University of Lisbon colleague who held the chair of psychiatry labeled them "pure cerebral mythology."[5]

In conducting a lobotomy, Egas Moniz first passed a long-needled syringe containing alcohol through drill holes in the skull and through the brain to destroy the fibers that connect the frontal lobes to the main body of the brain. Then he used a specially designed knife, called a leucotome, to cut away the frontal lobes. Whether or not the patient's anxiety truly abated as a result of the surgery was difficult to prove. What was apparent was that the patient simply did not care. Indifference, apathy, dullness, and disorientation came to dominate the patient's behavior. In a barrage of grandiose publications, Egas Moniz claimed a high success rate of "cure" or "improvement," but these were only very short-term observations and were poorly documented.[6] He performed several more operations before World War II began. Soon thereafter, psychosurgery — a term coined by Egas Moniz — was being performed all over the world.

PERSONALITY, CUT TO MEASURE

In the United States, the procedure was enthusiastically picked up by Walter Freeman, professor of neurology at George Washington University in Washington, D.C. Since he had no qualifications as a surgeon, he required a neurosurgeon, James Watts, as a collaborator. They practiced on brains from the morgue before selecting their first patient. Within only a few months, in the fall of 1936, they performed twenty lobotomies. They varied the technique in several ways and first adopted a narrow steel blade, blunt and flat like a butter knife. Indeed, in assessing its ease of use in arcing through nerve matter, Watts commented, "It goes through just like soft butter." Basically, however, the technique remained blind; surgeons couldn't see or monitor what they were cutting.

They then began to perform the procedures under local anesthesia so that the patient could be asked to sing a song or to perform arithmetic. As long as these functions remained intact, the scalpel chopped away at more and more "cores" of the brain tissue. Amazingly, an editorial in the influential *New England Journal of Medicine* proclaimed: "The operation is based . . . on sound physiological observations." Yet classical analysts were in an uproar. In 1939 an English psychiatrist named William Sargant attended the American Psychiatric Association's convention in St. Louis and witnessed the antagonism and outrage at Walter Freeman, who was in attendance: "They felt so insulted by this attempt to treat otherwise incurable mental disorders with the knife that some would almost have used their own on him at the least excuse."[7] Sadly, the reservations of psychiatrists and psychoanalysts were not voiced to the public.

Freeman was adept at self-promotion. He effectively publicized a widely read article, "Turning the Mind Inside Out," that appeared in the *Saturday Evening Post* in 1941. Freeman had explained the operation to the writer, the science editor of the *New York Times,* as one that separated the prefrontal lobes, "the rational brain," from the thalamic brain, "the emotional brain." The following year, Freeman and Watts's book *Psychosurgery* stimulated worldwide professional interest. Freeman himself wrote the jacket copy, and in a phrase that reveals not only his hubris but the anticipation of wider applications than those he first cautiously espoused, he claims: "This work reveals how personality can be cut to measure."[8] For many years Freeman had a "psychosurgery exhibit" at the AMA's annual meeting. There he buttonholed journalists and, according to his partner, Watts, often behaved "like a barker at a carnival," using "a clicker which made a sharp staccato noise" to attract attention.

In 1945 Freeman undertook a new surgical approach by himself: transorbital lobotomy, also known as "ice pick" lobotomy. After practicing on corpses, he developed a technique of perforating the back of the bony orbit (the eye socket) behind the eyes with a modified ice pick he found in his kitchen drawer,[9] to cut the nerves at the base of the frontal lobes. Now the procedure could be accomplished more widely by physicians without the need for a neurosurgeon. The

---

"I'd Like to Pick Your Brain"

Professor Walter Freeman became an indefatigable evangelist for the ice-pick lobotomy procedure. As Freeman's biographer Elliot Valenstein describes it, "On one five-week summer trip [in 1951], he drove 11,000 miles with a station wagon loaded, in addition to camping equipment, with an electroconvulsive shock box, a dictaphone, and a file cabinet filled with patient records, photographs, and correspondence; his surgical instruments were in his pocket." The most important of his surgical instruments was the ice pick. He performed the procedure at mental institutions across the country, demonstrating it before audiences, on as many as thirteen patients in one day.

popular press hailed it as a major advance, while many in the profession viewed it as "not an operation, but a mutilation."

The awarding of the Nobel Prize for Medicine to Egas Moniz in 1949 for prefrontal leucotomy broke down any barriers of resistance among superintendents of overcrowded, underfunded asylums for the mentally ill. "A new psychiatry may be said to have been born in 1935," proclaimed the *New England Journal of Medicine*, "when Moniz took his first bold step in the field of psychosurgery." Within eight months, 515 transorbital lobotomies were performed in Texas alone. More people were lobotomized in the three years after Egas Moniz received the prize than in the previous fourteen years.

Two lobotomy cases in particular reached national consciousness:

At the request of Joseph Kennedy, ambassador to Britain and father of JFK, his daughter Rose Marie (Rosemary) when in her early twenties was lobotomized by Freeman and Watts at the George Washington University Hospital in 1941. She was mildly retarded and very outgoing, and the family feared she might embarrass them by getting pregnant. After the procedure, she lost her personality and regressed into an infantlike state. She lived out the rest of her long life in a private sanitorium run by nuns.

Frances Farmer, a 1930s film star and radical political activist, had rebelled against authority and injustice all her life. The institutions of society, law, and psychiatry converged to restrain this behavior. Institutionalized at a state hospital, she was subjected to insulin shock and electroconvulsive treatments and then underwent transorbital lobotomy by Freeman in 1948.[10] A 1982 Hollywood movie, *Frances,* is graphic in its portrayal of these events.

In the severely psychotic, the operation did result in relief from compulsive and morbid anxieties. Although aberrant behavior often continued, many patients were able to live outside an institutional setting. Many, however, became the "walking wounded," with loss of imagination, foresight, sensitivity, and some loss of individual personality. Initially proposed as an operation of last resort, it had become the first step in creating a manageable personality.

By 1955, more than 40,000 people in the United States had undergone psychosurgery to treat psychotic and various other forms of abnormal behavior. Nearly twice as many women as men were lobot-omized.[11] Freeman himself was responsible for 3,500 operations.[12]

In March 1954 Thorazine (chlorpromazine) was approved by the Food and Drug Administration. This and the rapid succession of other psychoactive drugs led to the marked decrease and ultimate abandon-ment of lobotomy. Not only had simple, inexpensive, and effective al-ternatives arisen, but the awareness had grown that lobotomy had created many brain-damaged people. The procedure fell into disre-pute, replaced by psychopharmacology.

"THE WRONG END OF THE RIGHT PATH"

Looking back, it is remarkable that the theoretical bases for all these modes of somatic treatment were so ill-founded and yet so widely ac-cepted. Theories were proposed, sometimes forcefully, and were often so vaguely stated that they strain scientific plausibility. There may have been critics, but it is a fact that these radical treatments were quickly embraced by mainstream medicine. Psychotic conditions did not lend themselves to treatment by psychoanalysis, and many considered Freud's contributions to be greater in the field of literature than in medicine.

An English psychiatrist, speaking of his experiences at a mental institution in the period before the new organic treatments, is disarm-ingly frank:

> We developed a self-protective intellectual atmosphere of its own. . . . We improvised special formalities, such as recording very lengthy case-histories for every admission [sometimes running] to more than thirty pages of de-tailed information. Today such long screeds would gener-ally be laughed at, the simple treatments now available often making them superfluous, but then they gave us a feeling that we were doing something for the patient by learning so much about him, even if we could not yet find any relief for his suffering. We also compiled "social" his-tories. . . . This, too, was often a waste of time, but what else could one do? Nowadays [we have no] need for elab-

orate case-history or social investigation, still less for the former eternity of talk.[13]

While accepting that "no one really claimed to understand the cause, nature or cure of this dread malady [schizophrenia]," the *New England Journal of Medicine* in 1938 applauded the idea that these new therapies, including shock treatment and ice-pick surgery, seemed "to bring the field of psychiatry a little closer to that of general medicine."

Medicine has long accepted that successful results may be shown empirically before an understanding of a treatment's scientific basis is reached. Manfred Sakel offered an insight on the bounty of accidental discovery:

> I have a high regard for strict scientific procedure and would be glad if we could follow the accustomed path in solving this special problem: it would have been preferable to have been able to trace the cause of the disease first, and then to follow the path by looking for a suitable treatment. But since it has so happened that we by chance hit upon the wrong end of the right path, shall we undertake to leave it before better alternatives present themselves? . . . For it should perhaps now enable us to work backwards from it to the nature and cause of schizophrenia itself.[14]

33

Lithium

In 1948 the thirty-seven-year-old Australian John Cade was, by his own description, "an unknown psychiatrist, working alone in a small chronic hospital with no research training, primitive techniques and negligible equipment."[1] A survivor of three years in a Japanese prisoner-of-war camp, he emerged with his scientific curiosity intact and took a position at a mental institution in Victoria, Australia, focusing his attention on manic-depressive illness. On the basis of an unsound hypothesis, he conducted tests that led to one of the most significant discoveries in the history of pharmacotherapy, thanks to a chance selection of certain materials for his experiments.

The stereotype of a manic is that of a person talking very fast and becoming euphoric. In milder cases of acute mania, the patient can be easily annoyed and very hostile. There are also so-called mixed states, in which patients have such symptoms and are depressed as well. In more manic stages, thoughts rapidly flow, a surge in creativity is felt, social inhibitions are lost, and schizophrenic-like delusions may occur.

In thinking about the nature of manic-depressive illness, Cade drew an analogy to the behavior of patients afflicted with thyroid disorders. He had observed that extreme hyperactivity of the thyroid gland (thyrotoxicosis) seemed to cause a form of mania, while a marked depletion of thyroid function (myxedema) seemed to trigger depression. Cade asked whether mania might similarly be a state of intoxication produced by a circulating excess of some metabolite, while depression, where it was associated with mania, might correspondingly be due to its absence or relative lack. He reasoned that the manic patient would be expected to excrete the chemical or its breakdown product in greater magnitude than other categories of patients or normal individuals. Accordingly, he collected urine samples from manic patients, melancholics, and schizophrenics, as well as from normal individuals.

Working single-handedly in a pantry laboratory attached to a psychiatric hospital in Bundoora, a suburb of Melbourne, Cade conducted tests that were relatively crude but nonetheless effective. To test for the presence of a toxic substance, he concentrated the urines and then injected it into guinea pigs. If sufficient urine was injected, the guinea pigs developed severe toxic convulsions, fell unconscious, and died. All the urine samples proved fatal to the animals, but those from some of the manic patients were found to be by far the most toxic. He established that the toxicity was caused, unsurprisingly, by urea (the main substance in urine), but urea was not more abundant in the urine of manics. He postulated that the toxicity of urea might be heightened by the presence of uric acid and that this enhancement might occur to the greatest degree in manic patients.

It was at this point in carrying out tests to measure the toxicity of urea in the presence of varying concentrations of uric acid that Cade ran into difficulties in preparing solutions of the highly insoluble uric acid. The problem was overcome by using its most soluble salt, which was lithium urate. "And," Cade relates in a disarming admission, "that is how lithium came into the story."

The result took him aback.

Far from increasing the toxic effect of urea, the lithium urate exerted some kind of protection. Moreover, and what was equally unexpected, the normally jittery guinea pigs lost their natural timidity and became instead placid, tranquilized, and generally lacking in responsiveness to stimulation. Cade turned them on their backs, and instead of frantically trying to right themselves and scurry away, they lay still and gazed serenely back at him. Further experiments confirmed that this calming action was due not to the urate component but to the lithium ion itself.[2]

Lithium is a metallic element that is abundant. It is found in most rocks of volcanic origin. In 1817 it was discovered in a Swedish iron mine and quickly became the object of health claims. The British physician Sir A. B. Garrod advocated rubbing lithium and rosewater onto gouty joints to dissolve the crystal deposits of uric acid. Others promoted lithium preparations as nostrums for indigestion and gallstones. A St. Louis merchant, C. L. Grigg, marketed a fruity drink he named Bib-Label Lithiated Lemon-Lime Soda. A doctor's testimonial promised "an abundance of energy, enthusiasm, a clear complexion, lustrous hair, and shining eyes." Grigg later devised a punchier slogan — "You Like It, It Likes You" — and a new name, 7UP. Later, lithium was taken out of the soft drink, at about the time of reports of cardiac toxicity from its use as a salt substitute.

Bottled Lithium

In Lithia Springs, Georgia, Lithia Springs Mineral Water is still sold to devoted self-healers, one of whom is quoted as drinking a gallon a day because "it keeps my nerves steady." Other bottled curative waters, many of which are still on the market — Vichy, Apollinaire, Perrier, Lithée — were all promoted at one time for their high lithium content.

In the wake of the surprising results he saw in his becalmed guinea pigs, Cade then tested lithium salts on himself in the clinical doses contemplated. Aware that lithium salts had been used clinically during the nineteenth century to treat such diverse conditions as epilepsy, gout, cancer, and diabetes, Cade considered self-administration safe.

Finding no harmful effects, he then gave the lithium salt to a fifty-one-year-old male patient who had been in a state of manic excitement for five years. After five days, there was a clear improvement in the patient's condition, and within three weeks he was considered well enough to be transferred to a convalescence ward for the first time. In his fourth month of continuous treatment, he was sent home with instructions to take lithium carbonate daily. The man was even able to return to his former occupation. Unfortunately, neglecting to take his medication over several weeks led to irritability and erratic behavior and necessitated his readmission. Back on lithium, the patient strikingly improved and was discharged once again.

Cade achieved similar dramatic success with nine other manic cases. He watched, thunderstruck, as their raging moods subsided. All showed rapid and marked improvement.[3]

The serendipitous nature of the discovery of the therapeutic effect of lithium is stunning. Cade openly acknowledged: "My discovery of the specific anti-manic effect of the lithium ion was an unexpected . . . by-product of experimental work I was doing to test a hypothesis regarding the etiology of manic-depressive illness."[4] The decision to use lithium in the experiments rested solely upon its known solubility.

Doctors were slow to begin using lithium for manic patients. Cade's 1949 report appeared in an obscure regional publication, the *Medical Journal of Australia,* not usually read by the major international scientific and clinical centers.[5] Furthermore, Cade did not take an active role in proselytizing, with his next report not appearing until more than twenty years later, in 1970, at which time he was president of the Australian–New Zealand Psychiatric Association. His breakthrough discovery would have languished if not for the persistent efforts of a clinical researcher named Mogens Schou, professor of biological psychiatry at Århus University in Denmark, who undertook intense investigations of the biology and clinical actions of lithium salts, confirming Cade's claims.[6] Schou was drawn to the investigation because manic-depressive illness ran in his family. He and a number of his relatives benefited from lithium treatment.

It was not until the mid-1960s that anything like general awareness of lithium's dramatic effect on mania dawned on the psychiatric profession. By 1972 there were more than seventy reports on the treatment of mania with lithium, with improvement rates between 60 and 100 percent. It became accepted in European and English psychiatric practice as a highly effective and safe treatment for manic-depressive illness, both for the treatment of acute mania and for reducing the frequency and severity of recurrent mania and depression.

For a variety of reasons, lithium salts were not readily accepted into American medical practice. One reason was the notoriety attached to lithium salts following several cases of severe intoxication and even death during the 1940s among patients using large, uncontrolled amounts of lithium chloride as a salt substitute while on a sodium-restricted regimen for congestive heart failure or renal failure. (Such use was discontinued in 1950.) It is now known that lithium salts are contraindicated in these conditions. Another factor contributing to slow development of lithium therapy was the lack of commercial interest by pharmaceutical companies in this inexpensive, unpatentable mineral. However, evidence accumulated to support the usefulness and safety of lithium, and it finally appeared "as a public service" on the American market in 1970, more than twenty years after Cade's report.

An estimated 1 to 3 percent of Americans, or between 3 and 9 million people, have manic-depressive illness, now more commonly known as bipolar affective disorder. "Mania is sickness for one's friends," the poet Robert Lowell, a longtime sufferer declared, "depression for one's self." The affliction is unusually prevalent among poets: Lowell, Theodore Roethke, John Berryman, Anne Sexton, and Sylvia Plath. In this group, suicide claimed an unusually high number.

An Unquiet Mind

Kay Redfield Jamison, long afflicted with bipolar disorder and a professor of psychiatry at the Johns Hopkins University School of Medicine, brilliantly recorded the fluctuating pains and pleasures of the intense dichotomy in her 1995 memoir *An Unquiet Mind*: "I have often asked myself whether, given the choice, I would choose to have manic-depressive illness. If lithium were not available to me, the answer would be a simple no — and it would be an answer laced with terror. But lithium does work for me, and therefore I suppose I can afford to pose the question. Strangely enough, I think I would choose it. . . . Because I honestly believe that as a result of it I have felt things, more deeply; had more experiences, more intensely. . . . And I think much of this is related to my illness — the intensity it gives to things and the perspective it forces on me."

The simple lithium ion remains the first-line treatment for mania and the prevention of recurrent attacks of manic-depressive illness. Its dramatic surprise discovery marked the dawn of the modern era of psychotropic medicine.[7]

34

Thorazine

In the middle of the twentieth century, people believed that shock occurred through the release of histamine by the autonomic, or involuntary, nervous system. (Today, we know that this is not the case.) Military surgeons operating on soldiers in shock faced many problems. In 1949 Henri Laborit, a thirty-five-year-old French naval surgeon stationed at the Maritime Hospital in Bizerte, Tunisia, undertook a clinical investigation to see if an antihistamine could reduce the symptoms of shock. He obtained a synthetic antihistamine, promethazine, from the Rhône-Poulenc drug company and introduced it into surgery.

The origin of this drug dated back as far as 1883, when phenothiazine was first synthesized in the course of chemical analyses on a dye called methylene blue that was derived from coal tar, a by-product of coal gas production. Thereafter, a remarkable succession of fortunate happenings occurred. Paul Ehrlich, the eminent researcher, found uses for dyes in the staining of biological tissues, which prompted his idea that they might also be of value in selectively destroying microorganisms. Methylene blue was noted to have antimalarial activity. By the time of World War II, this led to Rhône-Poulenc's pursuit of synthesizing phenothiazine amines in the hope of uncovering a new antimalarial agent. (Fortunately, the French researchers did not see a scientific paper that appeared in the American literature around this time, showing that these products did not work against malaria, or they would have abandoned their research project.) In the fall of 1944, the investigators noted the antihistamine activity of some of these products and marketed one of these, promethazine, as Phenergan. Like many antihistamines, it had a strong sedative effect.

Laborit was immediately impressed by promethazine's tendency to promote what he termed a "euphoric quietude" in his patients. "Our patients are calm, relaxed, and euphoric even after major operations; they appear to really suffer less."[1] Promethazine and similar compounds replaced barbiturates preoperatively and morphine postoperatively. Surgeons found that it put patients into such a state of diminished arousal that they needed less anesthesia during surgery. This latter finding caused Rhône-Poulenc to shift its focus from antihistamines to drugs that affect the central nervous system, with the goal of developing even better drugs to enhance the effects of anesthesia. The potential market was enormous. The company's researchers synthesized chlorpromazine (CPZ), a phenothiazine derivative, in December 1950.

Laborit was transferred to Val-de-Grâce, the famed military hospital in Paris, in early 1951, where he studied the effects of CPZ upon his surgical patients. His results were groundbreaking. In a 1952 report summarizing his experience with CPZ in sixty patients, he made the first published suggestion that the drug could be used in psychiatry.[2]

Extensive clinical tests of CPZ were begun by Jean Delay, a highly respected and influential psychiatrist, and his associate, Pierre Deniker, at the Ste. Anne mental hospital in Paris, and the positive results of these trials were promptly embraced by other psychiatrists in France.[3]

"By May 1953," writes one historian, "the atmosphere in the disturbed wards of mental hospitals in Paris was transformed: straitjackets, psychohydraulic packs and noise were things of the past! Once more, Paris psychiatrists who long ago unchained the chained, became pioneers in liberating their patients, this time from inner torments, and with a drug: chlorpromazine. It accomplished the pharmacologic revolution of psychiatry."[4]

A SERENDIPITOUS OVERSIGHT

An accidental discovery over the next two years, as further clinical trials in England, Europe, and Canada were promoted by Rhône-Poulenc, provided another astonishing breakthrough. Dr. Heinz E. Lehmann in Montreal published the first North American report on CPZ in psychiatry in 1954, having found it most useful in manic patients. But he soon uncovered an inadvertent occurrence:

> About three months after the trial had ended, we discovered that some of the chronic, back-ward schizophrenics had been accidentally left on large doses of CPZ. And incredibly, to us, four or five of these back-ward patients were getting better. No one believed that a pill could cause remission in schizophrenia, and we seemed to be getting the best results with chronic paranoids, the group most refractory to treatment.[5]

Large clinical studies in Europe and the United States then confirmed Lehmann's observations: CPZ was effective in palliating and, in almost 75 percent, "socially" curing acute schizophrenia. In chronic schizophrenia, the drug improved 40 percent and "socially" cured 18 percent of patients.[6]

Institutionalized patients with aggressive, violent, and destructive behavior, kept in restraints, in locked wards, receiving repeated electroshock treatment to control explosive behavior, became calm, cooperative, and communicative. The improvement was so dramatic that most of them were able to go home as long as they received continuous outpatient management.

In May 1954, following a licensing agreement with Rhône-Poulenc, Smith Kline & French Laboratories of Philadelphia began to market CPZ in the United States under the trade name Thorazine. Like other antipsychotic agents, Thorazine acts by blocking dopamine receptors and inactivating certain sites of dopamine transmission in the forebrain. This strongly suggests that the underlying chemical defect in the brains of schizophrenics is either an excess of dopamine formation or a supersensitivity to dopamine receptors. Thorazine is easy to use and quick-acting, and its acute toxicity is practically zero, a useful feature in the treatment of acute schizophrenia and mania. Side effects are often extensions of the many pharmacological actions of these drugs. The major liability of Thorazine and similar drugs is a pernicious side effect of excessive dosage: tardive dyskinesia, a disorder that causes repetitive movements: chewing motions, lip smacking, and contortions of the arms and legs.

A revolution in psychiatry had occurred. Over the next eight months, Thorazine was given to an estimated 2 million patients. The impact on mental hospitals was astounding, particularly involving patients with schizophrenia, the most prevalent and severe of the major psychoses. Patients on antipsychotic drugs tend to show drowsiness and reduced initiative but retain intact intellectual functions. Typically, psychotic patients become less agitated, and withdrawn patients may become more responsive. Hallucinations, delusions, and disorganized thinking tend to gradually disappear.

Before the introduction of the psychotropic drugs, there were about 380 to 400 patients per 100,000 population in mental hospitals in the United States, a figure that had remained constant for almost a hundred years. After their introduction in the mid-1950s, this figure steadily and dramatically decreased, causing many state hospitals to close down for lack of patients. The number of mental patients in U.S. hospitals declined from 558,922 in 1955 to 61,772 in 1996. In the first decade of the twenty-first century, it is estimated that there are about 2.2 million American adults with schizophrenia.

35

Your Town, My Town, Miltown!

A humorous scene from a 1960s Hollywood film illustrates the public's embrace of tranquilizers. In a department store, a crowd of about thirty people gather around a woman who is having an anxiety attack. A man who has bent down to help her turns to the onlookers and shouts, "Anybody have a tranquilizer?" Twenty people promptly reach into their pockets and purses.

Anxiety is a universal human response to stress. The reaction may occur in either the presence or the absence of a definable threat, and its magnitude may be out of proportion to the threat perceived. Subjectively, it is experienced as a state of tension or vulnerability accompanied by feelings of dread. This psychological state is accompanied by various symptoms, including dilated pupils, muscular tension, a racing heart, hyperventilation, excessive perspiration, nausea, and flushing.

"MUCH TO MY SURPRISE . . ."

As miraculous as penicillin is in fighting infections, it does not work against certain bacteria, particularly gram-negative organisms (those that do not react to the microscopic staining technique). In 1945 Frank Berger, a Czechoslovakian pharmacologist working at British Drug Houses Ltd. in London, was trying to find antibiotics that would work against these other organisms.

At the time, there was a commercial disinfectant that was effective against gram-negative organisms. Berger hoped to devise some form of it that would be safe in the human body. He made chemical analogs of the disinfectant and began conducting toxicity tests on mice.

The results were totally unexpected. "Much to my surprise," these compounds produced paralysis of their limbs "unlike that I had ever seen before."[1] Depending on the size of the dose, complete recovery occurred within an hour. Notably, with doses too small to induce muscle impairment, the marked "quieting effect on the demeanor of the animals" remained evident, and this was described as "tranquilization." Many more compounds were synthesized, one of which had superior activity in acting directly on muscles rather than at the neuromuscular junction.

The new tranquilizer was brought to doctors under the name mephenesin (Myanesin) as an agent to produce muscular relaxation during light anesthesia, but its very short duration of action and lack of consistency were major drawbacks. A structural analog, Robaxin, was introduced later for the relief of muscle spasm by the AH Robins Company. Reports began to appear that mephenesin could ease symptoms of anxiety without clouding consciousness, bringing about a relaxed feeling in tense patients.

In 1947 Berger moved to New Jersey to work at Wallace Laboratories, where he eventually became president. Seeking a longer-acting central muscle relaxant with more reliable absorption than mephenesin, he and his chemist, Bernard Ludwig, synthesized and screened more than 1,200 compounds before developing one that was effective for about eight times longer than mephenesin. It was given the generic name meprobamate. Almost eleven years elapsed between the discovery of mephenesin and the availability of meprobamate as a prescription drug in April 1955. Marketed as both Miltown and Equanil, it was found to dispel anxiety with less sedation than barbiturates, along with promoting general muscle relaxation.[2]

The drug quickly grew in popularity, and within two years it was widely prescribed. The demand for it was unprecedented.[3] It was not surprising to see signs in pharmacy windows that read "Out of Miltown" or "Miltown available tomorrow." By 1956 one American in twenty was taking tranquilizers within a given month.[4] In the late fifties, it became the most popular among all psychotropic agents, especially in the United States, and remained unchallenged until the development of Librium and other benzodiazepines, which caused less drowsiness, almost a decade later.

36

Conquering the "Beast" of Depression

Depressions are a group of disorders universally marked by low energy and a mood often described by the person as sad, "blue," discouraged, or "down in the dumps." Depressive syndromes are characterized by a wide array of symptoms. There is always a persistent change in behavior, with loss of interest or pleasure in nearly all activities, and this may be accompanied by feelings of worthlessness, hopelessness, and helplessness. Often depressions are marked by a pervasive sense of emptiness and deprivation, as well as anxiety. The individual may have an impaired ability to think, concentrate, or make decisions, and in severe cases there may be thoughts of death, suicide, or suicide attempts.[1]

Depression is the most common mood disorder that comes to the attention of physicians. In 2005, about 10 million Americans had a major depressive disorder.

Throughout history, fanciful terms were adopted in the description of different complaints we now call clinical depression. "Neurasthenia" bespoke "nervous exhaustion." From before the time of Hippocrates and up until the middle of the nineteenth century, physicians believed that health depended upon a balance between various body fluids, or humors. "Melancholia" was attributed to an excess of black bile (from the Greek *melan-*, meaning black, and *cholē*, bile). The humors theory finally crumbled with the advance of cellular pathology by Rudolf Virchow in the middle of the nineteenth century, but it lingers in the terms "ill humor" and "good humor."

Treatments toward the end of the nineteenth century and extending well into the twentieth century included electrotherapy, hydrotherapy, rest, diet, and massage.[2]

MAO INHIBITORS

The breakthrough in discovering antidepressants and ultimately understanding the chemical basis of depression arose from an interlocking series of chance experiences. First among these was the accidental discovery of what became known as MAO inhibitors.

On a spring Sunday in 1953, Nathan Kline, a psychiatrist who was director of the Rockland State Hospital in Orangeburg, N.Y., read an interesting news item from Bombay in the *New York Times*. At a provincial medical conference, "a special prize was awarded to Dr. R. A. Hakim for his paper on 'Indigenous drugs in the treatment of mental diseases.'" The indigenous drugs were from the plant *Rauwolfia serpentina,* and the special prize was a gold medal for treating more than seven hundred schizophrenics.

The drugs were derived from the snakelike roots of a small bush that is native to India. Its powdered extract had long been used in folk medicine to quiet babies and to treat insomnia. Indian holy men, including Mahatma Gandhi, were habitual users, as they found it helped them attain states of introspection and meditation.[3]

At around this time, the Swiss pharmaceutical firm Ciba isolated the active ingredient from the plant, an alkaloid called reserpine, to treat people with high blood pressure. Each purified crystal was equivalent in activity to more than 10,000 times its weight in the crude drug. A few months after Kline read the *New York Times* announcement, Ciba marketed the drug as Serpasil for high blood pressure.

Kline was a bold, energetic, and often controversial innovator in psychiatric research. He tested the drug on many of his psychiatric patients. The results were highly variable. Several showed dramatic improvement, and most became sedated to varying degrees and indifferent to events around them. He went on to transform Rockland

State Hospital from a regular state mental institution into a leading research center. Kline became an evangelist in promoting the drug to treat schizophrenics. For his pioneer efforts in introducing the use of tranquilizers to the practice of psychiatry in the United States, he received the prestigious Albert Lasker Medical Research Award in 1957.

On the recommendation of the commissioner of mental health in New York State, reserpine was routinely administered to all 94,000 patients in the state psychiatric hospitals. The drug, however, had serious side effects, particularly permanent movement disorders, and its use was discontinued in the early 1960s.

But it was the drug's effect on mood in nonpsychiatric patients that threw a spotlight on the chemical mix at the brain synapses. As millions of patients in the U.S. with high blood pressure were now taking Ciba's Serpasil, it soon became apparent that reserpine precipitated a severe depression in many people, with some even committing suicide. Investigations into reserpine showed by 1955 that it reduces the level of the three major monoamines — serotonin, dopamine, and noradrenaline (now called norepinephrine) — in the brain.[4] This was a major illumination. And the investigators were able to figure out how it happened: Reserpine causes these amines to leak out from little storage pouches (called synaptic vesicles) at the terminal endings of neurons at the synapse, and they are broken down by the enzyme monoamine oxidase (MAO). The resultant deficiency causes the depression. For an antidepressant to be uncovered, the key was to find a way to inhibit this MAO enzyme.

In the early 1950s, clinical studies were being done in Europe to test how effective a new drug called iproniazid would be in treating tuberculosis.[5] Scientists discovered that the drug was extremely effective, but they also noticed a surprising side effect. The patients taking it experienced euphoria.[6] Kline, upon learning about the odd side effect, saw the potential, and conducted a clinical trial to test iproniazid, a known MAO inhibitor, as an antidepressant. The patient response was astonishing. In April 1957 Kline's team reported the results of their experiments with twenty-six patients that revealed iproniazid to be the first drug of value in treating depression.

It is difficult today — in this era of institutional review boards, informed consent, and governmental regulation — not to marvel at the directness, simplicity, and lack of restraint possible at that time. Ethics, responsibility, and competence were matters that were, for better or worse, completely up to the clinician who undertook the trial.

Who Gets the Credit?

The pride of authorship and the credit for discovery are powerful dynamics among researchers. The circumstances surrounding the press announcements and the presentation at the regional meeting of the American Psychiatric Association in Syracuse, New York, of the evident value of iproniazid led to a long and bitter conflict between the participants.

Harry Loomer, the physician in charge of the clinical unit involved at Rockland State Hospital, read the paper at the meeting,[7] but the *New York Times* incorrectly credited the presentation to Kline. At a press conference in New York a few days earlier, details of the success of the trial had been provided principally by Kline. Loomer was infuriated, and the *New York Times* printed a retraction. Another member of the team, John Saunders, a physician and clinical pharmacologist who supervised the project and had correctly attributed the mode of action of iproniazid to monoamine oxidase activity, also felt exploited. The following year, under his single authorship, Kline published a paper on the clinical experience with iproniazid.[8] The spark of discontent burst into flame in 1964 when Kline, for work with iproniazid, became the single recipient of a second Lasker Award. (The accompanying citation said: "Literally hundreds of thousands of people are leading productive, normal lives who — but for Dr. Kline's work — would be leading lives of fruitless despair and frustration.") Loomer and Saunders objected, saying they had made the discovery, but the awards committee was unrelenting. Seventeen years of litigation followed. A jury ruled for Loomer and Saunders, and the State Court of Appeals upheld that ruling and finally laid the case to rest in 1981.

Because iproniazid (Marsalid) had already been marketed since 1952 as an antitubercular drug, psychiatrists were able to obtain supplies as soon as they heard of its antidepressant properties. Within one year of Kline's report, more than 400,000 patients had received the drug for treatment of depression. A gratified Kline observed that "probably no drug in history was so widely used so soon after announcement of its application in the treatment of a specific disease."[9]

The report of a small number of cases of jaundice — most likely a coincidental epidemic of infectious hepatitis — led to the withdrawal of iproniazid from the American market by the manufacturer, Hoffman–La Roche, in 1961. It was then replaced by more potent MAO inhibitors, such as Nardil and Parnate.

MAO inhibitors (MAOIs) increase brain levels of norepinephrine, serotonin, and dopamine. Iproniazid's antidepressant effect is due to this mechanism. The discovery that depression can be lifted by agents that inhibit the enzyme MAO was a major breakthrough in the recognition of depression as a common disease. Moreover, MAO inhibitors had an important impact on the development of modern biological psychiatry.

The use of MAO inhibitors in the treatment of depression became severely restricted in the face of unexpected consequences. Complex, sometimes severe, and often unpredictable interactions with other food-derived amines (particularly those in cheese) chemically related to adrenaline and thus liable to raise blood pressure, sometimes sufficient to precipitate strokes or heart failure, made their medical use difficult and potentially hazardous.[10]

Fortunately, only a few years later, the next important, and now dominant, class of antidepressant compounds — the tricyclics — was introduced.

TRICYCLIC ANTIDEPRESSANTS

Roland Kuhn at the Cantonal Psychiatric Clinic in Münsterlingen, Switzerland, was a distinguished and cultivated thirty-eight-year-old psychiatrist who had been trained in psychoanalysis. Despite this training, his clinical experiences had led him to seriously consider the possibility that mental illnesses were caused by biological factors. At the request of the Geigy pharmaceutical company in Basel, he began testing a drug called imipramine, one of the tricyclics (so named for their chemical structure, characterized by three interlinked carbon rings) that had originally been synthesized with antihistaminic activity in mind. Kuhn set out in 1950 to investigate it for use in treatment of schizophrenic patients. Over the next several years, Kuhn made an astute observation of an apparent paradoxical effect. Rather than exerting a sedative effect, imipramine did exactly the opposite — it served as a stimulant.

A few quiet, withdrawn schizophrenics became less withdrawn and began reacting to their hallucinations, hence harder to handle. Rather than concluding that the drug had made them "worse," Kuhn took the view that the drug had eased the depressed component in these patients, resulting in what might be considered a "normal" reaction to hallucinations. He therefore decided to reevaluate the drug by trying it in patients who were primarily depressed and not necessarily schizophrenic.

The results were dramatic and breathtaking. Deeply depressed patients were resurrected to activity, socialization, contentment, and accessibility. The patients themselves spoke of a "miracle cure." Kuhn was candid in acknowledging that "chance admittedly had something to do with the discovery of imipramine. . . . Good fortune, too, played a part." But in also declaring that "a discovery does not arise of its own accord out of a particular scientific solution,"[11] this experienced psychiatrist paid tribute as well to Pasteur's dictum: "Chance favors the prepared mind."

Thus was born a class of drugs that would have a profound effect on millions of patients suffering from depression, thanks to both chance and sagacity. Kuhn's sagacity, in this case, was in his ability to see the drug's potential for use as an antidepressant, even though that potential good use was masked by its failure as a sedative.

PROZAC NATION

Depression was recognized as having biological causes, and the search was on for even more effective drugs to treat it. It is remarkable that, over the next thirty years, so little progress was made. The most notable advance was the introduction of agents that have, in essence, the same mechanism of action as the original tricyclic antidepressants, but with fewer side effects. These are called selective serotonin reuptake inhibitors, or SSRIs, and they include the well-known brand-name drugs Prozac, Paxil, and Zoloft. (Prozac's generic name is fluoxetine; Paxil's is paroxetine; and Zoloft's is sertraline.) SSRIs prolong serotonin's neurotransmission by inhibiting its reuptake by the nerves that release it, which is the normal way that serotonin signaling is terminated.

Prozac was the first SSRI approved by the FDA, in December 1987, and the others soon followed. Prozac in particular became a buzzword for the new zeitgeist, built on society's acceptance of the chemical regulation of mood. By 1994, according to *Newsweek,* the drug had "attained the familiarity of Kleenex and the social status of spring water."[12] Organizations like the National Alliance for the Mentally Ill and the National Depression and Manic-Depressive Association assert that these disorders are illnesses like diabetes or heart disease, not weaknesses or failures of character. Americans are coming to think of depression as an illness like any other, a topic discussed on dates and at dinner parties.

As of 2006, about 40 million people had taken Prozac since it was introduced in 1988. In 2001 it had retail sales of $2.7 billion in the United States, with similar figures posted by Zoloft and other agents. In 2004 antidepressants became the best-selling category of prescription medicines in the United States. Global sales totaled $20 billion dollars that year. In 2006, at least 7 million Americans were on antidepressants.

The search continues for even better drugs that are more effective and have fewer side effects. Despite a better understanding of drug effects on neurotransmission, a trial-and-error approach endures in the development of new drugs. What physically initiates and sustains depression remains poorly understood, largely because an animal model for human depression does not exist. Animal tests do not mimic anything like human depression. Not only do antidepressants affect different parts of the brain in animals and humans, but in people depression has a strong emotional and cognitive component.

Yet even with more sophisticated laboratory investigations,

"Discouraging data on the antidepressant."

additional serendipitous discoveries will be needed for real progress to take place. Eric Nestler of the Division of Molecular Psychiatry at the Yale University School of Medicine acknowledged that reality when he wrote in 1998: "It is likely that truly novel antidepressants will be developed either by luck once again or by revealing more about the basic mechanisms that underlie depressive syndromes."[13]

37

Librium and Valium

Following the early reports of the medical usefulness of the initial psychotropics, much interest was generated to develop further new drugs. In the mid-1950s the pharmaceutical company Hoffman–La Roche in New Jersey, eager to enter this exploding market, launched a program to synthesize novel compounds in order to discover a new "psychosedative" drug.

Leo H. Sternbach, a Polish immigrant, was Roche's director of medicinal chemistry. He decided to take a "from the shelf" approach that might prove therapeutic. It led Sternbach to a self-described "chain of events that started with the haphazard synthesis of new chemical compounds"[1] and led to what was later acknowledged at Hoffman–La Roche as an "accidental discovery"[2] of new blockbuster drugs.

Sternbach decided to reinvestigate some tricyclic compounds he had synthesized eighteen years earlier at the University of Cracow, just before the Nazis invaded Poland, while doing studies on dyestuffs.[3] Nobody else had paid any attention to these compounds — which had turned out to be useless as dyes — in the long interval since he had first prepared them. Having in mind the tricyclic nature of the newly discovered tranquilizer Thorazine, Sternbach's idea was to modify the molecular structure of his compounds by introducing a basic side chain, known to be important for the biological activity of many drugs.

TIDYING UP THE LAB BENCH: "EUREKA!"

Sternbach prepared around forty new compounds, which he screened for muscle relaxant, sedative, and anticonvulsant properties, but all proved biologically inactive. Then the program was virtually suspended for almost two years. In April 1957 Sternbach, tidying up a lab bench that had become littered with dishes, flasks, and beakers, was about to discard a crystalline compound when a colleague pointed out that it had been overlooked during the testing process. He suggested it be sent for animal testing. Agreeing, Sternbach promised management that this compound would be his last product of the series. "We thought that the expected negative pharmacological result would complete our work with this series of compounds and yield at least some publishable material."[4]

A few days later, he got a call from Roche's director of pharmacology. This compound, later called Librium, had extraordinary qualities. It tamed a colony of vicious monkeys, without affecting their alertness. A mouse would hang limply when held by one ear, but unlike those given Miltown, it was able to walk when prodded. In headlining his description of the results of the animal tests, Sternbach was justified in using the term "Eureka!"[5] The compound was superior to Miltown for allaying anxiety and muscle tension. All this with a very low level of acute toxicity and no significant side effects.

The unexpected finding of a pharmacologically active compound after what had been a fruitless search generated considerable interest and enthusiasm. However, the question loomed: Why was only this compound active of the forty he had synthesized?

MOTHER'S LITTLE HELPER

Sternbach reinvestigated its chemistry and found that a key intermediate was not truly of the class of compound that he originally thought but was rather an entirely different one. This last compound had followed a different synthetic pathway and had undergone an "unusual transformation" to an unexpected molecular rearrangement that was not then well known. The product thus formed was now shown to be a benzodiazepine.

Sternbach was granted a U.S. patent for this new tranquilizer in July 1959. The drug was quickly found to calm patients' tension and anxiety with a minimum of side effects. Most important, the calming action was accomplished without clouding consciousness or interfering with intellectual activity. The interest of clinicians grew enormously and, within what today appears an incredibly short time, initial clinical studies were conducted on around 16,000 patients. The drug received FDA approval in February 1960 and was marketed as Librium. It was highly successful as a tranquilizer that was stronger than Miltown. A brief clinical note in the March 12, 1960, issue of the *Journal of the American Medical Association* was the first published announcement that the drug was therapeutically effective.

As certain features of Librium's chemical structure were not necessary in order to achieve the desired effects, Sternbach did additional work to find simpler analogs. The one he came up with, diazepam, was five to ten times more powerful than Librium and did not have its bitter taste. Marketed as Valium in December 1963, it became the country's best-selling drug, as well as an American cultural icon.

By 1970, one woman in five and one man in thirteen were using "minor tranquilizers and sedatives," meaning mainly the benzodiazepines.[6] Valium reached the height of its popularity in 1978, a year when Americans consumed 2.3 billion of the little yellow pills. In 1980 Roche was making 30 million Valium tablets a day. By the mid-1980s, the benzodiazepines accounted for up to 100 million prescriptions per year in the United States alone.[7]

Today about thirty benzodiazepines, of thousands synthesized since 1960 by many pharmaceutical companies, are in regular use throughout the world, principally as antianxiety agents but also as muscle relaxants, anesthetics, and epilepsy medication. They are among the most commercially successful drugs of all time.[8] Although they have been proven valuable in patients whose anxiety interferes with their work, leisure, and personal relationships, they are considered by some to be widely misused in the management of the most trivial symptoms of stress.[9]

38

"That's Funny, I Have the Same Bug!"

During World War II, Erik Jacobsen, the forty-four-year-old director of the biological laboratory at the Danish pharmaceutical firm Medicinalco in Copenhagen, thought he was coming down with the flu. After eating a sandwich at lunchtime, accompanied by a customary bottle of beer, he felt nauseated and his head throbbed. But strangely, no flu developed.

A few weeks later, he and the managing director went out to lunch and celebrated the company's good fortunes with a few glasses of aquavit. Shortly thereafter, Jacobsen had a reoccurrence of his symptoms. Colleagues told him he was strikingly red-faced.

Some days passed, then the nausea and headache struck again while he was sipping a beer and eating a sandwich at lunch. He left work early and bicycled home, wondering if his family were similarly affected, but they were fine.

Even though the attacks continued at work for several more days, sometimes with heart palpitations, he did not alter his habit of having beer with his lunch. Soon thereafter, he met with a colleague, Jens Hald, with whom he was investigating the effects of a drug on intestinal worms. "We were just chatting," Jacobsen recalled, "and I mentioned this curious affair that I had been through. Hald smiled and said, 'That's funny. I have had the same bug!'"[1]

At this very point, Jacobsen and Hald had a flash of insight and immediately recognized the cause of their symptoms. While they had been continuing other investigations in the laboratory, they had each been testing a chemical, disulfiram (tetraethylthiuramdisulphide), upon themselves. Used for a long time in the rubber industry as an antioxidant, it had recently been employed as an ointment by Swedish researchers in patients with scabies, a parasitic disease of the skin endemic at that period of privation throughout occupied Europe. The disease causes maddening itching that leads to incessant scratching and infections. Hald had come up with the idea based on what was known of the drug's action that it might be used to interfere with the metabolism of intestinal parasites.

Jacobsen and Hald prepared pills of disulfiram as a potential oral drug and were encouraged by preliminary tests in rabbits infected with intestinal worms. Next, they had to see if the pills would be safe for humans. Jacobsen had earlier undergone several self-experiments at Medicinalco with other research studies. "It never occurred to us to ask others to volunteer first. We would have regarded that as unethical."[2] They both began taking disulfiram daily.

At first there were no apparent ill effects. They had attributed their episodic symptoms of flushing, pulsating headache, palpitations, and nausea to the flu, but when they each learned that the other was experiencing the same symptoms, their trained scientific minds led them to understand that these circumstances were not coincidences. They then began to focus their attention on the drug they were both taking. Realizing that the drug had altered their response to alcohol, they began a series of studies, including some in which Jacobsen volunteered for further self-experiments. As it turned out, disulfiram by itself is innocuous, but combined with alcohol it breaks down to form a highly toxic product. At one point, Jacobsen's blood pressure fell to almost zero and he came close to death.

Studies on alcoholic patients were then undertaken by a psychiatrist specializing in aversion therapy. In other words, patients would take small doses of disulfiram specifically to make themselves feel sick if they consumed alcohol, in the hope that their behavior would be conditioned to avoid alcohol. These studies provided the basis for the use of disulfiram in the treatment of chronic alcoholism.

Disulfiram acts by altering the intermediary metabolism of alcohol, causing a concentration of acetaldehyde in the body.[3] Ordinarily, this does not accumulate in the tissues, but disulfiram interferes with its breakdown so that the blood acetaldehyde concentration rises to a toxicity level five to ten times higher than normal. Dramatic signs and symptoms accompany this effect, known as the Acetaldehyde Syndrome.

The first paper Jacobsen and Hald published, in 1948, established both Jacobsen's trade name for the drug, Antabuse, and its new usefulness.[4]

An interesting harbinger of the observations of Jacobsen and Hald had occurred a decade before.[5] In 1937 a physician to a chemical company in Connecticut noted that the workers, exposed to a chemical similar to disulfiram, had "become involuntary total abstainers" of alcohol. He even speculated in his report in the *Journal of the American Medical Association* "whether one has discovered the cure for alcoholism."[6] But no one took steps to investigate this further.

Jacobsen and Hald's experience leading to Antabuse is a clear example of a discovery that could have occurred only through the channel of serendipity. Researchers conducting animal experiments with the chemical never would have imagined to test with alcohol.

Antabuse has a highly specific usefulness in chronic alcoholism. It is probably most useful in well-motivated patients whose drinking is triggered largely by recurrent psychological stress and who want to remain in a state of enforced sobriety so that supportive and psychotherapeutic treatment may be applied to best advantage. Importantly, the patient is warned to avoid alcohol in disguised form — in sauces, vinegars, cough mixtures, and even preparations containing alcohol that can be absorbed through the skin, like aftershave lotions and massage oils.

39

LSD

The accidental discovery in 1943 of the hallucinogenic effects of LSD is one of the most dramatic episodes in the modern history of drug development.

Albert Hofmann, a thirty-seven-year-old research chemist at the Sandoz pharmaceutical company in Basel, Switzerland, was investigating ergot alkaloids derived from the fungus *Claviceps purpurea*. Alkaloids are nitrogenous organic compounds found in plants. This was an area of research in which Sandoz had long pioneered under the direction of the distinguished Swiss chemist Arthur Stoll. At Sandoz, Stoll had earlier isolated ergotamine, a widely used product to control bleeding after childbirth, but it took him another thirty-three years before he was able to establish its full complex chemical structure. The characteristic nucleus of the ergot alkaloids is lysergic acid.

Lysergic acid itself is not hallucinogenic. Hofmann's original purpose was to synthesize the basic components of ergot in hopes of developing new compounds that might be useful in other fields of medicine. In 1938 Hofmann had synthesized lysergic acid diethylamide with the chief aim of obtaining a stimulus to the circulatory and respiratory systems.

Lysergic acid diethylamide (in German, *Lysergische-Säure Diäthylamid,* or LSD) was given the laboratory code name LSD-25 because it was the twenty-fifth in a series of ergot derivatives of lysergic acid that he and his colleagues synthesized. Not unexpectedly, it had the effect of inducing uterine contractions. What was not anticipated was the excitation that was observed in some of the animals, but that was of no interest to Hofmann. He put aside the investigation and did not pick it up again until April 1943, when "a peculiar presentiment" overcame him that perhaps he had missed something five years earlier. He synthesized a new batch of LSD-25 one morning and by noon had a crystalline sample that was easily soluble in water.

Hofmann described his now-famous "accidental observation" in his laboratory notes:

> Last Friday, April 16, 1943, I was forced to stop my work in the laboratory in the middle of the afternoon and to go home, as I was seized by a peculiar restlessness associated with a sensation of mild dizziness. . . . As I lay in a dazed condition . . . there surged upon me an uninterrupted stream of fantastic images of extraordinary plasticity and vividness and accompanied by an intense, kaleidoscope-like play of colors. This condition gradually passed off after about two hours.

What Hofmann had experienced was the first acid trip.

This bizarre reaction baffled him. Over the weekend, he reviewed his preparation of LSD step by step. How could he have been exposed to the effects of this new compound? The possibility of his having unconsciously licked his fingers after some of the LSD-25 accidentally spilled on them seemed most unlikely. Was it possible that he had absorbed a minute amount of LSD solution through his skin? He had trained himself to be careful and meticulous in the laboratory, knowing full well that even small amounts of ergot preparations can be dangerous. Could he have inhaled and then absorbed some of the LSD-25?

As a scientist, Hofmann was compelled to pursue this serendipitous happening. Did this compound have mind-altering effects of greater potency than any previously seen? "In order to get to the root of this matter, I decided to conduct some experiments on myself." The following Monday, the nineteenth, using other ergot alkaloids as a guide, Hofmann took 0.25mg of LSD orally, believing that to be the lowest dose that might be expected to have any effect. In fact, this is five to ten times the average effective dose of LSD! The reaction he had to such a large dose was quite spectacular. After forty minutes, he noted a "slight dizziness, unrest, difficulty in concentration, visual disturbances, and a marked desire to laugh." At this point, the entries in the laboratory notebook end, and the last words were written only with the greatest difficulty. He was used to bicycling the four miles home from his lab, and on this day he asked his laboratory assistant to accompany him. He continued his accounts later:

> While we were cycling home, it became clear that the symptoms were much stronger than the first time. I had great difficulty in speaking coherently, my field of vision swayed before me, and objects appeared distorted like images in curved mirrors. I had the impression of being unable to move from the spot, although my assistant told me afterwards that we had cycled at a good pace. . . .
>
> By the time the doctor arrived, the peak of the crisis had already passed. As far as I remember, the following were the most outstanding symptoms: vertigo, visual disturbances; the faces of those around me appeared as grotesque, colored masks; marked motoric unrest, alternating with paralysis; an intermittent heavy feeling in the head, limbs and the entire body, as if they were filled with lead; dry, constricted sensation in the throat; feeling of choking; clear recognition of my condition, in which state I sometimes observed, in the manner of an independent, neutral observer, that I shouted half insanely or babbled incoherent words. Occasionally I felt as if I were out of my body. . . .
>
> Six hours after ingestion of the LSD my condition had already improved considerably. Only the visual disturbances were still pronounced . . . all objects appeared in unpleasant, constantly changing colors . . . fantastic images surged in upon me. A remarkable feature was the manner in which all acoustic perceptions (e.g., the noise of a passing car) were transformed into optical effects,

every sound evoking a corresponding colored hallucina-
tion constantly changing in shape and color like pictures
in a kaleidoscope. At about one o'clock I fell asleep and
awoke next morning feeling perfectly well.[1]

Hofmann had been propelled into visual hallucinations, out-of-
body experiences, cognitive distortions of time and space, a synes-
thetic swoon, and elements of paranoia. For a while, he feared he was
losing his mind, that the drug had precipitated a psychosis, and this
anxiety made his experience quite negative. Remarkably, despite his
response to the potent properties of LSD mimicking a psychosis, he
could later recall the experience in sharp detail.

Hofmann reported his self-experiment to Dr. Ernst Rothlin, Sandoz's chief pharmacologist, who initially doubted that such a tiny
amount of any substance could so affect the mind. Rothlin agreed to
try it on himself and two other colleagues, prudently taking only one-
quarter of the dose that Hofmann had taken. All had hallucinations
and, as Hofmann later ruefully recalled, "Rothlin believed it then."[2]

Dr. Werner Stoll, the son of Arthur Stoll, the Sandoz scientist
who had pioneered the investigations on ergot chemistry, also tested
on himself one-quarter of the dose that Hofmann had taken. Then in
1947 he reported the results of a systematic investigation of LSD in
sixteen normal volunteers and six schizophrenic patients at the Psychiatric Clinic of the University of Zurich, in which Hofmann's original findings were essentially confirmed.[3] LSD could cause profound
changes in human perception, reminiscent of those experienced by
schizophrenics. Stoll also discovered that in low doses LSD seemed to
facilitate the psychotherapeutic process by the release of repressed
material into consciousness.

LSD turned out to be about 5,000 to 10,000 times more potent
than an equivalent dose of mescaline, the substance it most closely resembled. Mescaline was first synthesized in 1919 and allowed the initiation in the 1920s of psychopharmacological studies of its
hallucinogenic effects, but interest in such research lapsed until the
discovery of LSD.[4]

Hofmann's discovery of the high potency of LSD led to the popularization of the concept that a toxic substance or product of metabolism might be a cause of mental illness. Shortly after Werner Stoll's 1947 article, Sandoz made LSD available to select researchers under the trade name Delysid. Its arrival in the United States in 1949 enjoyed particularly propitious timing for a new mind drug. It coincided with the ascent of the neurosciences, an exciting period in which the various brain centers were first mapped and the brain's neurotransmitters, its chemical messengers, discovered.[5]

It had long been accepted that the course of some mental disorders could be modified by certain drugs, such as barbiturates. Thorazine, the first major tranquilizer, appeared in 1954, the sedative Miltown a year later, and then Librium, Valium, Elavil, Tofranil, transforming psychiatry and Western culture itself. Now the pharmaceutical companies — aware of the accumulating evidence and sensitive to the market potential — actively supported the search for new mind drugs. The course of LSD over the next years followed dramatically disparate elements: scientific research on its psychiatric legitimacy, the flowering of the hippie subculture, and clandestine mind control by governmental agencies.

The psychiatric use of LSD remained sporadic until late in 1954, when Thorazine revitalized interest in hallucinogens. For both psychopharmacologic research and psychotherapy, LSD required a dependable antidote, and Thorazine became the drug of choice.

Two terms for such psychoactive drugs were introduced. "Psychedelic," derived from the Greek and meaning "mind-manifesting," became engraved in the public consciousness. The preferred scientific term is "hallucinogen," and the effect of such drugs has been eloquently described by Albert Hofmann:

> Hallucinogens distinguish themselves from all other psychoactive substances through their extremely profound effects upon the human psyche. They bring about radical psychological changes which are associated with altered experiences of space and time, the most basic categories

of human existence. Even the consciousness of one's own corporeality and one's own self may be changed dramatically. Hallucinogens take us to another world, to a type of dream world which is nevertheless experienced as completely real, as even more intense and consequently in some ways more real than the ordinary world of everyday reality. At the same time, if the dosage is not too high, consciousness and memory are retained completely. This is a key distinction between these substances and the opiates and other intoxicants, whose effects are associated with an obscuration of consciousness.[6]

LSD, replacing mescaline as the leading hallucinogenic drug in psychiatry, was used in various doses to study its effects on creative expression and perception of normal persons or to induce in them "model psychoses" with the goal of revealing what causes schizophrenia. The term *psychotomimetic* (mimicking a psychosis) was adopted for the latter purpose.

Sidney Cohen, a psychiatrist at the Los Angeles Veterans Administration Hospital, tried LSD on himself and subsequently described the results in a series of patients in *The Beyond Within.*[7] LSD is an immensely powerful psychochemical, one ounce of which would provide a psychedelic experience for 300,000 adults. That the whole brain is involved is clear. One of the most impressive experiences of the drug's influence is synesthesia — the response by one of the senses to a stimulus ordinarily responded to by another of the senses. Albert Hofmann had perceived sound as color. Others were able to taste colors or smell sounds. Some who took the drug felt that they gained striking "insightfulness into themselves, an awareness of their place in the environment, and a sense of order in life." Cohen recognized, however, that such therapeutic benefits might be short-lived, with a full return of the patient's original crippling neuroses.

In California, the use of small doses of LSD in psychotherapy spread across several private clinics and veterans, state, and county hospitals in the late 1950s and early 1960s. The publicized description of the value of LSD by the film star Cary Grant, who took the drug more than sixty times after a long period of depression and felt that he had been reborn, was taken as a ringing endorsement at least by the Hollywood community.

With the intensification of sensory perception, an observer could understandably raise questions: Was cognition impaired or was insight deepened? How "real" is the spiritual/mystical/religious effect? Is this what mystics and saints have experienced? Is this a supranormal or abnormal response? The claim of many patients that they had experienced "a higher power," or "ultimate reality," raised provocative questions. What of the distinction actress Lily Tomlin makes: "When we talk to God, we're praying. When God talks to us, we're schizophrenic"?

Many of the clinical researchers were themselves taking the drug. The conventional therapeutic community was aghast at what was seen as unprofessionalism, if not outright chicanery. Charges of "bad science" abounded.

According to Dr. Timothy Leary, a onetime Harvard psychologist who became the counterculture's leading guru, LSD would transform America into a spiritual utopia. Leary was the Pied Piper promoting LSD to the pop culture of the 1960s, and his slogan was "Tune in, turn on, drop out." Appealing to people in their late teens and twenties at a time of great social upheavals — anti–Vietnam War protests, the Free Speech movement, civil rights demonstrations — Leary promised an "ecstasy of revelation." This recreational drug would re-create. To its adherents, the ordinary became extraordinary.

LSD AND THE CIA

The potential use of new mind-control drugs did not escape the attention of the Central Intelligence Agency. Sidney Gottlieb, the director of the agency's chemical division from 1953 to 1972, has gone down in history as the chemist who brought LSD to the CIA in the name of national security. Gottlieb, a slight man with short gray hair and a clubfoot, was a sorcerer-scientist. Throughout the 1950s and early 1960s, under his direction, the agency gave mind-altering drugs to hundreds of unsuspecting Americans — mental patients, prisoners, drug addicts, prostitutes — to explore the possibilities of developing a truth serum and of controlling human consciousness. With the Cold War raging, the agency was fearful that the Soviet Union would use LSD as a chemical weapon to disorient enemy forces or that China would employ techniques of brainwashing.

Most of the experiments were administered by various psychology researchers throughout the country through an insidious operation called MKULTRA, funded through front foundations, often in clear violation of human rights.[8] In one case, a mental patient in Lexington, Kentucky, was dosed with LSD continuously for 174 days. Other experiments involved CIA employees, military personnel, and college students who either had limited knowledge or were totally unwitting about the tests. At least one participant, an army scientist, thinking he had lost his mind, committed suicide by jumping out of a New York hotel window. Others suffered psychological damage after participating in the project. Some subjects were not told what had been done to them until some twenty years later. Some soldiers sued the CIA, and the case went as far as the U.S. Supreme Court, which rejected their claims on the grounds that soldiers can't sue for injuries that "arise out of or are in the course of activity incident to service."

A Nice "Trip" from Langley, Virginia

As detailed in *Who Goes First?* by Lawrence K. Altman, self-experimentation is a noble tradition in the history of medicine. Many researchers feel ethically bound to initiate human experiments on themselves with a new drug or procedure. After Albert Hofmann stumbled on and confirmed the hallucinogenic effects of LSD by taking it himself, he did not ever take it again. Sidney Gottlieb, the CIA man who directed the program to give LSD to thousands of unsuspecting people, on the other hand, took it hundreds of times.

When Sandoz in Basel balked at continuing to provide large quantities of LSD for military purposes, the CIA pressured the Eli Lilly

Company of Indianapolis to synthesize the compound, in disregard of international patent accords. In the words of John Marks, author of *The Search for the "Manchurian Candidate,"* the definitive book on the experiments, Gottlieb "never did what he did for inhumane reasons. He thought he was doing exactly what was needed. And in the context of the time, who would argue?"[9]

By 1972, shortly before he retired, Gottlieb declared the experiments useless. On orders from Richard Helms, the CIA director, the agency in 1973 deliberately destroyed most of the MKULTRA records, according to Gottlieb, because of a "burgeoning paper problem." After he left the CIA, Gottlieb ran a leper hospital in India and, in his later years, cared for dying patients in a hospice in Virginia, while defending himself against lawsuits from survivors of his secret tests.

The Beatles' song "Lucy in the Sky with Diamonds" was obviously a lyrical tribute to an LSD trip. However, LSD in unstable personalities resulted in disturbing conditions, described by new terminology: "freaked" (an anxious or panicked state), "bad trip" (a fully psychotic reaction), and "flashback" (recurrence of symptoms a few days or weeks after the initial drug experience).

In June 1963, as established by congressional law, the FDA took control over all investigational drugs, requiring the submission of research projects to the National Institute of Mental Health for its approval. A 1965 report in the *New England Journal of Medicine* called attention to waves of highly disturbed LSD-users showing up in emergency rooms and requiring hospitalization. It urged an end to all LSD research, stating that "there is no published evidence that further experimentation is likely to yield invaluable data." A year later, America was in the midst of an LSD epidemic. By October 1966, possession of LSD was illegal. In 1970 the Nixon administration secured passage of the Controlled Substances Act, introducing the concept of scheduling, whereby the government establishes a limited range of categories for different levels of drugs (narcotics being one) with corresponding requirements for prescriptions and use. LSD and mescaline were categorized along with narcotics as Schedule I drugs, defined as having no medical use and with a high potential for misuse. However, a lucrative drug operation continued illicitly into the 1970s with laboratories chemically converting other, legal ergot compounds into LSD. Little wonder that Albert Hofmann called LSD "my problem child."[10]

As Jay Stevens, author of *Storming Heaven: LSD and the American Dream*, the definitive work on the political and social history of the drug, put it: "To discover, in the recesses of the mind, something that felt a lot like God, was not a situation that either organized science or organized religion wished to contemplate."[11]

Bewitching or Just Tripping?

Ergot is a fungus that sometimes infects rye and other cereals. It produces alkaloids that when ingested often cause neurological disturbances. They may also result in narrowing of the blood vessels, which sometimes leads to gangrene of the extremities.

Ergotism was the cause of terrible suffering in Europe, particularly during the Middle Ages. In an epidemic of ergotism in Aquitaine in the year 994, as many as 40,000 people are said to have died, most of them peasants. (The affluent were spared because they ate white wheat bread, whereas peasants ate rye and other dark breads.) Another decimated Paris in 1418, killing 50,000 people in one month.[12] The symptoms — terrifying hallucinations accompanied by twitching and convulsive contractions (St. Vitus's dance) and a "holy fire" afflicting the extremities with agonizing burning sensations and gangrene — were horrific. In severe cases, the tissues became dry and black, and the mummified limbs broke off without loss of blood. People assumed these were retributions for their sins.

Prayers were offered to St. Anthony, the patron saint with protective powers against fire, infection, and epilepsy. The Order of St. Anthony was founded in 1093, and Antonine monasteries in France became the focus of pilgrimages by sufferers of what came to be known as St. Anthony's Fire. Their hospitals hung the detached limbs above their entrance portals, and it was said that victims could retrieve them at the Last Judgment. The relief that followed visits to the shrine of St. Anthony in the Dauphiné, containing his relics brought back by Crusaders from the Holy Land, was probably real, since the sufferers received a diet free of contaminated grain during their journey to the shrine.[13]

It was not until 1670 that ergot was proved to be the cause of the epidemics that had raged for centuries. Oddly enough, the ergot baked in bread dough forms LSD. The action of other fungi may also produce LSD in the natural ergot. The historian Mary Matossian has correlated, over seven centuries, the consumption of rye bread and climatic conditions favorable to the growth of the ergot mold in Europe and America with periods of sporadic outbursts of bizarre behavior and the symptoms of bewitchment.[14] Her arguments are persuasive.

It now appears likely that the alleged bewitchings that occurred in Salem, Massachusetts, in 1692 were caused by a relatively mild outbreak of ergotism. From available meteorological records for 1691, the early rains and a warm spring followed by a hot, humid summer formed a weather pattern conducive to the growth of ergot mold on the local rye. In contrast, a drought in 1692 was accompanied by an abrupt cessation of the "epidemic of bewitchings." These Salem "witches" — nineteen were hanged, one was pressed to death with stones, and two died in prison — like some of the earlier French sufferers of ergot poisoning, were probably just high on acid.

THERAPEUTIC USES OF HALLUCINOGENS

After 1970 LSD research came to a virtual standstill for three decades. The stages LSD had gone through generated a profoundly nihilistic attitude among scientists and clinicians. Heralded initially as a scientific tool for unlocking the unconscious, viewed by converts such as Aldous Huxley as a force that could push mankind up the evolutionary ladder, propagated by Timothy Leary and others as leading to a spiritual utopia, then embraced for widespread recreational abuse, LSD use culminated in a mindless cultural revolt with mystical underpinnings.

Psychedelic research was grievously injured by this unwarranted messianism.

Nevertheless, by the turn of the century, researchers in several countries were once again beginning to explore potential therapeutic uses for hallucinogens. Several factors conspired to encourage the trend. Modern psychiatry had come to routinely use psychotropic drugs, which are known to affect the same brain neurotransmitters as do hallucinogens. Hallucinogens enhance the brain's serotonin system, which may be useful in the treatment of phobias, depression, obsessive-compulsive disorders, and substance abuse. Furthermore, techniques to evaluate brain function, neurochemistry, and response to drugs have become more sophisticated.

Hallucinogens were being used in the treatment of depression and schizophrenia at the University of Zurich. Members of the Native American Church of North America had volunteered to form the core of a study of the influence of mescaline on alcoholism. Psilocybin, a hallucinogen isolated from the Mexican "sacred mushroom" by Albert Hofmann in 1958 and subsequently synthesized, was undergoing FDA-approved trials at the University of Arizona to see if it could allay the symptoms of obsessive-compulsive disorder.[15] As of 2006 LSD was being reevaluated for use in the reduction of pain and fear in dying patients.[16]

A new era of biological psychiatry characterized by advances in the understanding of brain chemistry had dawned, bringing with it profound cultural changes. But the drugs are far from perfect. The claims made for psychotropic medications are sometimes overblown, with too little awareness of their limitations. For instance, about 30 percent of people who try antidepressants are not helped by them. Side effects include an increased risk of suicidal thoughts and behavior in children and adolescents. As research marches on, investigators are working to devise different treatments for the various subtypes of depression.

There are three main fronts in the current research on mental illness: the possibility of other neurotransmitters yet to be found (besides the four known ones: serotonin, dopamine, norepinephrine, and GABA); investigations into physical changes that are now known to occur in the brain (as revealed by modern brain-imaging techniques); and research into possible genetic causes or predispositions.

In an astonishingly rapid series of chance occurrences in the 1940s and 1950s, mood stabilizers, behavior-conditioning drugs, tranquilizers, antidepressants, and antianxiety drugs all became part of the human experience. There is no question that serendipity, not deductive reasoning, was the force behind virtually all these exciting discoveries in the psychotropics.

There is little doubt that the advances of the future, like those of the past, will come about under the good graces of both luck and sagacity.

Conclusion

Taking a Chance on Chance: Cultivating Serendipity

In his farewell address on January 17, 1961, President Dwight Eisenhower famously cautioned the nation about the influence of the "military-industrial complex," coining a phrase that became part of the political vernacular. However, in the same speech, he presciently warned that scientific and academic research might become too dependent on, and thus shaped by, government grants. He foresaw a situation in which "a government contract becomes virtually a substitute for intellectual curiosity."

As we have seen, many of the most essential medical discoveries in history came about not because someone came up with a hypothesis, tested it, and discovered that it was correct, but more typically because someone stumbled upon an answer and, after some creative thought, figured out what problem had been inadvertently solved. Such investigations were driven by curiosity, creativity, and often disregard for conventional wisdom. While one might assume that as technology continues to advance, so will major medical discoveries, the truth is that over the past two decades there has been a marked decline in major discoveries. The current system of medical research simply does not foster such serendipitous discoveries. Why is this so, and what can we do about it?

In earlier times, discoverers were more apt to be motivated in their pursuits by personal curiosity. In the nineteenth century, the typical discoverer was not a professional scientist but a well-educated gentleman of independent means (like Charles Darwin) tinkering around on his estate, designing experiments and conceptualizing patterns. There were no committees or granting agencies that had to be romanced and persuaded of the worth of a particular project. Such an individual, driven by nothing other than intellectual curiosity, was open to unexpected observations and would pursue them wherever they led with no restrictions.

Even in the early twentieth century, the climate was more conducive to serendipitous discovery. In the United States, for example, scientific research was funded by private foundations, notably the Rockefeller Institute for Medical Research in New York (established 1901) and the Rockefeller Foundation (1913). The Rockefeller Institute modeled itself on prestigious European organizations such as the Pasteur Institute in France and the Koch Institute in Germany, recruiting the world's best scientists and providing them with comfortable stipends, well-equipped laboratories, and freedom from teaching obligations and university politics, so that they could devote their energies to research. The Rockefeller Foundation, which was the most expansive supporter of basic research, especially in biology, between the two world wars, relied on successful programs to seek promising scientists to identify and accelerate burgeoning fields of interest. In Britain, too, the Medical Research Council believed in "picking the man, not the project," and nurturing successful results with progressive grants.

After World War II, everything about scientific research changed. The U.S. government — which previously had had little to do with funding research except for some agricultural projects — took on a major role. The National Institutes of Health (NIH) grew out of feeble beginnings in 1930 but became foremost among the granting agencies in the early 1940s at around the time they moved to Bethesda, Maryland. The government then established the National Science Foundation (NSF) in 1950 to promote progress in science and engineering.[1] Research in the United States became centralized and therefore suffused with bureaucracy. The lone scientist working independently was now a rarity. Research came to be characterized by large teams drawing upon multiple scientific disciplines and using highly technical methods in an environment that promoted the not-very-creative phenomenon known as "groupthink." Under this new regime, the competition among researchers for grant approvals fostered a kind of conformity with existing dogma. As the bureaucracy of granting agencies expanded, planning and justification became the order of the day, thwarting the climate in which imaginative thought and creative ideas flourish.

About 90 percent of NIH-funded research is carried out not by the NIH itself on its Bethesda campus but by other (mostly academic medical) organizations throughout the country. The NIH gets more than 43,000 applications a year. Through several stages of review, panels of experts in different fields evaluate the applications and choose which deserve to be funded. About 22 percent are approved for a period of three to five years. The typical grant recipient works at a university and does not draw a salary but is dependent on NIH funding for his or her livelihood. After the three- to five-year grant expires, the researcher has to renew the funding. The pressure is typically even greater the second time around, as the university has gotten used to "absorbing" up to 50 percent of the grant money to use for "overhead," and by now the scientist has a staff of paid postdocs and graduate students who depend on the funding, not to mention the fact that the continuation of the scientist's faculty position is at stake.

Inherent in the system is a mindset of conformity: one will tend to submit only proposals that are likely to be approved, which is to say, those that conform to the beliefs of most members on the committee of experts. Because of the intense competition for limited money, investigators are reluctant to submit novel or maverick proposals. Needless to say, this environment stifles the spirit of innovation. Taking risks, pioneering new paths, thwarting conventional wisdom — the very things one associates with the wild-eyed, wild-haired scientists of the past — don't much enter into the picture nowadays.

These realities of how science is practiced lead to a systemic problem of scientists working essentially with blinders on. Research is "targeted" toward a specifically defined goal with a carefully laid-out plan of procedures and experiments. There is no real room for significant deviation. A researcher can make a supplementary request if an unanticipated finding occurs, as long as its pursuit falls within the general theme — the presumptive goal — that was originally funded. In

2005 the NIH received a meager 170 such applications for supplementary grants, only 51 of which were funded. But if, for instance, someone is funded to study the effect of diet on ulcers, that person wouldn't go where the evidence of a bacterial cause might lead. Alexander Fleming would not have been funded. Nor would Barnet Rosenberg in his incidental observation that eventually led to the discovery of the chemotherapeutic agent cisplatin.

In the past, the real advances in medicine have often come not from research specifically directed against a target but rather from discoveries made in fields other than the one being studied. In cancer, chemotherapy arose from the development of instruments of chemical warfare, the study of nutritional disease, and the effect of electric current on bacterial growth. A major tumor-suppressor gene was discovered through research on polio vaccines. Stem cells were discovered through research on bone-marrow transplants for leukemia and for whole-body irradiation from nuclear weapons. Agents for alteration of mood and behavior came from attempts to combat surgical shock, the treatment of tuberculosis, and the search for antibiotics. When scientists were allowed to pursue whatever they found, serendipitous discovery flourished.

Today, targeted research is pretty much all there is. Yet, as Richard Feynman put it in his typical rough-hewn but insightful manner, giving more money "just increases the number of guys following the comet head."[2] Money doesn't foster new ideas, ideas that drive science; it only fosters applications of old ideas, most often enabling improvements but not discoveries.

PEER REVIEW

The government does not employ scientists to work at the NIH or the NSF to review grant applications. Rather, the task is accomplished through a system known as "peer review": independent qualified experts judge the work of their "peers" on a pro bono basis.[3] The system of peer review within government granting agencies developed after World War II as a noble attempt to keep science apolitical by keeping the decision-making regarding how research money is spent within the scientific community. The same system is also used by scientific journals to determine whether the results of a research study are solid enough to be published. However, questions regarding the peer review system's flaws began to arise in the 1970s and became more insistent by the mid-1990s.

In 2005 the NIH invested more than $28 billion in medical research. Competitive grants went to 9,600 grant projects at some 2,800 universities, medical schools, and other research institutions in the United States and around the world. An applicant for a research grant is expected to have a clearly defined program for a period of three to five years. Implicit is the assumption that nothing unforeseen will be discovered during that time and, even if something were, it would not cause distraction from the approved line of research. Yet the reality is that many medical discoveries were made by researchers working on the basis of a fallacious hypothesis that led them down an unexpected fortuitous path. Paul Ehrlich, for example, began with a false belief regarding the classification of the syphilis spirochete that nevertheless set him upon the right road to find a cure. Ladislaus von Meduna incorrectly hypothesized that schizophrenia was incompatible with epilepsy, but this led to a breakthrough convulsive therapy. John Cade, by pursuing a fanciful speculation down a false trail, was led to discover the value of lithium for treating mania. Today such people, if funded by the NIH or NSF, would not be allowed to stray from their original agenda.

The peer review system forces investigators to work on problems others think are important and to describe the work in a way that convinces the reviewers that results will be obtained. This is precisely what prevents funded work from being highly preliminary, speculative, or radical.[4] How can a venture into the unknown offer predictability of results? Biochemist Stanley Cohen, who, along with Herbert Boyer, was the first to splice and recombine genes, showed how little he thought of peer review when he speculated: "Herb and I didn't set out to invent genetic engineering. Were we to make the same proposals today, some peer review committees would come to the conclusion that the research had too small a likelihood of success to merit support."[5]

Peer review institutionalizes dogmatism by promoting ortho-doxy. Reviewers prefer applications that mesh with their own perspec-tive on how an issue should be conceptualized, and they favor individuals whom they know or whose reputations have already been established, making it harder for new people to break into the system.[6] Indeed, the basic process of peer review demands conformity of think-ing and disdains a maverick's approach. "We can hardly expect a com-mittee," said the biologist and historian of science, Garrett Hardin, "to acquiesce in the dethronement of tradition. Only an individual can do that."[7] Young investigators get the message loud and clear: Do not challenge existing beliefs and practices.

So enmeshed in the conventional wisdoms of the day, so-called "peers" have again and again failed to appreciate major break-throughs even when they were staring them in the face. This reality is evidenced by the fact that so many pioneering researchers were inap-propriately scheduled to present their findings at undesirable times when few people were in the audience to hear about them. Consider the following examples:

- The psychiatrist Pierre Deniker's presentation on the positive results of clinical trials of Thorazine at the Fiftieth French Congress of Psychiatry and Neurol-ogy in Paris in July 1952 was scheduled at the end of the last session of the week during the lunch hour and was delivered to not more than twenty regis-trants in a large auditorium.

- The psychiatrist Roland Kuhn presented his findings on a new tricyclic antidepressant (imipramine) at the Second International Congress of Psychiatry in Zurich in September 1957 to an audience of barely a dozen people.

- Robert Noble found himself presenting his findings on vinca alkaloids as cancer chemotherapy at midnight before a small cluster of scientists at a meeting of the New York Academy of Sciences in 1958.

- Australian Barry Marshall, rebuffed by gastroenterologists, sought an infectious disease conference in Brussels in 1983 to present his observations on bacterial association with stomach ulcers.

Nevertheless, each of these discoveries eventually transformed the world.

The same flawed peer review process also controls what is deemed worthy of publication in scientific journals. In a similar vein, there are some striking examples of prestigious journals rejecting for publication core papers outlining the results of groundbreaking studies:[8]

- Rosalyn Yalow was so bitter over her letter of rejection from a prestigious American journal regarding her path-breaking original submission, which would lead to radioimmunoassay, that she even reproduced it in her 1977 Nobel Prize lecture twenty-two years later.[9]

- Paul Lauterbur's landmark report in *Nature,* the leading British scientific journal, establishing a new principle of magnetic resonance image (MRI) formation was initially rejected. Lauterbur rewrote it, including a general statement that "these techniques should find many useful applications in studies of the internal structures," surely one of the great understatements in the scientific literature. After a few further revisions were required, it was finally published.[10]

- Barry Marshall persisted for ten years against long-held dogma regarding the causation of peptic ulcer disease before his revolutionary concepts were widely accepted. A report first submitted by him to the *New*

England Journal of Medicine was rejected and then published in the *Lancet*.

- Judah Folkman has persevered for more than forty years in delineating the role of angiogenesis in tumor growth and spread, finding difficulty in the early years in having his work published.

Hungarian Humor

Arbiters of the scientific literature demand declarative sentences of no uncertainty or ambiguity. A case in point is the experience of Albert Szent-Györgyi, a Hungarian biochemist working at Cambridge University in England. In 1928 he extracted a compound from a cow's adrenal gland, not yet recognizing it as vitamin C. Thinking he had isolated a new sugarlike hormone, he named it "ignose," the suffix *-ose* being used by chemists for sugars or carbohydrates (like gluc*ose* and fruct*ose*) and the *igno-* part indicating he was ignorant of the substance's structure. The editor of the *Biochemical Journal* did not share his humor and rejected the submitted manuscript. When Szent-Györgyi's second suggestion for a name, "Godnose," was similarly rejected, he settled upon the name hexuronic acid, based upon the known six carbon atoms in the formula. He subsequently identified it as ascorbic acid, or vitamin C, for which he was awarded the Nobel Prize in 1937.

Peers are, almost by definition, part of the established order and typically mired in traditional thinking. They are also human beings with their own agendas and priorities. What it comes down to is this: Who on a review committee is the peer of a maverick?[11]

THE DRUG PUSHERS

The dearth of scientific innovation is systemic, and government grants and peer reviews are not the only culprits. The other major offender is the pharmaceutical-industrial complex, a behemoth that has become more of a marketing machine than a fountainhead of research and development. Since the mid-1980s, the industry has shown a striking decline in innovation and productivity, even while its profits have soared. The number of new drugs approved by the Food and Drug Administration (FDA) fell especially sharply in the ten-year period from 1996 to 2005, from 53 in 1996 to only 20 in 2005, despite record-high research spending by the industry, averaging more than $38 billion a year during that time.

Why did the industry have so few major breakthroughs in such a free-spending period? Because it has progressively shifted the core of its business away from the unpredictable and increasingly expensive task of creating drugs and toward the steadier business of marketing them. Most drug makers now spend twice as much marketing medicines as they do researching them. They compensate for dramatically diminished productivity and loss of patent protection by raising prices, maneuvering to extend patents, engaging in direct-to-consumer advertising, and developing and marketing more and more "me-too" and lifestyle drugs that do not enhance health and longevity. Hope has been largely replaced by hype.

Since the FDA lifted limits on direct-to-consumer (DTC) advertising of drugs in 1997, this form of marketing grew into a $4.2 billion business. (Only one other country, New Zealand — with a population of less than 4 million — allows such advertising.) The ensuing onslaught of ads, mostly on television, that urged you to "ask your doctor about" the drug being advertised was so effective that patients began demanding prescriptions from their doctors, who often obliged even if the new drugs were considerably more expensive and no better than existing drugs.[12]

Indeed, this compliance on the part of doctors is part of the problem. The average number of prescriptions they wrote for each American rose from seven in 1993 to twelve in 2004.[13] American doctors wrote some 3 billion prescriptions in 2005. In the same year, prescription drug sales soared to about $250 billion in the U.S. and an astronomical $500 billion worldwide. In a typical year, the fifty most heavily advertised drugs account for nearly half the increase in spending on prescription drugs.[14] (In their better moments, most doctors express feelings of discomfort about their acquiescence. A 1997 survey reported that more than 80 percent of physicians had negative feelings about drug advertising to the consumer.[15] And at the American Medical Association meeting in June 2005, six separate resolutions were introduced advocating for limitations or outright bans on DTC marketing of prescription drugs.)

In an even more insidious practice that is part and parcel of consumer-oriented advertising, pharmaceutical companies are widening the very definition of "illness" to create new markets for drugs. As documented in the book *Selling Sickness,* old conditions are expanded, new ones are created, common complaints are labeled "medical conditions," and people's "unmet needs" are taken advantage of as markets for medications to grow ever larger.[16] Under such a system, premenstrual syndrome (PMS) became a psychiatric illness with the imposing name of premenstrual dysphoric disorder (PMDD), and for the very bad moods some women suffer before their periods, Eli Lilly repackaged its aging antidepressant Prozac in a lavender pill dubbed Sarafem. Hyperactive children all have attention deficit disorder (ADD), and there are now maladies called female sexual dysfunction (FSD) and social anxiety disorder (Glaxo advertised its antidepressant Paxil for this newly noted "disorder"). By 2006, nearly 4 million people — adults and children — were taking drugs to treat attention deficit disorder and hyperactivity.

Perhaps the biggest reason the pharmaceutical industry is so profitable and yet so unproductive is its current obsession with "me-too" drugs, the insider term for variants of existing drugs that contain virtually the same ingredients as previously approved ones. Instead of looking for truly innovative medicines to succeed those slated to lose patent protection, the industry focuses most of its research efforts on developing and producing drugs that are either minor variations or outright duplicates of drugs already on the market. Of the ten top-selling drugs in the world, half offer almost no benefit over drugs marketed previously.[17]

An especially egregious example is Nexium, the "purple pill" for heartburn that was the nation's most widely advertised prescription drug in 2006. When British company AstraZeneca's patent on Prilosec, a blockbuster prescription drug for heartburn that reached astronomical sales of $6 billion in 2000, was due to expire in 2001, the company simply extracted the active half of its molecule, which the Patent Office then recognized as a separate invention called Nexium. In clinical trials, it had a 90 percent healing rate in erosive esophagitis compared to Prilosec's 87 percent. Approved by the FDA, it became one of the best-selling drugs in America, retailing for more than $5 a pill. The company continued to market Prilosec over the counter for a fraction of the cost, a maneuver that knocked out the usage of its generic drug, omeprazole.[18] Furthermore, the industry continues to launch an increasing number of "lifestyle" drugs — such as the trinity of impotence relievers Viagra, Levitra, and Cialis; Pfizer's and Merck's baldness remedies; and Bristol-Myers's drug to eliminate women's facial hair — whose sales largely depend on inducing consumers to ask doctors for them. These are indicative of the fact that the drug companies are motivated purely by profit and not by any desire to find cures for serious illnesses for their own sake. Relatively rare diseases, for example, with a small potential market are not on their accountants' agenda. The story of the drug Gleevec is a case in point. Academic research uncovered a localized genetic flaw underlying a rare fatal type of leukemia called chronic myeloid leukemia. Novartis had patented the molecule that inactivated the gene's enzyme, but the company showed little interest in committing resources to the drug's development, even after Brian Druker, an NIH-funded university researcher, showed its usefulness. As Jerry Avorn tells it in his book *Powerful Medicines,* "cancer researchers had to resort to the bizarre tactic of sending a petition to the company's CEO, signed by scientists in the Leukemia and Lymphoma Society of America, imploring him to make more of the drug available for clinical studies." The FDA approved the drug within two years.[19] Novartis now not only charges $27,000 for a year's supply of Gleevec, it also uses it as the poster child for drug company innovation.

To take other examples, even after genetic fingerprinting by the Human Genome Project and Craig Venter's privately funded gene research of diseases such as sickle cell anemia, cystic fibrosis, and Huntington's disease, little progress toward curing or treating the diseases resulted. Researchers and Big Pharma generally disregard "niche" diseases, those affecting so few people that a breakthrough treatment would not lead to glory or profit.

In one major instance, it was the very pursuit of a cure for a rare disease that led to a conceptual breakthrough that eventually brought both glory and profit. Michael Brown and Joseph Goldstein's discovery of the mechanism of cholesterol metabolism through their study of the rare inherited disease known as familial hypercholesterolemia led to the widespread use of statins for the treatment of high cholesterol.

One oft-quoted estimate is that the industry as a whole invests up to 40 percent of its revenue in promotional efforts. Without a doubt, it spends far more on salesmen than on scientists. An army of more than 90,000 pharmaceutical sales representatives ("detailmen" and "drug reps") in the United States costs well over $7.5 billion a year. And, unfortunately, neither doctors nor patients have the time, knowledge, or inclination to pay close attention to the scientific sophistry behind many of the new drugs coming on the market.

Hair Today, Gone Tomorrow

In 1980 a Pennsylvania physician was treating a bald man for high blood pressure with the oral drug minoxidil. The doctor was surprised to note new follicles growing on the patient's scalp, and this hair-raising experience soon led to a commercial product. Approved by the FDA for men in 1988 and for women in 1991, it is now sold over the counter in the form of a lotion called Rogaine. It is believed to work by inducing dilation of scalp blood vessels. Balding American men spent $96 million on Rogaine in 2005.

Two other drugs related to remedying the problems of either not enough or too much hair were also discovered unexpectedly. Propecia was originally approved to treat benign prostatic hyperplasia, an enlargement of the prostate gland, by blocking a testosterone. This hormone has long been known to be associated with baldness. A drug in pill form was approved in 1997 only for men because it may cause birth defects in women of childbearing age.

The drug eflornithine, which in 1979 was found useful in treating African sleeping sickness, was later discovered to suppress the enzyme that causes facial hair to grow. It is now marketed as a cream, Vaniqa, for the removal of unwanted facial hair.

The first step in modern drug development is identifying the molecule a drug could potentially attach itself to, called the target. The raw statistics are daunting. For every 5,000 compounds the industry screens as potential new medicines, 250 promising ones make it to the next stage: testing in animals. This stage of discovery and preclinical research typically takes six to seven years. Of the 250, five compounds survive to be tested in people in clinical trials. Phase I, investigating the drug's safety on 20 to 100 volunteers, may take one and a half years. Phase II, testing efficacy and side effects in 100 to 500 patients who are candidates for the drug, may take up to two years. Phase III, confirming the drug's usefulness and studying its long-term toxicity in 1,000 to 5,000 patients, generally requires another three to four years. Then, over perhaps a year and a half, the FDA reviews the evidence and decides if the drug is safe and effective enough to approve. At the end of the pipeline, only one out of the five drugs that went into clinical trials eventually emerges as approved for sale.

The current system of screening an endless number of substances has certainly not been notably efficacious, despite the fact that a slew of new technologies — combinatorial chemistry, improved screening, rational drug design, and pharmacogenomics — developed since the late 1970s promised to streamline the process.[20] Even with these useful technologies, the time spent to develop a drug has lengthened, the cost has more than doubled, and the failure rate has not declined. Technology is merely a tool and does not by itself provide answers. Only ideas and creative thought can do that, and those are things our existing system sadly lacks and fails to nurture.

Yet, despite the paucity of innovative drugs, the drug companies consistently rank as the most profitable among Fortune 500 companies. A lofty 25 percent of pharmaceutical revenues are profits. In 2002, the combined profits of the top ten pharmaceutical companies in the Fortune 500 exceeded the combined profits of the other 490 companies![21]

Making matters worse, in recent years, the drug companies have engaged in a mania of mergers to bolster earnings by reducing costs. Mergers subordinate research to strategies determined by short-term market forces and not by a quest for the truly novel. The sad fact is that these deals, which centralize research staff, do nothing to increase overall ability to produce truly new medicine. The mergers are distracting and hinder researchers' creativity, as they result in the reduction of multiple pursuits down to only a few that are overly managed and targeted. When it comes to research and development, bigger is not better; in fact, it appears to be worse.

EDUCATION

Discovery is an essentially creative enterprise. Yet the education that future researchers receive does not, for the most part, foster the mindset or the skills that are needed for that sort of creativity. The focus in medical schools and higher education institutions in the field of science is on facts, not ideas; on outcomes, not process. Incessant multiple-choice examinations put a premium on a form of quick thinking that may neglect other qualities of intelligence and creativity.[22] Their curricula completely ignore the process of how discoveries and current concepts came to be accepted. Lacking is any sense of awe and wonder for the magic of discovery, and there is little attempt to teach, or even encourage, the kind of creativity and complex synthesizing of ideas that has enabled discoverers to connect the dots that led to major breakthroughs in understanding.

As is apparent in the many stories of serendipitous discoveries, it takes more than just good luck. Opportunities for discovery present themselves every day, but not everyone is able to take advantage of them. Failure to follow puzzling observations and unexpected results of experiments has resulted in many missed opportunities. For example, depletion of white blood cells was observed in victims of mustard gas during World War I, but the agent's chemistry was not seized upon then as a possible treatment for lymphomas and leukemias. Many researchers encountered the phenomenon of mold preventing bacterial growth without realizing its significance and usefulness before Alexander Fleming discovered penicillin in 1928; then it was twelve years before Ernst Chain and Howard Florey stumbled upon its importance and figured out how to exploit it. Long before Barry Marshall, bacterial colonies in the stomach were seen but shrugged off.

Why are particular people able to seize on such opportunities and say, "I've stumbled upon a solution. What's the problem?" Typically, such people are not constrained by an overly focused or dogmatic mindset. In contrast, those with a firmly held set of preconceptions are less likely to be distracted by an unexpected or contradictory observation, and yet it is exactly such things that lead to the blessing of serendipitous discovery.

Serendipitous discoverers have certain traits in common. They have a passionate intensity. They insist on trying to see beyond their own and others' expectations and resist any pressure that would close off investigation. Successful medical discoverers let nothing stand in their way. They break through, sidestep, or ignore any obstacle or objection to their chosen course, which is simply to follow the evidence wherever it leads. They have no patience with dogma of any kind.

The only things successful discoverers do not dismiss out of hand are contradictory — and perhaps serendipitously valuable — facts. They painstakingly examine every aspect of uncomfortable facts until they understand how they fit with other facts. Far from being cavalier about method, serendipitous discoverers subject their evidence and suppositions to the most rigorous methods they can find. They do not run from uncertainty, but see it as the raw material from which new scientific and medical certainties can be wrought. They generally share a fondness for pattern play and draw from many diverse — and seemingly unrelated — sources. An epiphany is often the result of pattern recognition. Alert to change but not made uneasy by it, discoverers demonstrate an almost aesthetic feel for the way order arises from the unpredictable odds of an infinitely varied universe."[23]

Today's researchers are not being educated to think in the ways described above — in fact, just the opposite. A medical education is not designed to foster creativity: rather it consists of accumulating facts without any recognition of how those facts were obtained. "Facts are the enemy of truth!" cried Don Quixote de la Mancha. Certainly, unprocessed facts, facts taken at face value, not only tell us nothing but frequently deceive us. Understanding comes from making connections between many disparate facts. Information is not knowledge. The overemphasis on memorization to the detriment of analysis, critical thinking, and overall understanding is a big problem.

We must consider whether our current educational obsessions and fashions are likely to help or hinder serendipity in the future and ask how we can nurture a penchant for serendipitous discovery in today's children and future generations.

FOSTERING SERENDIPITY

Despite all the examples given, mainstream medical research stubbornly continues to assume that new drugs and other advances will follow exclusively from a predetermined research path. Many, in fact, will. Others, if history is any indication, will not. They will come not from a committee or a research team but from an individual, a maverick who views a problem with fresh eyes. Serendipity will strike and be seized upon by a well-trained scientist or clinician who also dares to rely upon intuition, imagination, and creativity. Unbound by traditional theory, willing to suspend the usual set of beliefs, unconstrained by the requirement to obtain approval or funding for his or her pursuits, this outsider will persevere and lead the way to a dazzling breakthrough. Eventually, once the breakthrough becomes part of accepted medical wisdom, the insiders will pretend that the outsider was one of them all along.

Some scientists are beginning to recognize the value of serendipity and are even trying to foster it. In the fall of 2006 the Howard Hughes Medical Institute opened a $500 million research center, the Janelia Farm Research Campus, built on the banks of the Potomac River, forty miles northwest of Washington, D.C. The center will emulate Bell Labs and the British Medical Research Council's Laboratory of Molecular Biology by encouraging scientists to do hands-on creative work on their own or in small, close-knit teams, and to focus on original projects. In other words, the Janelia Farm Research Campus is being set up as a serendipity incubator.

"Crowdsourcing"

Another recent phenomenon that may help research move outside the box is the "crowdsourcing" of research and development made possible by the Internet. A prime example is the Web site InnoCentive. Launched by pharmaceutical maker Eli Lilly in 2001 as a way to connect with brainpower outside the company that could help develop drugs and speed them to market, InnoCentive promptly threw open the doors to other firms eager to access the network's trove of ad hoc experts. Companies like Boeing, DuPont, and Procter & Gamble post problems that corporate R&D people have been unable to solve in the hope that outside "tinkerers" — amateur scientists, inventors, and researchers — will come up with solutions in exchange for a fee. InnoCentive boasts a 30 percent success rate. This sort of decentralized mechanism that allows fresh ideas from outside the establishment to percolate up through a broad network of diverse thinkers and problem solvers is a welcome development and can only have a positive effect on research.

One of the main reasons for the success of InnoCentive's "crowdsourcing" is its ability to tap into the fresh perspectives of nonexperts. An analysis of the phenomenon by Karim Lakhani, a lecturer in technology and innovation at MIT, who surveyed 166 problems posted to InnoCentive from more than 25 different firms, actually showed that "the odds of a solver's success increased in fields in which they had no formal expertise."[24]

Discoveries are surprises. You can't plan surprises, but you can certainly create an environment in which they are apt to happen and are likely to be recognized and pursued when they do. Our current system is set up in such a way as to discourage the kind of curiosity-driven research that leads to serendipitous findings.

To turn around this situation, we as a society should make changes in several key areas:

- *General education:* Students, particularly in science, must be educated not only to know facts and the scientific method but also to be prepared to recognize and exploit departures from expected results. Essential are the tools related to thinking: pattern recognition and pattern formation, alertness to similar differences and different similarities, analogies, imagination, visual thinking, aesthetics of nature and of knowledge.

- *Medical and general science education:* Current methods of educating medical students overemphasize memorization at the expense of full understanding, reinforce compulsive behavior, and stifle creativity. The curricula of medical schools and graduate programs in science should teach students about the role of serendipity in past discoveries and be honest about the likelihood that it will play a major role in the future. Creative and critical thinking and open-mindedness to unanticipated observations, not only the acquisition of facts, are essential skills for the next generation of serendipitous discoverers that should be encouraged in such institutions.

- *Big Pharmaceuticals:* Restrictions should be placed on the pharmaceutical industry to shift the emphasis from "me-too" drugs to innovative drugs and to break the cozy link between drug companies and the thousands of doctors who take their gifts, consulting jobs, and trips, then turn around and write prescriptions for their benefactors' drugs, which may just mimic the action of easily available and less expensive generic medications.[25] The FDA should regulate the incessant din of DTC commercials that flood the airwaves, along with print ads, stoking demand for very expensive me-too drugs.

- *Research grants:* Agencies and foundations that fund research grants should allow curiosity-driven investigators to pursue promising findings that may deviate from the scientific question originally proposed. Some flexibility should be provided in the funding mechanisms to allow investigators to follow any unexpected findings wherever they may lead. For investigators to be forced to secretly divert funds for this purpose from their declared and approved missions incurs a sense of dishonesty and, in the end, is not the most productive path. Granting bodies should consider the qualities of creative people and the personae of the applicants as well as their proposals. The challenge is to have insight into whom to support before a track record exists.

- *Peer review process:* Whereas peer review is intended to protect the autonomy and self-governance of the sciences, it has become an agent for the defense of orthodoxy and a constraint on creativity. The review process should be modified to reduce the inherent bias toward prevailing concepts and to encourage mavericks and outsiders. A firm standard should be the degree to which a researcher's work threatens to disturb conventional beliefs.

- *Scientific journals:* Editors of medical and scientific journals should encourage researchers to be more forthcoming in fessing up to serendipity's contributions to their experimental results. The widespread contribution of chance is typically obscured, so other researchers remain ignorant of the important role serendipity plays.[26]

It would be too bold a statement to assert that pulling out the thread of serendipity would unravel the entire tapestry of modern scientific discoveries. Yet, while serendipity is not to be overemphasized, neither should it be denied its due credit. Even with increasingly sophisticated technology and team efforts inclusive of multiple disciplines, serendipity continues to be a major influence to which a resourceful investigator cannot afford to turn a blind eye. Serendipity teaches that it matters less *where* we start looking for something interesting, and more *how* we go about looking for it.

Certain leitmotifs underlie many of the individual medical advances of the past century. Published reports are usually cool and dispassionate, but as one delves below their surface, one begins to see recurring themes: the role of unconscious factors, the ironies of circumstance, the elements of surprise and wonder, the subtle variables that influence the discovery process, and the inadvertent observations leading to breakthroughs in understanding. These factors are as ubiquitous as they are hidden. Journeying among them lands you squarely in the pathways and patterns of creative thought. In today's overly managed, bureaucratized research environment, these elements have become submerged.

We must consider whether the industrialization of invention and discovery has fulfilled the promise of "the endless frontier" of medical science.[27] We must ask whether the yield so far amounts to as much as Robert Noble's chance discovery of the *Vinca* alkaloids for cancer chemotherapy or Barry Marshall's uncovering of the role of *H. pylori* in ulcers and stomach cancer or Frank Berger's stumbling upon Miltown or Mason Sones entering upon coronary arteriography or Baruch Blumberg's identification of the hepatitis B virus, or McCulloch and Till's unexpected discovery of stem cells. Unfortunately, the conditions that foster serendipitous discovery so rarely exist now.

In one of his famous aphorisms, Yogi Berra quite correctly

pointed out: "If you don't know where you're going, you will wind up somewhere else." We know that, in many cases, scientists are looking in all the wrong places, and that "somewhere else" is exactly where they need to go. In their attempts at targeted research guided so often by conventional wisdom, they are operating rather like the proverbial drunk looking for his keys beneath the streetlamp because that is where the light is best.

As John Barth wrote in *The Last Voyage of Somebody the Sailor,* "You don't reach Serendip by plotting a course for it. You have to set out in good faith for elsewhere and lose your bearings serendipitously."[28] The challenge for educational institutions, government policy, research centers, funding agencies, and, by extension, all modern medicine, will be how to encourage scientists to lose their bearings creatively. What they discover may just save our lives!

Acknowledgments

In the course of research for this book, I interviewed several Nobel laureates and winners of other major awards. At SUNY Stony Brook, Chen Ning Yang, Paul Lauterbur, and Arnold Levine provided intimate details of the creative process. Baruch Blumberg, Robert Furchgott, Barry Marshall, Ernest McCulloch, and James Till were candid in acknowledging the role of serendipity in their major discoveries.

I am grateful to others who have given me the benefit of their experience and knowledge: Karl Holubar, Helmut Wyklicky, and Manfred Skopec of the Institute of the History of Medicine and the Josephinum Museum of Medical History, University of Vienna; Kevin Brown, curator of the Alexander Fleming Laboratory Museum at St. Mary's Hospital, London; John Lesch, historian, UC Berkeley; Ernest Hook, professor of public health and pediatrics, UC Berkeley; Martin Blaser, chairman of the Department of Medicine, NYU School of Medicine; Paul Thagard, professor of philosophy and director of the cognitive science program at the University of Waterloo, Canada; Sven-Ivar Seldinger, the Swedish pioneer in vascular catheterization; Max Fink of Stony Brook's Department of Psychiatry; historian of medicine Marcia Meldrum of UCLA; historian of science Wilbur Applebaum of the Illinois Institute of Technology; Lawrence K. Altman for an understanding of the motives of self-experimentation in medicine;

Ivar Strang for information on applying for research funding; Robert Moore and Anthony Demsey for aspects of the peer review process and granting mechanisms at the National Institutes of Health; and Alexander Scheeline, professor of chemistry at the University of Illinois at Urbana, for a critical analysis of the flaws in funding programs for scientific research.

My warmest thanks to those who sustained me with encouragement and support: Michiel Feldberg, Nicholas Gourtsoyiannis, Robert Berk, Gerald Friedland, William Thompson, Michael Oliphant, Claus Pierach, Charles J. Hatem, Gerald Reminick, Carol Hochberg Holker, and, not least, Amy, Rich, and Karen Meyers. And of course, throughout, my muse, my wife, Bea.

Many thanks as well to the research librarians at the Wellcome Institute for the History of Science in London, Cambridge University, the Boerhaave Museum of the History of Science and Medicine in Leiden, the New York Academy of Medicine, and my local Emma S. Clark Memorial Library. Most of all in this regard, I am indebted to Colleen Kenefick, senior librarian at the Health Sciences Center at SUNY Stony Brook, who not only tracked the appropriate sources to all my queries but pointed me to other rewarding sources.

Leo Weinstein provided translations of foreign-language publications. Julia Jannen, my archivist, and Susan Simpson, my photo researcher, rendered useful services. Carrie Nichols Cantor helped shape the book's contents. My agent, Joëlle Delbourgo, was the indispensable catalyst for the entire process. My editor at Arcade, Casey Ebro, blended a cheerful personality with an incisive mind; her sharing my enthusiasm in the book's vision was constantly refreshing. My gratitude to my assistant, Alice Jimenez, for her consistent dependability and selfless cooperation.

Notes

INTRODUCTION: Serendipity, Science's Well-Guarded Secret

1. Quoted in Victor Weisskopf, *The Joy of Insight: Passions of a Physicist,* trans. Douglas Worth (New York: Basic Books, 1991), 296–97.

2. L. N. Gay and P. E. Carliner, "The prevention and treatment of motion sickness. I. Seasickness," *Johns Hopkins Medical Bulletin* 84 (1949): 470–87.

3. Kenneth Chang, "Two Americans Win Nobel for Chemistry," *New York Times,* October 9, 2003.

4. David Anderson, "The Alchemy of Stem Cell Research," *New York Times,* July 15, 2001.

5. E. C. Kendall, "The crystalline compound containing iodine which occurs in the thyroid," *Endocrinology* 1 (1917): 153–69. As is common with researchers who in time achieve success, it was not until fifty-five years later that Kendall told of his early frustrations with the use of metal tanks. Edward C. Kendall, *Cortisone* (New York: Scribner, 1971), 32–33.

6. Thomas S. Kuhn, *The Structure of Scientific Revolutions* (Chicago: University of Chicago Press, 1962/1970).

7. Quoted in J. G. Crowther, *Science in Modern Society* (New York: Schocken, 1968), 363.

8. A. Kantorovich and Y. Ne'eman, "Serendipity as a source of evolutionary progress in science," *Stud Hist Phil Sci* 20 (1989): 505–29. In his 1970 book *Chance and Necessity* (trans. Austryn Wainhouse; New York: Knopf, 1971), Jacques Monod showed by biochemical evidence how all life, including human, stems from the random chance of mutation and the necessity of Darwinian natural selection. Five years earlier, he

shared the Nobel Prize in Medicine with François Jacob and André Lwoff for their work on cellular genetic function.

9. R. S. Root-Bernstein, "Who discovers and who invents?" *Research and Technology Measurement* 32 (1989): 43–50.

10. Theodore G. Remer, ed., *Serendipity and the Three Princes: From the Peregrinaggio of 1557* (Norman: University of Oklahoma Press, 1965).

11. Walter Bradford Cannon, *The Way of an Investigator* (New York: W. W. Norton, 1945).

12. John Godfrey Saxe, *The Blind Men and the Elephant* (New York: McGraw-Hill, 1963).

13. Robert L. Park, *Voodoo Science: The Road from Foolishness to Fraud* (New York: Oxford University Press, 2000), 172–74.

14. Albert Szent-Györgyi, *Bioenergetics* (New York: Academic Press, 1957), 57.

15. Frank Herbert, *Heretics of Dune* (New York: G. P. Putnam's Sons, 1984), 368.

16. Author interview with Chen Ning Yang, December 6, 1995.

17. Albert Einstein, preface to *Where Is Science Going?* by Max Planck (London: Allen and Unwin, 1933).

18. Quoted in "Men of the Year," *Time* 77 (1) (1961): 40–46.

19. Quoted in Steven Levy, "Annals of Science: Dr. Edelman's Brain," *New Yorker,* May 2, 1994, 62–66.

20. Elkhonon Goldberg, *The Wisdom Paradox: How Your Mind Can Grow Stronger as Your Brain Grows Older* (New York: Gotham Books, 2005).

21. Timothy D. Wilson, *Strangers to Ourselves: Discovering the Adaptive Unconscious* (Cambridge, Mass.: Harvard University Press, 2002).

22. Quoted in W. D. Foster, *A History of Medical Bacteriology and Immunology* (London: Heinemann Medical, 1970).

23. Anne Fadiman, *Ex Libris* (New York: Farrar, Straus and Giroux, 1998), 90–91.

24. Arthur Koestler, *The Act of Creation* (London: Hutchinson, 1964).

25. Aser Rothstein, "Nonlogical factors in research: chance and serendipity," *Biochemical Cell Biology* 64 (1986): 1055–65.

26. Douglas R. Hofstadter, "Analogy as the Core of Cognition," in *The Best American Science Writing,* ed. James Gleick (New York: Ecco Press, 2000), 116.

27. David Bohm, *On Creativity,* ed. Lee Nichol (London: Routledge, 1998), 7, 15.

28. Francis Darwin, ed., *The Life and Letters of Charles Darwin* (London: John Murray, 1888).

29. Sharon Bertsch McGrayne, *Nobel Prize Women in Science: Their Lives, Struggles, and Momentous Discoveries* (New York: Birch Lane Press, 1993), 345.

30. Eugene Straus, *Rosalyn Yalow, Nobel Laureate* (New York: Plenum, 1998).

31. Lynn Gilbert and Gaylen Moore, *Particular Passions: Talks with Women Who Have Shaped Our Times* (New York: Crown, 1981), 44.

32. Richard P. Feynman, *"Surely You're Joking, Mr. Feynman!": Adventures of a Curious Character* (New York: W. W. Norton, 1985), 173–74.

33. Peter Medawar, *Pluto's Republic* (New York: Oxford University Press, 1982), 132.

34. Alan L. Hodgkin, "Chance and design in electrophysiology: An informal account of certain experiments on nerve carried out between 1934 and 1952," *J Physiol (Lond)* 263 (1976): 1–21; Alan Hodgkin, *Chance and Design: Reminiscences of Science in Peace and War* (Cambridge: Cambridge University Press, 1992), xi.

35. Quoted in Donald G. Mulder, "Serendipity in Surgery," *Pharos* 57, no. 3 (1994): 22–27.

36. Rothstein, "Nonlogical factors in research."

37. Peter Medawar, *The Limits of Science* (Oxford: Oxford University Press, 1987), 49.

38. J. H. Humphrey, "Serendipity in immunology," *Annu Rev Immunol* 2 (1984): 14.

39. J. H. Humphrey, "Serendipity and insight in immunology," *British Medical Journal* 293 (1986): 185.

40. Bernard Barber and Walter Hirsch, eds., *The Sociology of Science* (New York: Free Press, 1962), 525.

41. P. B. Medawar, "Is the scientific paper a fraud?" *Listener,* September 12, 1963, 377–78.

42. Rothstein, "Nonlogical factors in research."

43. Quoted in Philip Wheelwright, ed., *The Presocratics* (New York: Odyssey Press, 1966), 70.

44. Medawar, *Pluto's Republic,* 287.

Part I:
The Dawn of a New Era: Infectious Diseases and Antibiotics, the Miracle Drugs

CHAPTER 1: How Antony's Little Animals Led to the Development of Germ Theory

1. Leeuwenhoek used only a tiny bead of glass often less than 2 mm across as a single biconvex lens mounted between metal plates in a short tube. This was surprisingly small, measuring about one inch by three inches. Yet it achieved magnifying power as high as 500x with resolving power approaching 1 micron. To put that into perspective, compound microscopes used by others provided magnification of about 42x. They were limited optically because of the impure quality of the strong lenses and spherical and chromatic distortions. In addition he probably developed a method of backfield illumination that enabled him to darken the background of his specimen so that light-colored objects being studied stood out more clearly. L. E. Casida Jr., "Leeuwenhoek's Observation of Bacteria," *Science,* June 25, 1976, 193.

2. The *Transactions* of the Royal Society, which appeared in London in 1664, was the first scientific journal in Europe. Because there were few scientists as such, many of them amateurs, the journal served both professional and popular science. Today it is the oldest scientific journal in continuous publication.

3. Clifford Dobell, *Antony van Leeuwenhoek and His "Little Animals"* (New York: Harcourt, Brace, 1932), 245.

4. Leeuwenhoek's unquenchable drive to peer into the microscopic world and his unshakable faith in his observations can be likened to Galileo's exploration of the heavens with his telescope earlier in the century. In 1632, at the University of Padua, Galileo defended the heliocentric Copernican system in disregard of the Church's admonition. He was finally tried by the Inquisition and under threat of torture recanted. Legend has it that as he left the tribunal, he murmured under his breath, "Eppur si muove!" (And yet it moves). The sentence passed on Galileo by the Inquisition was formally retracted by Pope John Paul II on October 31, 1992. Leeuwenhoek shared a similar resolute certitude in his findings. During his lifetime, his observations became widely known and excited wonderment. The highest-born and most powerful people in Europe — Frederick the Great of Prussia, James II of England, and Peter the Great of Russia — came to his home in Delft to peer through his microscope and see for themselves. Some of his discoveries were doubted, in part be-

cause of the imperfect optical instruments used by others. However, any skepticism he faced, as he wrote in a letter to Herman Boerhaave in 1717, "does not bother me. I know I am in the right." William Bulloch, *The History of Bacteriology* (London: Oxford University Press, 1938), 29.

CHAPTER 2: The New Science of Bacteriology

1. Robert Koch, *Untersuchungen über die Aetiologie der Wundinfectionskrankheiten* (Leipzig: F. C. W. Vogel, 1878).

2. R. Koch, "Die aetiologie der Tuberkulose," *Berl Klin Wschr* 19 (1882): 221–30 (reprinted with translation in *Medical Classics* 2 [1938]: 821–80).

CHAPTER 3: Good Chemistry

1. Until the middle of the nineteenth century, only natural dyestuffs were generally employed. Their production was a huge industry, involving millions of acres of land and hundreds of thousands of persons. This was supplanted by coal-tar dyes which were less costly, frequently more brilliant, and usually simpler to use.

2. Georg Meyer-Thurow, "The Industrialization of Invention: A Case Study from the Germany Chemical Industry," *Isis* 73 (1982): 363–81.

3. Quoted in W. D. Foster, *A History of Medical Bacteriology and Immunology* (London: Heinemann Medical, 1970).

4. The principle relating biological activity to molecular chemical substances was not established until the late 1860s, by Alexander Crum Brown, professor of chemistry, and Thomas Fraser, professor of pharmacology, at the University of Edinburgh. They determined that certain arrangements of atoms could confer specific types of biological activity upon molecules. A. Crum Brown and T. R. Fraser, "On the physiological action of the ammonium bases derived from atropia and conia," *Trans Roy Soc Edin* 25 (1868): 693–739.

5. Paul Ehrlich and S. Hata, *Die experimentelle Chemotherapie der Spirillosen: Syphilis, Rückfallfieber, Hühnerspillose, Frambösie* (Berlin: J. Springer, 1910).

6. Martha Marquardt, "Paul Ehrlich: Some Reminiscences," *British Medical Journal* 1 (1954): 665–66.

7. W. I. B. Beveridge, *The Art of Scientific Investigation* (New York: Vintage, 1950), 61.

8. Quoted in Iago Galdston, *Behind the Sulfa Drugs* (New York: D. Appleton–Century, 1943), 127.

CHAPTER 4: The Art of Dyeing

1. Author interview with Ernõ Mako, M.D., Ph.D., Semmelweis University of Medical Sciences, Budapest, Hungary, September 14, 2003.

2. Years earlier, the need for the body to alter a chemical drug to render it effective had been come across by Ehrlich. To his surprise, Ehrlich found that the arsenic-containing compound Atoxyl didn't work in his cultures of trypanosomes but was converted into an active form by the infected animal. Was Domagk's team aware of this precedent?

3. Quoted in Frank Ryan, *The Forgotten Plague* (Boston: Little, Brown, 1992), 97.

4. G. Domagk, "Ein Beitrag zur Chemotherapie der bakteriellen Infektionen," *Dtsch Med Wschr* 61 (1935): 250–53.

5. Daniel Bovet, *Une chimie qui guérit: Histoire de la découverte des sulfamides* (Paris: Payot, 1988), 42–43.

6. J. Tréfouël, T. Tréfouël, F. Nitti, and D. Bovet, "Chimiothérapie des infections streptococciques par les dérivés du *p*-aminophénylsulfamide," *Ann Inst Pasteur* 58 (1937): 30–47.

7. Bovet, *Une chimie qui guérit*.

8. By the late 1930s IG Farben's successful diversification into strategic raw materials, such as synthetic rubber and oil, proved of great use to the Nazi war machine. As an unrivaled chemical syndicate, it came to employ 120,000 people, including a thousand chemists. Peter Hayes, *Industry and Ideology: IG Farben in the Nazi Era* (Cambridge: Cambridge University Press, 1987).

The chillingly deceptive phrase "Arbeit macht frei" (Work makes you free) mounted in time above the gates to the concentration camps originated from posters against trade unionism at IG Farben's factories for making Buna rubber. In 1941 the slave labor camp Monowitz-Buna was established as an outstation only a few miles from the main camp, Auschwitz, using about 40,000 of its prisoners, mostly Jews. One of these was Primo Levi, a skilled chemist. After liberation, resuming his life and career in Turin, Italy, Levi experienced a grim incident related in his 1975 book *The Periodic Table* (trans. Raymond Rosenthal; New York: Schocken, 1984). Recognizing an idiosyncratic phrase in a business correspondence with a chemical distributor in Germany, he realized that this was the same individual who was the chief of the laboratory in which he himself had worked as a starved and abused prisoner. In a written exchange, the German declared that he and Levi were both victims who should collaborate on "overcoming the past." Levi found the invitation to admit his own share in responsibility for Auschwitz puzzling.

IG Farben's notoriety did not end here. One of its subsidiaries, Degesch, manufactured Zyklon B, the poisonous gas initially marketed as an insecticide, used at Auschwitz and other camps for the mass murdering of Europe's Jews. IG Farben was dismantled by the allies in 1952 — its factories split up among Bayer, BASF, and other German chemical companies and its chief scientist condemned as a mass murderer by the Nuremberg war crimes tribunal.

9. Despite this ill-fated history, some have questioned why only Domagk was the recipient of the Nobel Prize. Of the two IG Farben chemists intimately involved in the synthesis of azo dyes for this type of drug research, one in particular expressed bitter dissatisfaction at his exclusion. Others thought that the director of Farben's medical division, as team organizer, should have been honored. And, most notably, it was the investigators at the Pasteur Institute who had determined that the metabolism in the body broke down Prontosil and unleashed its active ingredient, sulfanilamide. (In 1937 Domagk had been deceptive in explicitly dismissing the possibility of metabolic breakdown in the body even as IG Farben had synthesized the active constituent, sulfanilamide, under the trade name Prontosil Album.) Nevertheless, since Alfred Nobel's will stipulated that the original discoverer be recognized, the prize was given to Gerhard Domagk, "for the discovery of the antibacterial effects of Prontosil."

10. Quoted in Ryan, *The Forgotten Plague,* 119.

11. J. E. Lesch, "The Discovery of M & B 693 (sulfapyridine)," *American Institute of the History of Pharmacy* 16 (1997): 106.

12. One reason was the development, particularly in the United States, of a number of well-equipped university hospitals with full-time teachers and investigators and sophisticated laboratories for clinical research. Close relationships between clinical and basic science departments were established, and research by even junior staff was encouraged and facilitated. This was in distinct contrast to the rigid hierarchal system in Germany, where the professor dictated to his staff the type of clinical research to be done on his patients; there, the science departments remained compartmentalized as separate institutes. In France, Great Britain, Switzerland, and the United States, pharmaceutical companies developed and heavily invested in their own research and development activities.

13. Richard Lovell, *Churchill's Doctor: A Biography of Lord Moran* (London: Royal Society of Medicine, 1992), 228–35.

14. Sulfanilamide and the family of sulfa drugs work by preventing bacteria from utilizing an essential growth requirement called para-aminobenzoic acid, but generally known by its acronym PABA. (Many people are familiar with PABA as an active ingredient in sunscreen products.

It absorbs ultraviolet light at the very wavelengths that have been found to be most damaging to skin cells). PABA is a component of the vitamin folic acid and is chemically similar to sulfanilamide and the family of sulfa drugs. These drugs function as antimetabolites, competitively inhibiting the uptake of PABA for the synthesis of folic acid, causing the bacteria to die. D. D. Woods, "The relation of p-amino benzoic acid to the mechanism of the action of sulphonamide," *Brit J Exp Pathol* 21 (1940): 74–90.

Humans, relying on folic acid absorbed from our food, are not negatively affected by the action of sulfanilamide. Researchers in other fields subsequently seized upon this new concept. Reasoning that since folic acid is needed for the production of blood cells and its action might be blocked by means of a "folic acid antagonist," Dr. Sidney Farber of the Children's Hospital in Boston put it to work in treating children with leukemia. The concept of antimetabolites provided a springboard in the search for other chemotherapeutic agents, psychotropic drugs, antihistamines, and antihypertensives. It also led to the discovery of PAS, para-aminosalicylic acid, a drug effective against tuberculosis. The tubercle bacillus depends upon salicylates, a substance similar to ordinary aspirin, for growth and nutrition. Over a period of six years in the 1940s, a Swedish investigator tested a large number of derivatives of salicylates before coming upon PAS as an effective competitive inhibitor. Another major drug unanticipated as a spinoff from the sulfa drugs was chlorothiazide. It forces the kidneys to eliminate salt resulting in diuresis, or an increased secretion of urine. Marketed as Diuril in January 1958 and still widely used, it revolutionized the treatment of heart disease and high blood pressure. K. H. Beyer Jr., "Discovery of the Thiazides: Where Biology and Chemistry Meet," *Perspectives in Biology and Chemistry* 20 (1977): 410–20.

CHAPTER 5: Mold, Glorious Mold

1. In addition to his scientific background, Wright was a brilliant linguist and was prone to quote sections from the Bible, Milton, Dante, Goethe, Browning, and Kipling. George Bernard Shaw used Wright as the model for his main character, Sir Colenso Ridgeon, in *The Doctor's Dilemma,* a brilliant satire on the medical profession but also one introducing the necessity of ethical choices engendered by limited resources. Little wonder that the doctor's detractors took delight in calling him "Almost Wright."

2. The outcome of this tragedy was that by the time World War I broke out in 1914, medical authorities bowed to the inevitable. For the first time, the whole of the British army and navy were vaccinated against typhoid.

3. André Maurois, *The Life of Sir Alexander Fleming, Discoverer of Penicillin,* trans. Gerard Hopkins (New York: E. P. Dutton, 1959), 97, 93.

4. V. D. Allison, "Personal Recollections of Sir Almroth Wright and Alexander Fleming," *Ulster Medical Journal* 43 (1974): 89–98.

5. Gwyn Macfarlane, *Alexander Fleming: The Man and the Myth* (Cambridge, Mass.: Harvard University Press, 1984), 103.

6. Hare convincingly dispels the myth concerning an open window by Fleming's bench. Besides the difficulty of reaching the window to open it, with the risk of flasks and dishes kept on the windowsill clattering down on the heads of passing pedestrians, only to admit the unbearable roar of traffic, Fleming was too skilled to run the high risk of contamination of his cultures. Contamination is the bane of bacteriologic work, spoiling a crop of colonies. Bacteriologists must then just throw the cultures down the drain.

7. Alexander Fleming, "History and Development of Penicillin," in *Penicillin, Its Practical Application* (London: Butterworth, 1946).

8. Allison, "Personal Recollections," 89.

9. Macfarlane, *Alexander Fleming,* 116, 248.

10. In the 1880s a fundamental method of differential staining of bacteria was stumbled upon by Christian Gram, a Danish physician working in Berlin. Trying to develop a double stain for kidney sections, he used gentian violet and iodine, then placed the preparations in alcohol. He observed that some bacteria in the sections retained the color and others not. Christian Gram, "Über die isolierte Färbung der Schizomyceten in Schnitt-und Trockenpräparaten," *Fortschr Med* 2 (1884): 1985–89. Henceforth, the world of bacteriology has been divided into gram-positive and gram-negative organisms. In the 1930s, bacteriologists discovered that these two groups react differently to antibiotics, making Gram's stain more useful in diagnosis.

11. Maurois, *Life of Sir Alexander Fleming,* 137.

12. Alexander Fleming, "On the Antibacterial Action of Cultures of a Penicillium, with Special Reference to Their Use in the Isolation of *B. influenzae,*" *Brit J Exper Path* 10 (1929): 226–36.

13. The continuous subculture of this mold by Fleming and others was the principal source of penicillin for approximately fourteen years until the large-scale commercial production of penicillin during World War II.

14. Nor was its therapeutic importance recognized by others during this period. Indeed, similar observations on the destructive effects of mold on bacteria could be found in the literature dating back to the 1870s. John Tyndall, one of England's most distinguished physicists, inspired by Pasteur's discovery of microorganisms in "fresh air," wondered if they were evenly distributed through the atmosphere. He set up a series of open test tubes containing broth and found settling on the surface of the broth in some an "exquisitely beautiful" mold, called penicillium. Of the more than three hundred strains of penicillium, some had been used by the French cheese industry for centuries. The blue veins of Roquefort cheese are due to the mold *Penicillium roqueforti,* and the distinctive taste of Camembert is imparted by *Penicillium camemberti.* Tyndall made another interesting observation: "in every case where the mold was thick and coherent, the bacteria died or became dormant, and fell to the bottom as a sediment." John Tyndall, *Essays on the Floating Matter of the Air, in Relation to Putrefaction and Infection* (New York: D. Appleton, 1882). This effect was noted in a few brief sentences buried in a seventy-four-page article describing his interest as a physicist in the scattering of light by unseen particles floating in air, and furthermore preceded by seven years the proof by Robert Koch in 1882 that bacteria could cause disease. Also about the same time, Lister noted in his laboratory book that in a sample of urine containing bacteria as well as some filaments of mold, the bacteria seemed unable to grow. It's further startling to note that a medical scientist, John Burdon Sanderson, in the early years of St. Mary's Hospital in 1871, had observed that penicillium molds can inhibit the growth of bacteria in culture. Just a year before Fleming's publication, in 1928, a book published in France dealt extensively with bacterial inhibition by molds and by other bacteria, citing in a sixty-page chapter the concept of "antibiosis." Georges Papacostas and Jean Gaté, *Les associations microbiennes* (Paris: Doin, 1928). The word had been coined in 1889.

15. Lewis Thomas, *The Youngest Science: Notes of a Medicine-Watcher* (New York: Viking, 1983), 29.

16. Gwyn Macfarlane, *Howard Florey: The Making of a Great Scientist* (Oxford: Oxford University Press, 1979).

17. Ronald W. Clark, *The Life of Ernst Chain: Penicillin and Beyond* (New York: St. Martin's Press, 1985), 1.

18. Ibid., 33.

19. David Wilson, *In Search of Penicillin* (New York: Knopf, 1976).

20. The Oxford team had quickly found out that penicillin was not, in fact, an enzyme but a relatively small molecule. This surprised and dis-

appointed Chain, but he assumed that it would be easy to determine its molecular structure and to synthesize it. He was proved wrong on both counts. It was not until 1945, through the X-ray crystallography work of Oxford University's Dorothy Hodgkin, that the molecule was found to contain thirty-nine atoms in an unusual four-membered ring called a beta-lactam structure to which was attached a side chain. X-ray diffraction crystallography involves shining a beam of X-rays through a pure crystal of a substance. The crystal causes the beam to split up into a complex pattern, which can then be analyzed mathematically to show the positions of individual atoms in a molecule. Such a chemical arrangement had never before been found in nature and only rarely in the laboratory.

21. E. B. Chain, "Thirty years of penicillin therapy," *J R Coll Physicians Lond* 6 (1972): 103–31.

22. Clark, *Life of Ernst Chain,* 37.

23. E. B. Chain, H. W. Florey, A. D. Gardner, et al., "Penicillin as a chemotherapeutic agent," *Lancet* 2 (1940): 226–28.

24. The paper caught the eye of Alexander Fleming. His visit to the Oxford laboratory was a surprise to some of the staff, especially Chain, who remarked, "Good God! I thought he was dead!" But there he was — a modest man with shaggy eyebrows, pale blue eyes behind wire-rimmed spectacles, and his customary blue suit and bowtie inquiring, "What have you been doing with my old penicillin?" Given a tour by Florey and Chain, Fleming typically said little else. This left Chain with the impression that Fleming had not understood anything that was shown him.

25. Macfarlane, *Howard Florey,* 331.

26. E. P. Abraham, E. Chain, C. M. Fletcher, et al., "Further observations on penicillin," *Lancet* 2 (1941): 177–89.

27. Agriculture was the one area of scientific research for which Congress generously appropriated money. From the 1890s to the 1930s the Department of Agriculture was the leading agency of the federal government with scientific interests and health-related scientific work.

28. Lennard Bickel, *Rise Up to Life* (London: Angus and Robertson, 1972), 147.

29. Penicillin functions by inhibiting an enzyme necessary for chemical construction of the bacterium's cell wall. Without the wall, the cell eventually bursts and dies. Variations in the composition of the side chain determine the various characteristics of the different penicillins. Most commercial penicillins today are produced from a strain of *P. chrysogenum.* The main form produced is called penicillin G. Penicillin is effective against most gram-positive bacteria, including those that cause gonor-

rhea, syphilis, meningococcal meningitis, pneumococcal pneumonia, and some staphylococcal and streptococcal infections. Most gram-negative bacteria are resistant to penicillin. Two synthetic penicillins, ampicillin and amoxicillin, are active against both gram-positive and gram-negative bacteria. Bacterial strains that become penicillin-resistant are mutants that produce a penicillin-destroying enzyme, penicillinase. This acts by opening the beta-lactam ring of the penicillin molecule. Semisynthetically produced penicillins, such as oxacillin and methicillin, that are not degraded by penicillinase have been developed.

30. Penicillin was not successfully synthesized until 1957, but it remains cheaper to produce the drug from the mold. It has been found that exposing the culture to X-rays further raises the yield.

31. Remarkably, the first people to note this phenomenon of microbial antagonism were Pasteur and Joubert in 1877, observing in a famous phrase that "life hinders life." They speculated on the clinical potential of microbial products as therapeutic agents.

32. Alexander Fleming, "Penicillin, Nobel Lecture, December 11, 1945," in *Nobel Lectures, Physiology or Medicine, 1942–1962* (Amsterdam: Elsevier, 1964), 83–93.

33. D. C. Balfour, N. M. Keith, J. Cameron, and A. Fleming, "Remarks made at the dinner for Sir Alexander Fleming held at Mayo Foundation House, July 16, 1945, Rochester, MN." Quoted in J. W. Henderson, "The yellow brick road to penicillin: A story of serendipity," *Mayo Clinic Proceedings* 72 (1997): 683–87.

34. Alexander Fleming, "Réponse." In "Séance solonnelle et extraordinaire du 4 septembre 1945. Séance tenue en l'honnneur de la visite à l'Académie de Sir Alexander Fleming," *Bull Acad Natl Méd* (1945): 537–44.

35. Macfarlane, *Howard Florey,* 364.

36. Clark, *Life of Ernst Chain,* 31.

37. Macfarlane, *Howard Florey,* 304.

38. E. B. Chain, "The quest for new biodynamic substances," in *Reflections on Research and the Future of Medicine,* ed. Charles E. Lyght (New York: McGraw-Hill, 1967), 167–68.

CHAPTER 6: The Next Wonder Drug: Streptomycin

1. Selman A. Waksman, *The Antibiotic Era* (Tokyo: Waksman Foundation of Japan, 1975).

2. Quoted in J. Comroe, "Pay dirt: The story of streptomycin," *Am Rev Resp Dis* (1978): 778.

3. S. A. Waksman and H. B. Woodruff, "The soil as a source of micro-organisms antagonistic to disease producing bacteria," *J Bact* 40 (1940): 581.

4. Years earlier, researchers in Belgium had used some actinomycetes to dissolve bacterial cultures, but their aim related to immunity. With the antigens released from the bacterial contents, they hoped to induce formation of antibodies against the specific bacterium. A. Gratia and S. Dath, "Moisissures et Microbes Bacteriophages," *C R Soc Biol Paris* 92 (1925): 461–62. Waksman's quest was to find a bacteriolytic substance safe enough to be used as an antibiotic.

5. A strain of actinomycin was later found useful in Hodgkin's disease and thus crossed the line from antibiotic to chemotherapy.

6. Ryan, *Forgotten Plague*, 218.

7. The first test tube culture is now in the collection of the Smithsonian Institution.

8. A. Schatz and S. A. Waksman, "Effect of streptomycin and other antibiotic substances upon *Mycobacterium tuberculosis* and related organisms," *Proc Soc Exp Biol Med* 57 (1944): 244–48.

9. Antibiotics work by interfering with a bacterium's reproduction, its metabolic exchange through its cell envelope, or its source of energy. They can interfere with cell-wall synthesis, as penicillin and vancomycin do. Or, like the tetracyclines and erythromycin, they can inhibit the vital process of DNA synthesis, leading to cell death. A class of antibiotics known as polymyxins affect the permeability of the cell membrane, leading to leakage of intracellular components. Finally, an antibiotic may block specific metabolic steps that are essential, as the sulfa drugs do in competing with the vitamin substance PABA.

10. Another bounty of soil and mold research is a useful agent for lowering blood cholesterol. Lovastatin (marketed as Mevacor by Merck) was isolated from a strain of *Aspergillus terreus* and approved for use in the United States by the FDA in 1987. A. W. Alberts, J. S. MacDonald, A. E. Till, and J. A. Tobert, "Lovastatin," *Cardiol Drug Rev* 7 (1989): 89–109.

11. "Streptomycin Suit Is Labeled 'Baseless,'" *New York Times,* March 13, 1950.

12. Burton Feldman, *The Nobel Prize: A History of Genius, Controversy, and Prestige* (New York: Arcade, 2000), 276.

13. Selman A. Waksman, *My Life with the Microbes* (New York: Simon and Schuster, 1954), 285.

14. Selman A. Waksman, *The Conquest of Tuberculosis* (Berkeley and Los Angeles: University of California Press, 1964), 90.

CHAPTER 7: The Mysterious Protein from Down Under

1. Baruch S. Blumberg, "Australia antigen and the biology of hepatitis B," *Nobel Lectures in Physiology or Medicine, 1971–1980,* ed. Jan Lindsten (Singapore: World Scientific Publishing, 1992), 275–96.

2. Baruch S. Blumberg, B. J. S. Gerstley, D. A. Hungerford, W. T. London, and A. I. Sutnick, "A serum antigen (Australia antigen) in Down's syndrome, leukemia and hepatitis," *Ann Intern Med* 66 (1967): 924–31.

3. Baruch S. Blumberg, *Hepatitis B: The Hunt for a Killer Virus* (Princeton: Princeton University Press, 2002), 102.

4. W. T. London, A. I. Sutnick, and B. S. Blumberg, "Australia antigen and acute viral hepatitis," *Ann Intern Med* 70 (1969): 55–59.

5. Blumberg, *Hepatitis B,* 65.

6. S. Krugman, J. P. Giles, and J. Hammond, "Infectious hepatitis: Evidence for two distinctive clinical, epidemiological, and immunological types of infection," *Journal of the American Medical Association* 200 (1967): 365–73. Reprinted as a landmark article, *JAMA* 252 (1984): 393–401.

7. J. P. Giles, R. W. MacCallum, L. W. Berndtson Jr., and S. Krugman, "Viral hepatitis: Relation of Australia/SH antigen to the Willowbrook MS-2 strain," *N Engl J Med* 281 (1969): 119–22.

8. David J. Rothman and Sheila M. Rothman, *The Willowbrook Wars* (New York: Harper and Row, 1984), 260.

9. Two HAV vaccines, HAVRIX and VAQTA, are FDA-approved and commercially available and provide long-term protection. In a period of just over two decades, knowledge of viral hepatitis grew from the belief in the early 1970s that there were only two viruses, hepatitis A and hepatitis B, to the recognition that there are at least five — A through E. The hepatitis C virus was identified in 1989 by investigators at the Chiron Corporation in California working in collaboration with investigators at the Centers for Disease Control (CDC) in Atlanta. It is responsible for most chronic viral hepatitis cases in the United States, with an estimated 170,000 new cases annually. Parenteral transmission of HVC is well established as occurring via needles and through sex between males. CDC investigators estimated that in 1990 there were 3.5 million HCV carriers in the United States. Chronic HBV and HCV infections appear to be the link to the development of primary hepatocellular carcinoma. This is the most common cancer worldwide and accounts for approximately 1 million deaths annually. It is relatively rare in the United States, but many parts of Africa and Asia have an extremely high incidence. The hepatitis D virus (HDV), identified in 1977, can exist only in the presence of hepatitis B infection. It is estimated that there are at least 70,000 HDV carri-

ers in the United States. Hepatitis E, while presumably an old and perhaps even an ancient disease, was first recognized in 1980 through the study of serum and fecal samples from an extensive waterborne hepatitis outbreak in India.

10. E. Norby, "Introduction to Baruch Blumberg," in *Nobel Lectures, Physiology or Medicine 1971–1980,* ed. J. Lindsten (Singapore: World Scientific, 1992).

11. M. L. M. Luy, "Investigative coup: Discovery of the Australia antigen. An interview with Dr. Baruch S. Blumberg," *Modern Medicine* 42 (1974): 41–44.

12. Author interview with Baruch Blumberg, January 25, 2004.

CHAPTER 8: "This Ulcer 'Bugs' Me!"

1. K. Schwartz, "Über penetrierende Magen — und Jejunal-geschwüre," *Beitr Klin Chir* 67 (1910): 96–128.

2. B. W. Sippy, "Gastric duodenal ulcer: Medical cure by an efficient removal of gastric juice corrosion," *JAMA* 64 (1915): 1625–30.

3. E. Palmer, "Investigation of the gastric *spirochaetes* of the human," *Gastroenterology* 27 (1954): 218–20.

4. Terence Monmaney, "Marshall's Hunch," *New Yorker,* September 20, 1993, 64–72.

5. W. I. B. Beveridge, *The Art of Scientific Investigation* (New York: Vintage, 1950), 6.

6. B. J. Marshall, "History of the discovery of C. *pylori*," in *Campylobacter pylori in Gastritis and Peptic Ulcer Disease,* ed. Martin J. Blaser (New York: Igaku-Shoin, 1989), 7–22.

7. Claude Bernard, *An Introduction to the Study of Experimental Medicine,* trans. Henry Copley Greene (New York: Macmillan, 1927).

8. Author interview with Martin Blaser, M.D., professor and chairman of the Department of Medicine, New York University School of Medicine, December 16, 2003.

9. Quoted in Monmaney, "Marshall's Hunch."

10. J. R. Warren and B. J. Marshall, "Unidentified curved bacilli on gastric epithelium in active chronic gastritis," *Lancet* 1 (1983): 1273–75.

11. B. J. Marshall and J. R. Warren, "Unidentified curved bacilli in the stomach of patients with gastritis and peptic ulceration," *Lancet* 1 (1984): 1311–15.

12. B. J. Marshall, J. A. Armstrong, D. B. McGechie, and R. J. Clancy, "Attempt to fulfill Koch's postulates for pyloric *Campylobacter,*" *Medical Journal of Australia* 142 (1985): 436–39.

13. E. J. S. Boyd and K. G. Worsley, "Etiology and pathogenesis of peptic ulcer," in *Bockus Gastroenterology,* 4th ed., ed. J. Edward Berk (Philadelphia: W. B. Saunders, 1985).

14. Author interview with Barry Marshall, March 14, 2000.

15. B. J. Marshall, C. S. Goodwin, J. R. Warren, et al., "A prospective double-blind trial of duodenal ulcer relapse after eradication of *Campylobacter pylori,*" *Lancet* 2 (1988): 1437–41.

16. Quoted in S. Chazan, "The Doctor Who Wouldn't Accept No," *Reader's Digest,* October 1993, 118–23.

17. D. Y. Graham and M. F. Go, "*Helicobacter pylori:* Current status," *Gastroenterology* 105 (1993): 279–82.

18. Thomas S. Kuhn, *The Structure of Scientific Revolutions,* 2nd ed. (Chicago: University of Chicago Press, 1970), 150.

19. Quoted in Monmaney, "Marshall's Hunch."

Part II:
The Smell of Garlic Launches the War on Cancer

CHAPTER 9: Tragedy at Bari

1. Quoted in Glenn B. Infield, *Disaster at Bari* (New York: Macmillan, 1971), 121.

2. As in World War I, Germany's mustard gas was manufactured by the IG Farben chemical conglomerate. It was also used by the Spanish army in Morocco and the Italians in Abyssinia, now called Ethiopia.

3. S. L. A. Marshall, *World War I* (Boston: Houghton Mifflin, 1985), 167–69, 322.

4. In World War I as many as fifty different toxic agents were employed on the battlefields by at least one of the combatants, most of them proving relatively ineffective. Mustard, the most effective chemical incapacitator, was delivered in artillery and mortar shells, which on detonation sent out a liquid spray within a twenty-yard radius. Augustin M. Prentiss, *Chemicals in War: A Treatise on Chemical Warfare* (New York: McGraw-Hill, 1937). Large quantities of the agent were delivered on a wide front against enemy forces. In one offensive drive, the Germans used 1,700 batteries firing chemical shells on a seventy-mile front, an average of one battery every seventy-two yards. In a three-day German assault against British lines in early 1918, the Germans fired 120,000 to 150,000 mustard shells. D. K. Clark, *Effectiveness of Chemical Weapons in WWI*

(Bethesda, Md.: Johns Hopkins University, 1959). U.S. troops were not involved in gas attacks until February 25, 1918, when they were hit by German shells containing the choking gas, phosgene. A similar agent, diphosgene (German "green cross"), conveyed a chocolate-like smell across the battlefield. Four months later, the American forces initiated their own offensive use of toxic gas.

5. Evidence of this general nature had been noted as early as 1919 from the effects of chemical warfare in World War I. Autopsies had been conducted by the University of Pennsylvania's department of research medicine on seventy-five soldiers who had been killed in an accident with mustard gas. The examiners were surprised to find extreme reductions in the numbers of white blood cells and noted it to be caused by direct depression of blood cell formation. These observations were obscured by the debilitating and deadly blistering and respiratory consequences, and the potential significance of the blood changes was overlooked. E. B. Krumbhaar and H. D. Krumbhaar, "The blood and bone marrow in yellow cross mustard (mustard gas) poisoning," *Journal of Medical Research* 40 (1919): 497–506.

6. Alfred Gilman and Louis Goodman are known today to every medical student for their classic textbook *The Pharmacologic Basis of Therapeutics*. When it made its first appearance in 1938, not only was no section of the book devoted to cancer, but the word "cancer" did not even appear in the index. They went on to become the first to discover an effective drug against cancer.

7. The bizarre constellation of symptoms in the casualties at Bari is explained by the cascade of poisonous breakdown products. An initial action led to the excessive tearing. This was followed by a paralytic effect, which caused the profound apathy and weakness. A successive blistering action produced the brawny edema of the skin exposed to the dilute solution of mustard in the fuel oil. Several actions led to the virtual disappearance of white blood cells.

8. The capability to transplant cancerous tumors in mice had been known since the work of Carl Jensen, a Danish veterinarian in 1903. This was a significant discovery because it enabled cancer research in the laboratory using animals. Now a researcher could observe not only the growth of cancer but also the effects of agents upon it. In animals such as mice, which have a short life span, tumors grow much more rapidly than in human beings.

9. Alfred Gilman, "The initial clinical trial of nitrogen mustard," *Am J Surg* 105 (1963): 574–78.

10. Ibid.

11. A. Gilman and F. Phillips, "The Biological Actions and Therapeutic Applications of the B-chloranthyamines and Sulfides," *Science* 103 (1946): 409–15.

12. Cornelius P. Rhoads, "Report on a cooperative study of nitrogen mustard (HN2) therapy of neoplastic disease," *Trans Assoc Am Physicians* 60 (1947): 110–17.

13. Similar government suppression was later seen with the consequences of atom bomb testing in the 1950s, Agent Orange in Vietnam in the 1960s, and neurotoxins in the first Gulf War in 1991.

14. Arthur I. Holleb and Michael Braun Randers-Pehrson, *Classics in Oncology* (New York: American Cancer Society, 1987), 383.

CHAPTER 10: Antagonists to Cancer

1. L. Wills, P. W. Clutterbuck, and P. D. F. Evans, "A new factor in the production and cure of macrocytic anaemias and its relation to other haemopoietic principles curative in pernicious anaemia," *Biochem J* 31 (1937): 2136–47.

2. R. Leuchtenberger, C. Leuchtenberger, D. Laszlo, and R. Lewisohn, "The Influence of 'Folic Acid' on Spontaneous Breast Cancers in Mice," *Science* 101 (1945): 46.

3. S. Farber, L. K. Diamond, R. D. Mercer, R. F. Sylvester Jr., and J. A. Wolf, "Temporary remissions in acute leukemia in children produced by folic acid antagonist, 4-aminopteroylglutamic acid (Aminopterin)," *N Engl J Med* 238 (1948): 787–93.

4. M. C. Li, R. Hertz, and D. B. Spencer, "Effect of methotrexate upon choriocarcinoma and chorioadenoma," *Proc Soc Exp Biol Med* 93 (1956): 361–66.

5. In 1952 Gertrude Elion and colleagues introduced 6-mercaptopurine, a potent antagonist of a factor essential to a cell's nucleus.

6. Other antimetabolites currently in use, in addition to methotrexate and 6-mercaptopurine, include 6-thioguanine, 5-fluorouracil (5-FU), and arabinosyl cytosine. All antimetabolites act primarily during the DNA synthetic phase of the cell cycle.

7. Several of the drugs used for cancer therapy have become important components as well as treatments for other diseases: immunosuppressive regimens for rheumatoid arthritis (methotrexate and cyclophosphamide), multiple sclerosis (Novantrone), organ transplantation (methotrexate and azathioprine), and psoriasis (methotrexate), as well as for sickle cell anemia (hydroxyurea) and the parasitic disease leishmaniasis (miltefosine).

CHAPTER 11: Veni, Vidi, Vinca: The Healing Power of Periwinkle

1. John Mann, *Murder, Magic, and Medicine* (Oxford: Oxford University Press, 1992), 213.

2. R. L. Noble, C. T. Beer, and J. H. Cutts, "Role of chance observations in chemotherapy: *Vinca rosea*," *Ann N Y Acad Sci* 76 (1958): 882–94.

3. R. L. Noble, "The discovery of the vinca alkaloids — chemotherapeutic agents against cancer," *Biochem Cell Biol* 68 (1990): 1344–51.

4. G. H. Svoboda, "A note on several alkaloids from *Vinca rosea* Linn. I. Leurosine, Virsine and Perivine," *J Pharm Sci* 47 (1958): 834.

5. The *Vinca* alkaloids do this by inhibiting the action of microtubules, structures that are required for separation of chromosomes at mitosis.

6. Mann, *Murder, Magic, and Medicine,* 214.

CHAPTER 12: A Heavy Metal Rocks: The Value of Platinum

1. B. Rosenberg, L. Van Camp, and T. Krigas, "Inhibition of cell division in *Escherichia coli* by electrolysis products from a platinum electrode," *Nature* 2906 (1965): 698–99.

2. Stephen J. Lippard, ed., *Platinum, Gold, and Other Metal Chemotherapeutic Agents: Chemistry and Biochemistry* (Washington, D.C.: American Chemical Society, 1983).

3. B. Rosenberg, "Platinum coordination complexes in cancer chemotherapy," *Naturwissenchaften* 60 (1973): 399–406.

CHAPTER 13: Sex Hormones

1. C. Huggins and P. J. Clark, "Quantitative studies on prostatic secretion. II. The effect of castration and of estrogen injection on the normal and on the hyperplastic prostate gland of dogs," *J Exp Med* 72 (1940): 747–62.

2. A. Klopper and M. Hall, "New synthetic agent for the induction of ovulation: Preliminary trials in women," *British Medical Journal* 1 (1971): 152–54.

3. Early Breast Cancer Trialists' Collaborative Group, "Effects of chemotherapy and hormonal therapy for early breast cancer on recurrence and 15-year survival: An overview of the randomised trials," *Lancet* 365 (2005): 1687–717.

CHAPTER 14: Angiogenesis: The Birth of Blood Vessels

1. G. H. Algire and H. W. Chalkley, "Vascular reactions of normal and malignant tissues in vivo; vascular reactions of mice to wounds and to normal and neoplastic transplants," *J Natl Cancer Inst* 6 (1945): 73–85.

2. Judah Folkman, "How is blood vessel growth regulated in normal and neoplastic tissue?" GHA Clowes Memorial Award lecture, *Cancer Research* 46 (1986): 467–73.

3. Robert Cooke, *Dr. Folkman's War: Angiogenesis and the Struggle to Defeat Cancer* (New York: Random House, 2001).

4. Cooke, *Dr. Folkman's War,* 83.

5. Cooke, *Dr. Folkman's War,* 126.

6. J. Folkman and C. Haudenschild, "Angiogenesis in vitro," *Nature* 288 (1980): 551–56.

7. R. Langer, H. Brem, K. Falterman, M. Klein, and J. Folkman, "Isolation of a Cartilage Factor That Inhibits Tumor Neovascularization," *Science* 193 (1976): 70–72.

8. I. William Lane and Linda Comac, *Sharks Don't Get Cancer* (Garden City Park, N.Y.: Avery, 1992).

9. V. D. Herbert, "Laetrile: The Cult of Cyanide," *Am J Clin Nutr* 32 (1979): 1121–58.

10. C. G. Moertel, T. R. Fleming, J. Rubin, et al., "A clinical trial of amygdalin (Laetrile) in the treatment of human cancer," *N Engl J Med* 306 (1982): 201–6.

11. F. Rastinejad, P. J. Polverini, and N. P. Bouck, "Regulation of the activity of a new inhibitor of angiogenesis by a cancer suppressor gene," *Cell* 56 (1989): 345–55.

12. M. S. O'Reilly, L. Holmgren, Y. Shing, et al., "Angiostatin — a novel angiogenesis inhibitor that mediates the suppression of metastases by a Lewis lung-carcinoma," *Cell* 79 (1994): 315–28.

13. A new way of looking at many conditions on the basis of the mechanisms that underlie abnormal vessel growth stimulated research and clinical applications to a variety of diseases in many specialties besides oncology, including cardiology and vascular surgery, ophthalmology, hematology, and dermatology.

CHAPTER 15: Aspirin Kills More than Pain

1. How cancers of the colon arise from its lining had been a matter of controversy for a long time. Many felt that it developed directly from one mutant cell dividing without restraint, to perhaps invade the struc-

tures of its wall. But the answer came in the 1960s and 1970s from a dedicated pathologist, Basil Morson, working alone at St. Mark's Hospital in London. This small hospital uniquely specialized in diseases of the colon and rectum and retained records of its patients dating back for generations. Morson, with keen diligence, established a fundamental truth: carcinomas arise within adenomatous polyps (benign tumors).

2. W. R. Waddell and R. W. Loughry, "Sulindac for polyposis of the colon," *J Surg Oncol* 24 (1983): 83–87.

3. F. M. Giardiello, S. R. Hamilton, A. J. Krush, S. Piantadosi, L. M. Hylind, P. Celano, S. V. Booker, C. R. Robinson, and G. J. A. Offerhaus, "Treatment of colonic and rectal adenomas with sulindac in familial adenomatous polyposis," *N Engl J Med* 328 (1993): 1313–16.

4. C. S. Williams, M. Tsujii, J. Reese, et al., "Host cyclooxygenase-2 modulates carcinoma growth," *J Clin Invest* 105 (2000): 1589–94.

CHAPTER 16: Thalidomide: From Tragedy to Hope

1. Much of the thalidomide saga is told by Trent Stephens and Rock Brynner in *Dark Remedy: The Impact of Thalidomide and Its Revival as a Vital Medicine* (Cambridge, Mass.: Perseus, 2001).

2. W. G. McBride, "Thalidomide and Congenital Abnormalities," *Lancet* 2 (1961): 1358.

3. G. Rogerson, "Thalidomide and Congenital Abnormalities," *Lancet* 1 (1962): 691.

4. In the early 1960s a physician treating patients with Hansen's disease (leprosy) in Jerusalem, Dr. Jacob Sheskin, was caring for a patient with a severe inflammatory and agonizing complication known as ENL (erythema nodosum leprosum), characterized by large weeping boils all over the body and excruciating and unremitting pain. The patient had been sleepless for months, despite taking every existing sedative. As a last resort, Sheskin gave him thalidomide because he knew that certain mental patients, whom no other sleep aid had helped, had been effectively treated with it. Remarkably, not only did it allow the patient to sleep, but his pain vanished and his sores began to heal. Thalidomide is now the drug of choice for his condition. In 1998 the FDA approved it for treatment of ENL, which affects up to a couple hundred people in the United States. It is now known that it reduces inflammatory reactions by inhibiting the production of one of the body's self-destructive factors that provokes inflammation — tumor necrosis factor-alpha (TNF-∝).

5. S. Singhal et al., "Antitumor activity of thalidomide in refractory multiple myeloma," *N Engl J Med* 341 (1999): 1565–71.

6. Another new drug found useful in multiple myeloma is Velcade. It was initially developed as a treatment for muscle-wasting conditions because it could prevent the destruction of proteins needed for healthy cell growth.

CHAPTER 17: A Sick Chicken Leads to the Discovery of Cancer-
            Accelerating Genes

1. *Virus* means "poison" in Latin.

2. Peyton Rous, "A sarcoma of the fowl transmissible by an agent separable from the tumor cells," *J Exp Med* 13 (1911): 397–411.

3. In his 1925 novel about science, *Arrowsmith,* Sinclair Lewis based his character Rippleton Holabird upon Peyton Rous.

4. Some rejuvenation of interest was sparked when Burkitt's lymphoma, an unusual tumor of the lymph glands described in 1960, was found in 1982 to be caused by a retrovirus dubbed the Epstein-Barr virus.

5. J. Michael Bishop, *How to Win the Nobel Prize: An Unexpected Life in Science* (Cambridge, Mass.: Harvard University Press, 2003), 162.

6. D. Stehelin, H. E. Varmus, J. M. Bishop, and P. K. Vogt, "DNA Related to the Transforming Gene(s) of Rous Sarcoma Virus Is Present in Normal Avian DNA," *Nature* 260 (1976): 170–73.

7. Harold E. Varmus, "Retroviruses and Oncogenes I (Nobel Lecture)," *Angewandte Chemie* 29 (1990): 710.

8. Controversy erupted when Dominique Stehelin, Bishop's postdoctoral fellow, demanded a share of the prize for the important experiments he had done with the two laureates, but the committee believed that the fundamental intellectual creativity belonged to Bishop and Varmus. Manifold contributions by others in the fields of medical genetics, tumor virology, and cell biology helped to pave the way for their research pursuits. Varmus's Nobel Prize acceptance speech was a model of attributions, as he cited more than forty other researchers whose work had led him and Bishop to their basis discoveries.

9. Quoted in profile of Harold Varmus by Natalie Angier, *New York Times,* November 21, 1993.

10. Quoted in Mariana Cook, *Faces of Science* (New York: W. W. Norton, 2005), 156.

11. Bishop, who felt that "one lifetime as a scientist is enough — great fun, but enough" (quoted in Paula McGuire, ed., *Nobel Prize Winners: Supplement 1987–1991* [New York: H. W. Wilson, 1992]), became chancellor of the University of California at San Francisco.

CHAPTER 18: A Contaminated Vaccine Leads to Cancer-Braking Genes

1. Richard Carter, *Breakthrough: The Saga of Jonas Salk* (New York: Trident, 1966), 1.

2. A. J. Levine, "The Road to the Discovery of the p53 protein," *Int J Cancer* 56 (1994): 775–76; A. J. Levine, "P53, the cellular gatekeeper for growth and division," *Cell* 88 (1997): 323–31; B. H. Sweet and M. R. Hilleman, "The vacuolating virus, SV40," *Proc Soc Exp Biol* (New York) 105 (1960): 420–21.

3. S. J. Baker, R. White, E. Fearon, and B. Vogelstein, "Chromosome 17 Deletions and p53 Gene Mutations in Colorectal Carcinomas," *Science* 244 (1989): 217–21.

4. C. A. Finlay, P. W. Hinds, and A. J. Levine, "The p53 proto-oncogene can act as a suppressor of transformation," *Cell* 57 (1989): 1083–93.

5. Author interview with Arnold Levine, August 10, 1994.

6. Andrew Pollack, "Drugs May Turn Cancer Into Manageable Disease," *New York Times,* June 6, 2004.

7. Robert A. Weinberg, *One Renegade Cell: How Cancer Begins* (New York: Basic Books, 1998), 161–64.

CHAPTER 19: From Where It All Stems

1. E. A. McCulloch, *The Ontario Cancer Institute: Successes and Reverses at Sherbourne Street* (Montreal: McGill–Queen's University Press, 2003).

2. Author interview with Ernest McCulloch, January 25, 2006, and with James Till, February 21, 2006.

3. A. J. Becker, E. A. McCulloch, and J. E. Till, "Cytological demonstration of the clonal nature of spleen colonies derived from transplanted mouse marrow cells," *Nature* 197 (1963): 452–54.

4. Author interview with Ernest McCulloch, January 25, 2006.

5. Quoted in B. M. Kuehn and T. Hampton, "2005 Lasker Awards Honor Groundbreaking Biomedical Research Public Service," *JAMA* 294 (2005): 1327–30.

6. D. Rubio et al., "Spontaneous human adult cell transformation," *Cancer Research* 65 (2005): 3035.

7. E. A. McCulloch and J. E. Till, "Perspectives on the properties of stem cells," *Nature Medicine* 11 (2005): 1026–28.

8. In 1988 Lewis Thomas made a similar assessment of the scientific

method: "I have never been quite clear in my mind about what this means. 'Method' has the sound of an orderly, preordained, step-by-step process. . . . I do not believe it really works that way most of the time. . . . More often than not, the step-by-step process begins to come apart, because of what almost always seemed a piece of luck, good or bad, for the scientist; something unpredicted and surprising turned up, forcing the work to veer off in a different direction. Surprise is what scientists live for. . . . The very best ones revel in surprise, dance in the presence of astonishment." Lewis Thomas, foreword to *Natural Obsessions: The Search for the Oncogene,* by Natalie Angier (Boston: Houghton Mifflin, 1988), xiii–xiv.

9. Author interview with Ernest McCulloch, January 25, 2006.

CHAPTER 20: The Industrialization of Research and the War on Cancer

1. James T. Patterson, *The Dread Disease: Cancer and Modern American Culture* (Cambridge, Mass.: Harvard University Press, 1987).

2. George G. Crile Jr., *Cancer and Common Sense* (New York: Viking, 1955), 15–16.

3. Patterson, *The Dread Disease,* 182.

4. Vannevar Bush, *Science, the Endless Frontier: A Report to the President* (1945; reprinted Washington, D.C.: National Science Foundation, 1960).

5. C. G. Zubrod, S. Schepartz, J. Leiter, et al., "The chemotherapy program of the National Cancer Institute: history, analysis and plans," *Cancer Chemotherapy Reports* 50 (1966): 349–540.

6. Quoted in *Time,* June 25, 1949, 66–73.

7. Recent disclosures of events dating back to the early 1930s cast a dark shadow upon Rhoads's character. While conducting research in Puerto Rico funded by the Rockefeller Institute, he allegedly injected cancer cells into dozens of individuals without their knowledge or consent as part of an experiment designed to see how humans develop cancer. At least thirteen of his subjects eventually died of the cancer. G. Goliszek, *In the Name of Science: A History of Secret Programs, Medical Research, and Human Experimentation* (New York: St. Martin's Press, 2003), 221–22. In a letter to a friend in 1931, he explained why he chose Puerto Ricans for his research: "The Puerto Ricans are the dirtiest, laziest, most degenerate and thievish race of men ever to inhabit this sphere. . . . What the island needs is not public health work, but a tidal wave or something to totally exterminate the population." Carmelo Ruiz-Marrero, "Puerto Ricans Outraged Over Secret Medical Experiments," *Puerto Rican Herald,* October 21,

2002. These recent allegations regarding Rhoads's human experimentation became a matter of enough concern to the American Association for Cancer Research (AACR) that they changed the name of their Cornelius P. Rhoads Scientific Achievement Award, which they had been presenting annually to outstanding young cancer researchers from 1980 to 2002, to the AACR Award for Outstanding Achievement in Cancer Research.

8. R. and E. Brecher, "They Volunteered for Cancer: Inmates of Ohio State Penitentiary," *Reader's Digest,* April 1958, 62–66.

9. C. G. Zubrod, S. A. Schepartz, and S. K. Carter, "Historical Background of the National Cancer Institute's Drug Development Thrust," in *Methods of Development of New Anticancer Drugs: USA-USSR Monograph* (Washington, D.C.: NCI, 1977), 7–11.

10. Quoted in Patterson, *The Dread Disease,* 196.

11. Cited in Barbara J. Culliton, "National Cancer Plan: The Wheel and the Issues Go Round," *Science,* March 30, 1973, 1305–309.

12. Richard A. Rettig, *Cancer Crusade: The Story of the National Cancer Act of 1971* (Princeton, N.J.: Princeton University Press, 1977).

13. Farber's advocacy was perhaps further impelled by his own personal experience. He had had colorectal cancer and would have to use a colostomy bag for the rest of his life.

14. In 1869 Paul Langerhans first described scattered clusters of cells in the pancreas, subsequently identified as the source of insulin and classically known as the islets of Langerhans. See Paul Langerhans, *Beiträge zur mikroskopischen Anatomie der Bauchspeicheldrüse* (Berlin: Gustave Lange, 1869), translated by H. Morrison as *Contributions to the Microscopic Anatomy of the Pancreas* (Baltimore, Md.: Johns Hopkins Press, 1937).

15. Rettig, *Cancer Crusade,* 255–56.

16. Ted Kennedy was chairman of the Senate Health Subcommittee. One powerful motive on Nixon's part was to seize the health issue and impair Kennedy's potential candidacy as the Democratic presidential nominee in 1972.

17. D. S. Greenberg and J. E. Randall, "Waging the Wrong War on Cancer," *Washington Post,* May 1, 1977.

18. The current screening system uses human cell lines grown in culture, allowing automation of the testing of candidate drugs for high-volume screening.

19. Public acceptance of the concept of clinical trials had been galvanized by the voluntary participation of millions of American families in the 1954 double-blind trials of the Salk polio vaccine.

20. C. G. Zubrod, "The cure of cancer by chemotherapy — reflections on how it happened," *Med Pediatr Oncol* 8 (1980): 107–14.

21. Michael B. Shimkin, *As Memory Serves: Six Essays on a Personal Involvement with the National Cancer Institute, 1938 to 1978* (Washington, D.C.: NIH, 1983).

22. J. C. Bailar III and E. Smith, "Progress against cancer," *N Engl J Med* 314 (1986): 1226–32.

23. P. B. Chowka, "Cancer," *East West,* December 1987.

24. J. C. Bailar III and H. L. Gornik, "Cancer undefeated," *N Engl J Med* 336 (1997): 1569–74.

25. Jordan Goodman and Vivien Walsh, *The Story of Taxol: Nature and Politics in the Pursuit of an Anti-Cancer Drug* (Cambridge: Cambridge University Press, 2001).

26. Jerome Groopman, "The Thirty Years' War: Have We Been Fighting Cancer the Wrong Way?" *New Yorker,* June 4, 2001, 52–63.

27. Clifton Leaf, "Why We're Losing the War on Cancer (And How to Win It)," *Fortune,* March 22, 2004.

Part III:
A Quivering Quartz String Penetrates
the Mystery of the Heart

The part epigraph is from *Introduction à l'étude de la médecine expérimentale* (Paris: Baillière, 1865). Bernard is known as the founder of experimental medicine.

CHAPTER 22: An Unexpected Phenomenon: It's Electric!

1. A. Kölliker and H. Müller, "Nachweis der negativen Schwankung des Muskelstromes am naturlich sich contrahierenden Muskel," *Verhandl Phys Med Gesellsch* 6 (1856): 528–33.

2. This episode is reminiscent of an accidental observation made at the end of the eighteenth century by Luigi Galvani, professor of anatomy at the University of Bologna: "When one of my assistants by chance lightly applied the point of a scalpel to the inner crural nerves [main nerves of the legs of a dissected frog] suddenly all the muscles of the limbs were seen so to contract.... Another assistant ... observed that this phenomenon occurred when a spark was discharged from the conductor of the [nearby] electrical machine." L. Galvani, "De viribus electricitatis in motu musculari commentarius," *De Bononiensi Scientarium et Artium Instituto atque Academia Commentarii* 7 (1791): 363–418. Galvani was

placed on the right path by this chance finding and went on to confirm that electrical stimulation can induce muscle contraction.

3. E. Besterman and R. Creese, "Waller — pioneer of electrocardiography," *British Heart Journal* 42 (1979): 61–64.

4. A. D. Waller, "The electrocardiogram of man and of the dog as shown by Einthoven's string galvanometer," *Lancet* 1 (1909): 1448–50.

5. W. Einthoven, "Enregistreur galvanométrique de l'électrocardiogramme humain et controle des résultats obtenus par l'emploi de l'électromètre capillaire en physiologie," *Arch Néerland Sci Exactes Naturelles* 9 (1904): 202–209.

6. I. Erschler, "Willem Einthoven — the man. The string galvanometer electrocardiograph," *Arch Intern Med* 148 (1988): 453–55.

7. Einthoven's instrument could not have been designed without recent technological advances. The critical ultrathin quartz filament was first made by another investigator less than twenty years earlier, and Einthoven cleverly adopted it for his use. Its tiny movements were made visible with a projecting microscope lens requiring modern kinds of glass. The brightest available point source of light, the carbon arc, had been developed as a reliable device only in the previous two decades. And the movements of the filament were recorded on a photographic plate that was coated with a recently developed sensitive emulsion. J. Burnett, "The origins of the electrocardiograph as a clinical instrument," *Med Hist Suppl* 5 (1985): 53–76.

8. Einthoven successfully recorded ventricular extra systoles (contractions), heart block (defects in the conduction system between the atria and the ventricles), atrial flutter and fibrillation, and ventricular muscle hypertrophy. His classic article in 1908 illustrating electrocardiograms from a wide variety of cardiac diseases documented the string galvanometer's great practical as well as theoretical importance. Willem Einthoven, "Weiteres über das Elektrokardiogramm," *Pflügers Arch ges Physiol* 122 (1908): 517–84. His three lead combinations for ECG recordings remain standard today.

9. In May 1908, Thomas Lewis, a London physician, had written to Einthoven to ask for a reprint of his landmark 1906 paper "Le télécardiogramme." This was the beginning of a nearly twenty-year correspondence that resulted in major contributions to the development of clinical echocardiography. Lewis and Einthoven were ideal complements. Despite his medical training, Einthoven was essentially a physicist employing his talents in physiology. Lewis displayed great ingenuity in devising and carrying out experiments to clarify fundamental clinical problems.

10. W. Murrell, "Nitro-glycerine as a remedy for angina pectoris," *Lancet* 1 (1879): 80–81, 113–15, 151–52, 225–27.

11. James B. Herrick, "Historical Note," in *Diseases of the Coronary Arteries and Cardiac Pain,* ed. Robert L. Levy (New York: Macmillan, 1936), 17.

12. S. A. Levine, "Willem Einthoven: Some historical notes on the occasion of the centenary celebration of his birth," *Am Heart J* 61 (1961): 422–23.

CHAPTER 23: What a Catheter Can Do

1. Werner Forssmann, *Experiments on Myself: Memoirs of a Surgeon in Germany* (New York: St. Martin's Press, 1974), 85.

2. F. Splittgerber and D. E. Harken, "Catheterization of the right heart, by Werner Forssmann," *Cardiac Chronicle* 4 (1991): 13–15.

3. W. Forssmann, "Über Kontrastdarstellung der Höhlen des lebenden rechten Herzens und der Lungenschlagader," *Münch Med Wochenschr* 1 (1931): 489–92.

4. For studying the metabolism of the heart muscle itself, "the key in the lock" came about through a catheter tip misplaced during cardiac catheterization by Richard Bing at Johns Hopkins.

The heart muscle itself receives nourishing oxygen-rich blood through the coronary arteries and, like any other organ, drains the oxygen-depleted darker blood through a venous outlet. In this case, the outlet is termed the coronary sinus and it drains into the right side of the heart so that the blood can recirculate through the lungs. Bing's account thirty-five years after the chance occurrence is straightforward: "I pushed the catheter into what I thought was the right ventricle but when I drew the blood from the catheter it was black. This was a surprise because when we pulled the catheter back a bit, it was of the color which one expects to find in the right ventricle. We soon realized, working on cadavers, that we had entered the coronary sinus. This opened up a whole new field for me — that of cardiac metabolism." Richard J. Bing, "Personal Memories of Cardiac Catheterization and Metabolism of the Heart," in *History and Perspectives of Cardiology: Catheterization, Angiography, Surgery, and Concepts of Circular Control,* ed. H. A. Snellen, A. J. Dunning, and A. C. Arntzenius (The Hague: Leiden University Press, 1981), 43–45. The unintentional catheter probing of the coronary sinus enabled Bing to conduct new studies on the functional needs of the heart itself. Of all the muscles in the body, the heart has unique properties. It generally contracts sixty to one hundred times per minute, twenty-four hours a day, with only a brief rest between each cardiac cycle. Its nutritional demands are great. It extracts the most oxygen from the blood flowing through it.

5. G. Liljestrand, "1956 Nobel Prize Presentation Speech," in *Nobel Lectures, Physiology or Medicine,* vol. 3 (Amsterdam: Elsevier, 1964), 502.

6. S. I. Seldinger, "Catheter replacement of the needle in percutaneous arteriography," *Acta Radiol* 39 (1953): 368–76.

7. F. M. Sones Jr., E. K. Shirey, W. L. Proudfit, and R. N. Westcott, "Cine-coronary arteriography" (abstract), *Circulation* 20 (1959): 773.

8. F. M. Sones Jr., "Coronary Angiography," in *Beiträge zur Geschichte der Kardiologie,* ed. Gerhard Blümchen (Leichlingen: Klink Roderbirken, 1979).

9. J. Willis Hurst, "History of Cardiac Catheterization," in *Coronary Arteriography and Angioplasty,* ed. Spencer B. King III and John S. Douglas Jr. (New York: McGraw-Hill, 1985), 6.

10. F. M. Sones Jr. and E. K. Shirey, "Cine coronary arteriography," *Modern Concepts of Cardiovascular Disease* 31 (1962): 735–38.

11. Sones was an inveterate smoker who, even during a coronary catheterization on a patient, would use a long sterile forceps placed on a nearby tray in the pit to pick up a cigarette, have a nurse light it for him, and, after taking a few puffs, return the forceps with its tip overhanging the tray to keep everything sterile. He eventually died of lung cancer in 1985 at the age of sixty-six.

12. Donald B. Effler, introduction to *Surgical Treatment of Coronary Arteriosclerosis,* ed. René G. Favaloro (Baltimore: Williams and Wilkins, 1970), xi.

13. René G. Favaloro, "Saphenous vein autograft replacement of severe segmental coronary artery occlusion: Operative technique," *Ann Thorac Surg* 5 (1968): 334–39.

14. René G. Favaloro, *The Challenging Dream of Heart Surgery: From the Pampas to Cleveland* (Boston: Little, Brown, 1994), 98.

15. Quoted in Julius H. Comroe Jr., *Retrospectroscope: Insights into Medical Discovery* (Menlo Park, Calif.: Von Gehr, 1977), 94–95.

16. J. H. Jacobson II and E. L. Suarez, "Microsurgery in anastomoses of small vessels," *Surgical Forum* 11 (1960): 243–45.

CHAPTER 24: "Dottering"

1. C. T. Dotter and M. P. Judkins, "Transluminal treatment of arteriosclerotic obstruction: Description of a new technique and a preliminary report of its application," *Circulation* 30 (1964): 654–70.

2. William C. Sheldon and F. Mason Sones Jr., "Stormy Petrel of Cardiology," *Clinical Cardiology* 17 (1994): 405–407.

3. A. Gruentzig, "Transluminal dilatation of coronary-artery stenosis" (letter to editor), *Lancet* 1 (1978): 263.

4. A. Gruentzig, A. Sennings, and W. E. Siegenthaler, "Nonoperative dilatation of coronary artery stenosis: Percutaneous transluminal coronary angioplasty," *N Engl J Med* 301 (1979): 61–68.

5. J. Willis Hurst, "Tribute: Andreas Roland Gruentzig (1939–1985) — a private perspective," *Circulation* 73 (1986): 606–10.

6. Further modifications to the procedure involved atherectomy devices, lasers, and ultrasound for the removal, or ablation, of coronary plaque. As an alternative to bypass surgery, stents are being increasingly used to prop open coronary blood vessels to overcome the constrictions that cause heart attacks. Stents are metal mesh cylinders between 8 and 33 millimeters long and weighing around 27 thousandths of a gram. They are now drug-coated to prevent reclogging from scar tissue and generate more than $66 billion a year in sales.

CHAPTER 25: A Stitch in Time

1. In his early years, Carrel had taken sewing lessons from a local embroiderer in Lyon to improve his surgical dexterity. After moving to the United States, he used his technique of vascular anastomosis to pursue research on organ transplantation, for which he was awarded the Nobel Prize in 1912.

2. S. M. Levin, "Reminiscences and ruminations: Vascular surgery then and now," *Am J Surg* 154 (1987): 158–62.

3. A. B. Voorhees Jr., "The origin of the permeable arterial prosthesis: a personal reminiscence," *Surgical Rounds* 11 (1988): 79–84.

4. A. B. Voorhees Jr., A. Jaretzki, and A. H. Blakemore, "The use of tubes constructed from Vinyon 'N' cloth in bridging arterial defects: A preliminary report," *Ann Surg* 135 (1952): 332–36.

5. A. H. Blakemore and A. B. Voorhees, "The use of tubes constructed from Vinyon 'N' cloth in bridging arterial defects: Experimental and clinical," *Ann Surg* 140 (1954): 325–34.

6. Steven G. Friedman, *A History of Vascular Surgery* (Mount Kisco, N.Y.: Futura, 1989), 131–39.

7. A. B. Voorhees Jr., "The development of arterial prostheses: A personal view," *Arch Surg* 120 (1985): 289–95.

CHAPTER 26: The Nobel Committee Says Yes to NO

1. Author interview with Robert Furchgott, May 2, 2002.

2. Strips of intestine and later of aorta were routinely used to measure the effects of various agents on smooth muscle. Once, in using such strips, a laboratory technician's oversight in forgetting to add glucose to the solution led to an understanding of the energy needs for smooth muscle to contract. Furchgott also acknowledged that it was an "accidental discovery" that blood vessels undergo reversible relaxation when exposed to light. He observed this phenomenon of "photo relaxation" by pure chance. Strips of rabbit aorta placed near a window were observed oscillating. Contractions and dilations were apparent as clouds alternately blocked the sun or moved on to allow bright sunlight through the window. When a technician stood near the laboratory bench, casting a shadow on the preparation, Furchgott saw the strips contract. When she stepped to the side, he saw them relax. It was clearly the shadow that was producing the effect. When he closed the shade on the window, the strips of vessel contracted, and when he opened the shade, they relaxed. Overhead fluorescent lights did not have any effect. Author interview with Furchgott, May 2, 2002.

Years later, Furchgott hypothesized that ultraviolet light activates the release of nitric oxide from the vascular smooth muscle cells.

3. R. F. Furchgott, "A research trail over half a century," *Annu Rev Pharmacol Toxicol* 35 (1995): 1–27.

4. R. F. Furchgott, "The discovery of endothelium-dependent relaxation," *Circulation* 87 suppl. V (1993): V3–8.

5. R. F. Furchgott, D. Davidson, and C. I. Lin, "Conditions which determine whether muscarinic agonists contract or relax rabbit aortic rings and strips," *Blood Vessels* 16 (1979): 213–14.

6. R. F. Furchgott and J. V. Zawadzki, "The obligatory role of endothelial cells in the relaxation of arterial smooth muscle to acetylcholine," *Nature* 288 (1980): 373–76.

7. Ironically, Alfred Nobel, who discovered how to safely formulate dynamite from nitroglycerin, was prescribed nitroglycerin for his angina pectoris late in life, but he refused to take it because of the known vascular headaches of his factory workers.

8. In a further acknowledgment of serendipity, Furchgott's conclusion was based on recalling another "accidental finding" fifteen years earlier when a postdoctoral fellow in his laboratory mistakenly acidified a solution of sodium nitrite. Reaching for a bottle of physiological saline

from a laboratory shelf, he had instead grabbed a bottle of acidified saline solution. In acidifying the nitrite, NO was released, and transient vessel relaxation resulted. At the time, Furchgott did not realize the significance of this observation. Now he cites it as a "tantalizing near-miss." Author interview with Robert Furchgott, May 13, 2002.

9. Robert F. Furchgott, "Endothelium-derived relaxing factor: Discovery, early studies, and identification as nitric oxide," Nobel Lecture, December 8, 1998, *Les Prix Nobel*, 226–43.

CHAPTER 27: "It's Not You, Honey, It's NO"

1. C. E. Brown-Séquard, "Expérience démonstrant la puissance dynamogénique chez l'homme d'un liquide extrait de testicules d'animaux," *Archives de Physiologie Normale et Pathologique* 5 sér 1 (1889): 651–58, 739–46.

2. E. Laqueur, K. David, E. Dingemanse, J. Freud, and S. E. de Jongh, "Über männliches Hormon: Unterschied von Androsteron aus Harn und Testosteron aus Testis," *Acta Brev Neerland* 4 (1935): 5.

3. L. L. Stanley, "An analysis of one thousand testicular substance implantations," *Endocrinology* 6 (1922): 787–94.

4. V. Pruitt, "John R. Brinkley, Kansas physician, and the goat gland rejuvenation fad," *Pharos,* Summer 2002, 33–39.

5. NIH Consensus Conference Development Panel on Impotence, "Impotence," *JAMA* 270 (1993): 83–90.

6. H. A. Feldman, I. Goldstein, D. G. Hatzichristou, et al., "Impotence and its medical and psychological correlates: Results of the Massachusetts Male Aging Study," *Journal of Urology* 151 (1994): 54–61.

7. B. Handy, "The Viagra Craze," *Time,* May 4, 1998, 50–57.

8. "A Stampede Is On for Impotence Pill," *Wall Street Journal,* April 20, 1998.

9. Pfizer chose the name Viagra, perhaps because its marketing executives thought it evoked potency by suggesting "Niagara" and "vigor." Levitra's name came from the word "elevate." Furthermore, the prefix *le* implies masculinity — at least to the French — and *vitra* suggests the word "vitality." Donald G. McNeil Jr., "The Science of Naming Drugs (Sorry, 'Z' Is Already Taken)," *New York Times,* December 28, 2003.

CHAPTER 28: What's Your Number?

1. N. Anichkov and S. Chalatov, "Über experimentelle Cholesterinsteatose: Ihre Bedeutung für die Entstehung einiger pathologischer

Prozessen," *Centralblatt für Allgemeine Pathologie und Pathologische Anatomie* 1 (1913): 1.

2. J. W. Gofman, F. Lindgren, H. Elliott, et al., "The Role of Lipids and Lipoproteins in Arteriosclerosis," *Science* 111 (1950): 166–71.

3. L. I. Dublin and M. Spiegelman, "Factors in the higher mortality of our older age groups," *Am J Public Health* 42 (1952), 422–29.

4. Goldstein had attended medical school there, and his brilliance was recognized by the chairman of the department of internal medicine, who offered him the opportunity to return to establish the division of medical genetics after he completed his fellowship training elsewhere. Six years after graduation, he returned to his alma mater as an assistant professor to work with Michael Brown, whom he had encouraged to join the faculty a year earlier.

5. M. S. Brown and J. L. Goldstein, "Familial hypercholesterolemia: Biochemical, genetic, and pathophysiological considerations," *Adv Intern Med* 20 (1975): 78–96.

6. The idea of cell receptors was known, but it had never been studied in relationship to fat and cholesterol in the blood.

7. M. S. Brown and J. L. Goldstein, "Binding and degradation of low density lipoproteins by cultured human fibroblasts: Comparison of cells from normal subjects and from a patient with homozygous familial hypercholesterolemia," *J Biol Chem* 249 (1974): 5153–62.

8. Y. Watanabe, "Serial inbreeding of rabbits with hereditary hyperlipidemia (WHHL-rabbit): Incidence and development of atherosclerosis and xanthoma," *Atherosclerosis* 36 (1980): 261–68.

9. Lipid Research Clinics Program. *JAMA* 251 (1984): 351–64, 365–74.

10. Bob Banta, "Nobel Pair Trace Solution of Cholesterol Puzzle," *Austin American-Statesman,* October 27, 1985.

11. In yet another example of unintended discovery, lovastatin (marketed as Mevacor by Merck) was uncovered in a search of fungal metabolites for a possible new sulfa drug. A. W. Alberts, "Discovery, biochemistry and biology of lovastatin," *Am J Cardiol* 62, no. 15 (1988): 10J–15J.

12. The units are expressed in milligrams per deciliter.

CHAPTER 29: Thinning the Blood

1. J. McLean, "The thromboplastic action of cephalin," *Am J Physiol* 41 (1916): 250–57.

2. W. H. Howell and E. Holt, "Two new factors in blood coagulation — heparin and pro-antithrombin," *Am J Physiol* 47 (1918): 328–41.

3. J. McLean, "The discovery of heparin," *Circulation* 19 (1959): 75–78.

4. S. Sherry, "The origin of thrombolytic therapy," *American College of Cardiology* 14 (1989): 1085–92.

5. In another example of prematurity in discovery, dicumarol had been previously synthesized in 1903, albeit in impure form, and then promptly ignored. K. P. Link, "The anticoagulant from spoiled sweet clover hay," *Harvey Lectures* 34 (1943–44): 162–207.

6. Warfarin is from the acronym of the Wisconsin Alumni Research Foundation, which funded his research, and "arin" from coumarin. K. P. Link, "The discovery of dicumarol and its sequels," *Circulation* 19 (1959): 97–107.

7. L. Craven, "Acetyl salicylic acid: Possible preventive of coronary thrombosis," *Annals of Western Medicine and Surgery* 4 (1950): 95–99.

8. J. R. Vane, "Inhibition of prostaglandin synthesis as a mechanism of action for aspirin-like drugs," *Nature: New Biology* 231 (1971): 230–35.

Part IV:
The Flaw Lies in the Chemistry, Not the Character: Mood-Stabilizing Drugs, Antidepressants, and Other Psychotropics

CHAPTER 30: It Began with a Dream

1. Benedict Carey, "Most Will Be Mentally Ill at Some Point, Study Says," *New York Times,* June 7, 2005. Article based on a series of four papers in the June 2005 issue of *Archives of General Psychiatry.*

2. Ken Johnson, senior vice president of Pharmaceutical Research and Manufacturers of America (PHARMA), letter to the editor, *New York Times,* September 23, 2005.

3. Almost at the same time as Loewi's investigations began to become known, Walter Bradford Cannon in Harvard was carrying out experiments about a possible transmitter at sympathetic nerve endings. Eventually this was identified as noradrenaline.

4. Otto Loewi, "An autobiographical sketch," *Perspect Bio Med* 4 (1960): 17.

CHAPTER 31: Mental Straitjackets: Shocking Approaches

1. Elliot S. Valenstein, *Great and Desperate Cures: The Rise and De-*

cline of Psychosurgery and Other Radical Treatments for Mental Illness (New York: Basic Books, 1986).

2. J. Wagner von Jauregg, "Über den Einwirkung fieberhafter Erkrankungen auf Psychosen," *Jahrbücher für Psychiatrie und Neurologie* 7 (1887): 94–131.

3. The name of this clinical investigator changed with the times. He was born Julius Wagner, but when in 1883 his father, an Austrian government official, was granted the title "Ritter von Jauregg," he became Wagner von Jauregg. He took the hyphenated name "Wagner-Jauregg" when titles of nobility were abolished after the dissolution of the Hapsburg Empire in 1918. It was with this name that in 1927 he was awarded the Nobel Prize in Medicine, the only psychiatrist to be honored with this achievement. His work laid the foundation for the belief in an organic physical view of mental disorders.

4. A curious episode brought this physician and Sigmund Freud into a professional encounter. After World War I, it was alleged that Austrian military doctors in Wagner von Jauregg's psychiatric clinic had used cruel treatments on soldiers who had been suspected of malingering. The condition known as "shell shock" was not well understood, and the idea was to make electric shock treatments so disagreeable that the soldiers would prefer to return to the front. The strength of the current used on readmissions was increased to such an unbearable point that death and suicides resulted. In 1920 the Austrian military authorities appointed a commission to investigate, one of whose members was Freud, a former schoolmate of Wagner von Jauregg's. The latter was personally exonerated. According to Freud's biographer, Ernest Jones, Freud emphasized the distinction between a doctor's primary responsibility to the patient and the demand of the military authorities that the doctor's chief duty is to return soldiers to active duty promptly. Freud's observation was condemned as unpatriotic by the commission. Ernest Jones, *The Last Phase: 1919–1939*, vol. 3 of the *Life and Work of Sigmund Freud* (New York: Basic Books, 1953–57), 21–22.

5. Manfred J. Sakel, "A new treatment of schizophrenia," *Am J Psychiatry* 84 (1934): 1211–14.

6. Ladislaus J. Meduna, "Versuche über die biologische Beeinflussung des Ablaufes der Schizophrenie: Camphor und Cardiazolkrämpfe," *Zeitschrift für die gesamte Neurologie und Psychiatrie* 152 (1935): 235–62.

7. Ugo Cerletti, "Electroshock Therapy," in *The Great Physiodynamic Therapies in Psychiatry,* ed. Arthur M. Sackler et al. (New York: Hoeber, 1956), 91–120.

8. Sylvia Plath, *The Bell Jar* (1963; London: Faber and Faber, 1966), 151.

CHAPTER 32: Ice-Pick Psychiatry

1. Andrew Scull, *Madhouse: A Tragic Tale of Megalomania and Modern Medicine* (New Haven: Yale University Press, 2005).
2. Baptized António Caetano de Abreu Freire, he was also christened with the name Egas Moniz, after a twelfth-century Portuguese patriot who had fought against the Moors and from whom his family descended. Later, this was the name he adopted.
3. J. M. Harlow, "Recovery from the passage of an iron bar through the head," *Publication of the Massachusetts Medical Society* 2 (1868): 327–47.
4. P. P. Broca, "Sur le volume et la forme du cerveau suivant les individus et suivant les races," *Bulletin de la Société Anthropologique* 2 (1861): 301–21.
5. Egas Moniz had been nominated twice for a Nobel Prize for cerebral angiography, but did not win. He was a proud and ambitious man, "dominated," as he said, "by the desire to reach something new in the scientific world." He prided himself on his persistence, and in this regard a historian's assessment is illuminating: "He had the ability to pick important problems but did not attack them with great originality or creativity. His problem solving was characterized by trial and error, persistence and a willingness to take risks." Valenstein, *Great and Desperate Cures,* 82–83.
6. Egas Moniz, "How I Succeeded in Performing the Prefrontal Leucotomy," in *The Great Physiodynamic Therapies in Psychiatry,* ed. Arthur M. Sackler et al. (New York: Hoeber, 1956), 131–37.
7. William Sargant, *The Unquiet Mind* (London: Heinemann, 1967), 71.
8. Walter Freeman and James W. Watts, *Psychosurgery: Intelligence, Emotion, and Social Behavior Following Prefrontal Lobotomy for Mental Disorders* (Springfield, Ill.: C. C. Thomas, 1942).
9. Jack El-Hai, *The Lobotomist* (Hoboken, N.J.: John Wiley, 2005).
10. William Arnold, *Shadowland* (New York: McGraw-Hill, 1978).
11. Most lobotomies were done in state hospitals, and many others in medical school hospitals and Veterans Administration hospitals. Only about 6 percent were done in private hospitals.
12. A full-length feature film on lobotomy, *A Hole in One,* was made in 2004 by Richard Ledes, a first-time director. It is based on the

archives of Walter Freeman, kept at George Washington University, and the 1953 Annual Report of the New York State Department of Mental Hygiene. Randy Kennedy, "A Filmmaker Inspired by Lobotomy," *New York Times,* April 29, 2004.

13. Sargant, *The Unquiet Mind,* 36.

14. Quoted in J. Wortis, "In Memoriam, Manfred Sakel, M.D., 1900–1957," *Am J Psychiatry* 115 (1958–59): 287–88.

CHAPTER 33: Lithium

1. John F. J. Cade, "The Story of Lithium," in *Discoveries in Biological Psychiatry,* ed. Frank J. Ayd and Barry Blackwell (Philadelphia: Lippincott, 1970), 219.

2. A missed opportunity for the earlier discovery of the antimanic properties of lithium occurred in the 1870s. S. Weir Mitchell, a neurologist, introduced lithium bromide, straight from photography, in the treatment of epilepsy or mental exertion with insomnia from overwork. He found the lithium salts superior to other bromides, and came breathtakingly close to attributing its "more rapid and intense" hypnotic effects to lithium itself: "This result must have been due to some more sure or speedy action of this salt (lithium) because the other bromides failed." Yet he fixed on the value of the bromide of lithium. Anne E. Caldwell, "History of Psychopharmacology," in *Principles of Psychopharmacology,* ed. William G. Clark and Joseph del Giudice, 2nd ed. (New York: Academic Press, 1978), 22–23.

3. Questions about the large doses of lithium salts used in the guinea pigs have since been raised, and some have even conjectured that the lassitude and docility of the animals may have been due to lithium toxicity. The question as to why manic urine appeared to be "more toxic" was never answered.

4. Cade, "The Story of Lithium," 220.

5. John F. J. Cade, "Lithium salts in the treatment of psychotic excitement," *Med J Aust* 3 (1949): 349–52.

6. M. Schou, N. Juel-Nielsen, E. Stromgren, and H. Voldby, "The treatment of manic psychoses by the administration of lithium salts," *J Neurol Neurosurg Psychiatry* 17 (1954): 250–60.

7. Lithium, the lightest of all solid elements, has an atomic number of three. Its atomic structure is relatively simple. The nucleus comprises three or four neutrons and three protons. Two of three electrons occur in an inner shell, while one remaining electron occupies a distant outer shell from which it is readily lost. It is this that is the major determinant of the

element's chemical properties. Nevertheless, the precise mode of its action as a mood-stabilizing agent remains unknown. It is thought likely that it inhibits norepinephrine uptake into nerve endings.

CHAPTER 34: Thorazine

1. H. Laborit, "Sur l'utilisation de certains agents pharmacody-namiques à action neuro-végétative en periode pré- et postopératoire," *Acta Chirurgica Belgica* 48 (1949): 485–92.

2. H. Laborit, P. Huguenard, and R. Alluaume, "Un nouveau stabil-isateur végétatif (le 4560 RP)," *Presse Méd* 60 (1952): 206–8.

3. Deniker's presentation of these findings at the Fiftieth French Congress of Psychiatry and Neurology in Luxembourg at the end of July 1952 was not particularly auspicious. It was scheduled at the end of the last session of the week.

4. Caldwell, "History of Psychopharmacology," 30.

5. Quoted in Judith P. Swazey, *Chlorpromazine in Psychiatry* (Cambridge, Mass.: MIT Press, 1974), 157.

6. F. Labhardt, "Die Largactiltherapie bei Schizophrenien und an-deren psychotischen Zustanden," *Schweiz Arch Neurol Psychiat* 73 (1954): 309–38.

CHAPTER 35: Your Town, My Town, Miltown!

1. Frank M. Berger, "Anxiety and the Discovery of the Tranquiliz-ers," in *Discoveries in Biological Psychiatry,* ed. Frank J. Ayd and Barry Blackwell (Philadelphia: Lippincott, 1970), 115–29.

2. F. M. Berger, "The pharmacological properties of 2-methyl-2-*n*-propyl-1.3-propanediol dicarbamate (Miltown), a new interneuronal blocking agent," *J Pharmacol Exp Ther* 112 (1954): 413–23.

3. Frank M. Berger (interview), "The Social Chemistry of Pharma-cological Discovery: The Miltown Story," *Social Pharmacology* 2 (1988), 189–204.

4. " 'Ideal' in Tranquility," *Newsweek,* October 29, 1956, 63.

CHAPTER 36: Conquering the "Beast" of Depression

1. These depths have been movingly related in the novelist William Styron's true account of his own descent into a crippling and suicidal de-pression: "I felt an immense and aching solitude. I could no longer con-centrate during those afternoon hours, which for years had been my

working time, and the act of writing itself, becoming more and more difficult and exhausting, stalled, then finally ceased. There were also dreadful, pouncing seizures of anxiety. One bright day on a walk through the woods with my dog I heard a flock of Canada geese honking high above trees ablaze with foliage; ordinarily a sight and sound that would have exhilarated me, the flight of birds caused me to stop, riveted with fear, and I stood stranded there, helpless, shivering, aware for the first time that I had been stricken by no more pangs of withdrawal but by a serious illness whose name and actuality I was able finally to acknowledge. Going home, I couldn't rid my mind of the line of Baudelaire's dredged up from the distant past, that for several days had been skittering around at the edge of my consciousness: 'I have felt the wind of the wing of madness.'" William Styron, *Darkness Visible: A Memoir of Madness* (New York: Random House, 1990), 46.

2. Considerations of physical causes of mental disorders were given support by the finding in 1907 by Alois Alzheimer in Munich of brain abnormalities — "senile plaques," he called them — in some cases of premature dementia (Alzheimer's disease).

3. John Mann, *Murder, Magic, and Medicine* (Oxford: Oxford University Press, 1992), 186.

4. Alfred Pletscher, Parkhurst A. Shore, and Bernard B. Brodie, "Serotonin Release as a Possible Mechanism of Reserpine Action," *Science* 122 (1955): 374–75.

5. How this came to be found is replete with instances of unexpected findings. In 1945, Ernest Huant was administering nicotinamide to patients with tumors undergoing radiation therapy to lessen their postradiation nausea and vomiting. He was surprised to find clearing of densities in the chest films of some patients who also had pulmonary tuberculosis. E. Huant, "Note sur l'action de très fortes doses d'amide nicotinique dans les lésions bacillaires," *Gaz Hôp* 118 (1945): 259–60. The same year, Chorine reported that nicotinamide possessed tuberculostatic activity. V. Chorine, "Action de l'amide nicotinique sur les bacilles du genre *Mycobacterium*," *Comp Ren Acad Sci* 220 (1945): 150–51. In 1952, within a period of two weeks, two different pharmaceutical companies announced their independent discovery that isoniazid, a synthetic derivative, was a highly effective antitubercular drug. Yet again, remarkably, it was a serendipitous discovery in each case. Walter Sneader, *Drug Discovery: The Evolution of Modern Medicines* (New York: John Wiley, 1985), 293.

6. J. C. Saunders, D. Z. Rochlin, N. Radinger, et al., "Iproniazid in Depressed and Regressed Patients," in *Psychopharmacology Frontiers,* ed. Nathan S. Kline (Boston: Little, Brown, 1959), 177–94.

7. H. P. Loomer, J. C. Saunders, and N. S. Kline, "A clinical and pharmacodynamic evaluation of iproniazid as a psychic energizer," *Psychiatr Res Rep, Am Psychiatr Assoc* 8 (1958): 129–41.

8. Nathan S. Kline, "Clinical experience with iproniazid (marsilid)," *J Clin Exp Psychopathol* 19, no. 2, suppl. 1 (1958): 72–78.

9. Nathan S. Kline, "Monoamine Oxidase Inhibitors: An Unfinished Picaresque Tale," in Ayd and Blackwell, *Discoveries in Biological Psychiatry,* 194–204.

10. The British psychiatrist Barry Blackwell recalled the "chain of coincidence" that led him to the discovery of hypertensive crises provided by amines in cheese and other foodstuffs in patients prescribed MAO inhibitors; see "The Process of Discovery," in Ayd and Blackwell, *Discoveries in Biological Psychiatry,* 11–29. It is the equal of a Sherlock Holmes mystery. While a resident in psychiatry at the famous Maudsley Hospital, he published an initial letter in the *Lancet* in 1963 citing isolated case reports in the literature regarding the occasional association of hypertension in patients taking MAO inhibitors. The letter was read by a hospital pharmacist in Nottingham who noted that his wife had twice experienced identical symptoms after eating cheese and inquired regarding the possible link with the cheese's amino acids. Blackwell was amused by the suggestion and wrote a brief reply dismissing the notion. When he heard that the pharmaceutical company had received a few similar reports, including one death that occurred during treatment with an MAO inhibitor and an amino acid, he reviewed the dietary record of a previous incident at Maudsley, which revealed that the patient had eaten a cheese pie that evening. Soon another patient developed headache and a hypertensive crisis, and inquiry showed that she had eaten cheese sandwiches for supper. Blackwell tried to provoke the reaction in himself without success. However, the reaction was positive in a carefully monitored patient volunteer. "Coincidence continued to play a part." Two patients in a ward in the hospital simultaneously complained of headache. Both had recently returned from the hospital cafeteria, where cheese had just made its weekly appearance on the menu. It was at this point that certainty "dawned." The pharmacology of the reaction is now understood. The amine responsible is tyramine. An average portion of natural or particularly aged cheeses contains enough tyramine to provoke a marked rise in blood pressure and other cardiovascular changes. As a result of monoamine oxidase, tyramine brings this about by releasing catecholamines in nerve endings and the adrenal medulla. Another source of tyramine is yeast products used as food supplements.

11. Roland Kuhn, "The Imipramine Story," in Ayd and Blackwell, *Discoveries in Biological Psychiatry,* 205–17.

12. "The Culture of Prozac," *Newsweek,* February 7, 1994, 41.

13. E. J. Nestler, "Antidepressant treatments in the 21st century," *Biol Psychiatry* 44 (1998): 526–33.

Chapter 37: Librium and Valium

1. Leo H. Sternbach, "The benzodiazepine story," *Drug Res* 22 (1978): 230.

2. E. Kyburz, "Serendipity in drug discovery: The case of BZR ligands," *Il Farmaco* 44, no. 4 (1989): 345–82.

3. When World War II broke out, Sternbach was working as a research fellow at the Swiss Federal Institute of Technology in Zurich and could not return to Poland. He moved to Basel as a research chemist for Hoffman–La Roche, which, fearing a German invasion and wanting to get its Jewish scientists to safety, sent him to its subsidiary in New Jersey in the early 1940s. He became an American citizen in 1946.

4. Sternbach, "The benzodiazepine story," 235.

5. Ibid., 236.

6. H. J. Parry, M. B. Balter, G. D. Mellinger, I. H. Cisin, and D. I. Manheimer, "National patterns of psychotherapeutic drug use," *Arch Gen Psychiatry* 28, no. 6 (1973): 18–74.

7. Sternbach sold the patents to Librium and Valium (and some 230 others) for one dollar each to Hoffman–La Roche. When asked if he thought the company should be paying him royalties on the drugs' staggering sales, he replied, "I don't think that I was not treated appropriately. Yet, sometimes I find myself wondering why I didn't ask for more. I suppose I didn't have the chutzpah. Besides, what would I do with the money?" L. Kent, "Leo Sternbach: The tranquil chemist," *Sciquest* 53 (1980): 22–24.

8. Valium and similar products amplify the effects of the brain neurotransmitter gamma-aminobutyric acid (GABA). GABA is present in neurons that exert an inhibitory effect on other neurons in the central nervous system, and is active at up to 40 percent of brain synapses.

9. With the widespread use of these drugs, circumstances are commonly encountered in which an "antidote" would be helpful to reverse or reduce the drug's effects. These include cases of accidental or intentional overdosage and to shorten their effects when used in anesthesia. In another example of a "truly serendipitous discovery," the continued search for new improved anxiety-relieving drugs by researchers at Hoffman–La

Roche in Basel led to a specific benzodiazepine antagonist. Their candid description of stumbles in the late 1970s and early 1980s leading to an epiphany is a dramatic episode. "In contrast to our expectations" and "to our great surprise," they had stumbled upon the antagonist flumazenil, marketed as Romazicon in the United States and as Anexate in Europe.

Chapter 38: "That's Funny, I Have the Same Bug!"

1. Lawrence K. Altman, *Who Goes First? The Story of Self-Experimentation in Medicine* (New York: Random House, 1987), 100.
2. Ibid., 99.
3. D. I. Eneanya, J. R. Bianchine, D. O. Duran, and B. D. Andresen, "The actions and metabolic fate of disulfiram," *Annu Rev Pharmacol Toxicol* 21 (1981): 575–96.
4. J. Hald, E. Jacobsen, and V. Larsen, "The sensitizing effect of tetraethylthiuramdisulphide (Antabuse) to ethyl alcohol," *Acta Pharmacologica et Toxicologica* (Copenhagen) 4 (1948): 285–96.
5. Historically, hypersensitivity to alcohol brought about by other substances had been noted. The first was observed in 1914 from industrial exposure to cyanamide. The French cited *mal rouge* among the workers, referring to the intense facial blushing after consuming alcohol. This was accompanied by headache, accelerated and deepened respiration, accelerated pulse, and a feeling of giddiness. These attacks lasted from half an hour to two hours and were followed by sleepiness. Similar sensitization to alcohol is produced by eating the fungus *Coprinus atramentarius*.
6. E. E. Williams, "Effects of alcohol on workers with carbon disulfide," *JAMA* 109 (1937): 1472–73.

CHAPTER 39: LSD

1. Albert Hofmann, "The Discovery of LSD and Subsequent Investigation on Naturally Occurring Hallucinogens," in Ayd and Blackwell, *Discoveries in Biological Psychiatry*.
2. Quoted in Altman, *Who Goes First?* 73.
3. W. A. Stoll, "Lysergsäure-diäthyl-amid, ein Phantastikum aus der Mutterkorngruppe," *Schweiz Arch Neurol Psychiat* 60 (1947): 297–323.
4. Mescaline was isolated as the active ingredient from the peyote cactus in 1897 by the German chemist Arthur Heffter. This small spineless cactus grows in the desert regions of Texas and Mexico, and is used for its psychoactive properties by a number of Native American peoples.

5. In a series of experiments at UCLA, cats and monkeys with electrodes implanted in their brains could press a lever and stimulate their pleasure centers with little electrical discharges. A state of ecstasy must have been reached because some pressed the lever eight thousand times an hour, finally collapsing from exhaustion and lack of food. Observing this, Aldous Huxley wrote to a friend, "We are obviously very close to reproducing the Moslem paradise where every orgasm lasts six hundred years." Aldous Huxley, *Moksha* (Los Angeles: Tarcher, 1982). On the other hand, with the early discoveries of cerebral chemical neurotransmitters in the mid-1940s, new searching questions were raised: Could psychopathologies be due to depletion or overabundance of such chemicals?

6. From an unpublished lecture by Hofmann quoted in Christian Rätsch, *The Dictionary of Sacred and Magical Plants*, trans. John Baker (Bridgeport: Prism Press, 1992).

7. Sidney Cohen, *The Beyond Within: The LSD Story* (New York: Atheneum, 1965).

8. John Marks, *The Search for the "Manchurian Candidate": The CIA and Mind Control* (New York: Times Books, 1978).

9. Quoted in "Sidney Gottlieb, 80, Dies; Took LSD to CIA," *New York Times*, March 10, 1999.

10. Albert Hofmann, *LSD, My Problem Child*, trans. Jonathan Ott (New York: McGraw-Hill, 1980).

11. Jay Stevens, *Storming Heaven: LSD and the American Dream* (New York: Atlantic Monthly Press, 1987).

12. George Barger, *Ergot and Ergotism* (London: Gurney and Jackson, 1931).

13. This scenario has been imaginatively illustrated in one of Hieronymus Bosch's bizarre paintings, *St. Anthony Triptych*, from the early sixteenth century. Art historian Laurinda S. Dixon has recently interpreted many of the images; see "Bosch's 'St. Anthony Triptych' — An Apothecary's Apotheosis," *Art Journal* 44(2) (1984): 119–31. Many of the scenes depicted are as viewed through hallucinating eyes. Yet the image of an amputated and mummified human foot is evident. A strange figure — half human, half vegetable — is painted in the shape of a mandrake root. The mandrake root, with its forked shape resembling two human legs, was carried as a talisman against the "holy fire." A giant red fruit in the painting is the mandrake berry, the juice of which contains a belladonna-like substance, an alkaloid that has a narcotic effect similar to that of opium. It was effective as an anesthetic and used as an aid in the brutal amputations performed in the Antonine *hôpitals des démembrés*. A building is in the shape of an apothecary's retort, the distillery used to reduce medicinal herbs.

14. Mary Kilbourne Matossian, *Poisons of the Past: Molds, Epidemics, and History* (New Haven: Yale University Press, 1989).

15. P. L. Delgado and P. A. Moreno, "Hallucinogens, serotonin, and obsessive-compulsive disorder," *J Psychoactive Drugs* 30 (1998): 359–66.

16. T. J. Reedlinger and J. E. Reedlinger, "Psychedelic and entactogenic drugs in the treatment of depression," *J Psychoactive Drugs* 26 (1994): 41–55.

Conclusion: Taking a Chance on Chance: Cultivating Serendipity

1. The NSF's 2006 research budget for the biological sciences was $586 million. Along with the Department of Energy, the NSF also funds much of the physical sciences research in the United States, a fertile source for tools and software that enable biomedical research. Examples range from functional MRI instruments and electron microscopes to DNA microarrays and genome sequence. These critical advances were all the products of work by physicists, engineers, chemists, and computer scientists.

2. Quoted in James Gleick, *Genius: The Life and Science of Richard Feynman* (New York: Pantheon, 1992), 382.

3. The mechanism for the assessment via peer review of the competence, significance, and originality of work submitted for publication was instituted by the first scientific journal, *Philosophic Transactions,* published by the Royal Society of London, as early as the 1600s. Robert Hooke, the Royal Society's first curator of experimenters and himself a pioneering microscopist, had the distinction of confirming Leeuwenhoek's discovery of "animalcules" or bacteria. The motto of the society, *Nullius in verba* (On the word of no one), expresses the society's insistence on verification by observation or experiment, rather than by the voice of authority or tradition.

4. E. M. Allen, "Why Are Research Grant Applications Disapproved?" *Science* 132 (1960): 1532–34; D. J. Lisk, "Why Research Grant Applications Are Turned Down," *Science* 21 (1971): 1025–26; R. Smith, "Peer Review: Reform or Revolution?" *British Medical Journal* 315 (1997): 759–60.

5. In 2004 Cohen and Boyer won the Albany Medical Center Prize in Medicine and Biomedical Research, the nation's largest award in the field, for their pioneering work. J. C. McKinley Jr., "Big Medical Research Prize Goes to 2 Pioneers in Genetics Work," *New York Times,* April 24, 2004.

6. In 1994 the General Accounting Office of the U.S. Congress studied the use of peer review in government scientific grants and found that

reviewers often know applicants and tend to give preferential treatment to those they know. "Peer Review: Reforms Needed to Ensure Fairness in Federal Agency Grant Selection," General Accounting Office, June 24, 1994, GAO/PEMD-94-1. Similar evidence of cronyism as well as gender bias was found in an analysis of the peer review system of the Swedish Medical Research Council, a major funding agency for biomedical research in Sweden. C. Wennerås and A. Wold, "Nepotism and sexism in peer-review," *Nature* 387 (1997): 341–43.

7. Garrett James Hardin, *Nature and Man's Fate* (New York: Rinehart, 1959).

8. Juan Miguel Campanario, "Consolation for the Scientist: Sometimes It Is Hard to Publish Papers That Are Later Highly Cited," *Social Studies of Science* 23 (1993): 342–62; J. M. Campanario, "Commentary on Influential Books and Journal Articles Initially Rejected Because of Negative Referees' Evaluations," *Science Communication* 16 (1995): 304–25; F. Godlee, "The Ethics of Peer Review," in *Ethical Issues in Biomedical Publication,* ed. Anne Hudson Jones and Faith McLellan (Baltimore: Johns Hopkins University Press, 2000), 59–64.

9. Yalow, a feminist to her soul before feminism reached society's consciousness, suffered another indignity when her Nobel Prize received extensive coverage in women's magazines. In its June 1978 issue *Family Health,* a magazine that reached more than 5 million readers, headlined an article "She Cooks, She Cleans, She Wins the Nobel Prize" and introduced Yalow as a "Bronx housewife."

10. P. C. Lauterbur, "Image Formation by Induced Local Interactions: Examples Employing Nuclear Magnetic Resonance," *Nature* 242 (1973): 190–91.

11. For a perspective on the NIH and NSF process of peer review of grant applications, see chapter 6, "The Problems of Peer Review," in *The Great Betrayal: Fraud in Science* by Horace Freeland Judson (Orlando: Harcourt, 2004), 244–86.

12. Jerry Avorn, *Powerful Medicines: The Benefits, Risks, and Costs of Prescription Drugs* (New York: Knopf, 2004).

13. Gregg Critser, *Generation Rx: How Prescription Drugs Are Altering American Lives, Minds, and Bodies* (Boston: Houghton Mifflin, 2005), 2.

14. D. Young, "Studies show drug ads influence prescription decisions, drug costs," *Am J Health Syst Pharm* 59 (2002): 14–16.

15. M. S. Lipsky and C. A. Taylor, "The opinions and experiences of family physicians regarding direct-to-consumer advertising," *J Fam Pract* 45 (1997): 495–99.

16. Ray Moynihan and Alan Cassels, *Selling Sickness: How the World's Biggest Pharmaceutical Companies Are Turning Us All into Patients* (New York: Nation Books, 2005).

17. The power of marketing "me-too" drugs and a recognition of its unsavory characteristics was first demonstrated with the introduction of the broad-spectrum antibiotic tetracycline in the 1950s. Pfizer produced it by chemically manipulating another antibiotic called Aureomycin, manufactured by Lederle. Both companies, as well as three other firms, claimed patent rights. Collusion led to the circumvention of the Patent Office, and the problems were resolved, in one description, in "backrooms, boardrooms and courtrooms." The five firms reached agreement that they all should manufacture and market tetracycline. Each extensively promoted its product, drawing distinctions not on efficacy but on such characteristics as one's more stable liquid suspensions, another's more palatable liquid forms, or another's having combined injectable forms with a superior local anesthetic. J. Goodman, "Pharmaceutical Industry," in *Medicine in the Twentieth Century*, ed. Roger Cooter and John Pickstone (Amsterdam: Harwood Academic Publishers, 2000), 148.

18. Marcia Angell, *The Truth About the Drug Companies: How They Deceive Us and What to Do About It* (New York: Random House, 2004), 76–79.

19. Avorn, *Powerful Medicines*, 208; Angell, *Truth About the Drug Companies*, 62–64.

20. In *combinatorial chemistry*, small chemical building blocks are combined in every possible variation into larger molecules that together form a huge synthetic-chemical "library." These are screened by the thousands with robotic equipment for the ability to attach to or disturb the function of validated targets in a cell-culture system. *Rational drug design* is the process of designing new compounds based on the known structure or function of the target molecule. As drug-resistant variants arise in cancer chemotherapy, for example, DNA sequencing can identify the mutations responsible and provide a rational basis for further drug development. *Pharmacogenomics*, the newest technology, could increase the number of targets by examining the genetic underpinnings of disease in order to find new drugs and tailor them to the individual patients. Enthusiasts refer to this as "personalized medicine." Companies are finding thousands of genes that produce previously unknown proteins that might be involved in the disease.

21. Neal Pattison and Luke Warren, "2002 Drug Industry Profits: Hefty Pharmaceutical Company Margins Dwarf Other Industries," *Public Citizen's Congress Watch*, June 2003.

22. P. Rodenhauser, "On creativity and medicine," *Pharos* 59, no. 4 (1996): 2–5.

23. Carl Sagan eloquently put the issue of unpredictability into a human dimension: "For myself, I like a universe that includes much that is unknown and, at the same time, much that is knowable. A universe in which everything is known would be static and dull. . . . A universe that is unknowable is not a fit place for a thinking being. The ideal universe for us is one very much like the universe we inhabit. And I would guess that is not really much of a coincidence." *Broca's Brain: Reflections on the Romance of Science* (New York: Random House, 1979), 18.

24. Jeff Howe, "The Rise of Crowdsourcing," *Wired,* June 2006, 176–83.

25. Jerome P. Kassirer, *On the Take: How Medicine's Complicity with Big Business Can Endanger Your Health* (New York: Oxford University Press, 2005).

26. In 1996 Juan Miguel Campanario of the University of Alcalá near Madrid attempted a systematic analysis to identify classic papers in the scientific literature whose authors reported that unsought factors played a key role in their research; see "Using *Citation Classics* to study the incidence of serendipity in scientific discovery," *Scientometrics* 37 (1996): 3–24. Of those authors in Campanario's series who retrospectively commented on some kind of accident or oversight in performing the research, in only less than a third did a careful reading of the original paper reveal hints of the involvement of unplanned factors. Even then, such hints were often indirect and written in the passive voice. Examples include: "the mixture was allowed to stand overnight," "it was found that," and "with the intention, originally."

27. The phrase "the endless frontier" was coined by Vannevar Bush, who headed the federal government's Office of Scientific Research and Development in World War II. Vannevar Bush, *Science, the Endless Frontier: A Report to the President* (1945; reprinted Washington, D.C.: National Science Foundation, 1960).

28. John Barth, *The Last Voyage of Somebody the Sailor* (Boston: Little, Brown, 1991).

Selected Bibliography

Abramson, John. *Overdo$ed America: The Broken Promise of American Medicine.* New York: HarperCollins, 2004.

Altman, Lawrence K. *Who Goes First? The Story of Self-Experimentation in Medicine.* New York: Random House, 1987.

Angell, Marcia. *The Truth About the Drug Companies: How They Deceive Us and What to Do About It.* New York: Random House, 2004.

Avorn, Jerry. *Powerful Medicines: The Benefits, Risks and Costs of Prescription Drugs.* New York: Knopf, 2004.

Ayd, Frank J., and Barry Blackwell, eds. *Discoveries in Biological Psychiatry.* Philadelphia: Lippincott, 1970.

Beveridge, W. I. B. *The Art of Scientific Investigation.* New York: Vintage, 1950.

Bishop, J. Michael. *How to Win the Nobel Prize: An Unexpected Life in Science.* Cambridge, Mass.: Harvard University Press, 2003.

Blumberg, Baruch S. *Hepatitis B: The Hunt for a Killer Virus.* Princeton: Princeton University Press, 2002.

Boden, Margaret A. *The Creative Mind: Myths & Mechanisms.* New York: Basic Books, 1991.

Bohm, David. *On Creativity.* Edited by Lee Nichol. London: Routledge, 1998.

Caldwell, Anne E. "History of Psychopharmacology." In *Principles of Psychopharmacology,* edited by William G. Clark and Joseph del Giudice, 2nd ed., 9–40. New York: Academic Press, 1978.

Comroe, Julius H., Jr. *Retrospectroscope: Insights into Medical Discovery.* Menlo Park, Calif.: Von Gehr, 1977.

Cooke, Robert. *Dr. Folkman's War: Angiogenesis and the Struggle to Defeat Cancer.* New York: Random House, 2001.

Dobell, Clifford. *Antony van Leeuwenhoek and His "Little Animals."* New York: Harcourt, Brace, 1932.

Dowling, Harry F. *Fighting Infection: Conquests of the Twentieth Century.* Cambridge, Mass.: Harvard University Press, 1977.

El-Hai, Jack. *The Lobotomist.* Hoboken, N.J.: John Wiley, 2005.

Forssmann, Werner. *Experiments on Myself: Memoirs of a Surgeon in Germany.* Translated by Hilary Davies. New York: St. Martin's Press, 1974.

Friedman, Meyer, and Gerald W. Friedland. *Medicine's 10 Greatest Discoveries.* New Haven: Yale University Press, 1998.

Goozner, Merrill. *The $800 Million Pill: The Truth Behind the Cost of New Drugs.* Berkeley and Los Angeles: University of California Press, 2004.

Hare, Ronald. *The Birth of Penicillin, and the Disarming of Microbes.* London: Allen and Unwin, 1970.

Infield, Glenn B. *Disaster at Bari.* New York: Macmillan, 1971.

Koestler, Arthur. *The Act of Creation.* London: Hutchinson, 1964.

Koffka, Kurt. *Principles of Gestalt Psychology.* New York: Harcourt, Brace, 1935.

Kohn, Alexander. *Fortune or Failure: Missed Opportunities and Chance Discoveries in Science.* Oxford: Blackwell, 1989.

Kuhn, Thomas S. *The Structure of Scientific Revolutions.* 2nd ed. Chicago: University of Chicago Press, 1970.

Le Fanu, James. *The Rise and Fall of Modern Medicine.* New York: Carroll and Graf, 2000.

Macfarlane, Gwyn. *Alexander Fleming: The Man and the Myth.* Cambridge, Mass.: Harvard University Press, 1984.

———. *Howard Florey: The Making of a Great Scientist.* Oxford: Oxford University Press, 1979.

Maurois, André. *The Life of Sir Alexander Fleming, Discoverer of Penicillin.* Translated by Gerard Hopkins. New York: E. P. Dutton, 1959.

May, Rollo. *The Courage to Create.* New York: W. W. Norton, 1975.

Merton, Robert K., and Elinor Barber. *The Travels and Adventures of Serendipity.* Princeton: Princeton University Press, 2004.

Moynihan, Ray, and Alan Cassels. *Selling Sickness: How the World's*

Biggest Pharmaceutical Companies Are Turning Us All into Patients. New York: Nation Books, 2005.

Nuland, Sherwin B. *Doctors: The Biography of Medicine*. New York: Knopf, 1988.

Patterson, James T. *The Dread Disease: Cancer and Modern American Culture*. Cambridge, Mass.: Harvard University Press, 1987.

Popper, Karl R. *The Logic of Scientific Discovery*. London: Hutchinson, 1959.

Porter, Roy. *The Greatest Benefit to Mankind*. New York: W. W. Norton, 1997.

Roberts, Royston M. *Serendipity: Accidental Discoveries in Science*. New York: John Wiley, 1989.

Root-Bernstein, Robert Scott. *Discovering: Inventing and Solving Problems at the Frontiers of Scientific Knowledge*. Cambridge, Mass.: Harvard University Press, 1989.

Rothman, David J., and Sheila M. Rothman. *The Willowbrook Wars*. New York: Harper and Row, 1984.

Ryan, Frank. *The Forgotten Plague: How the Battle Against Tuberculosis Was Won — and Lost*. Boston: Little, Brown, 1992.

Sackler, Arthur M., Mortimer D. Sackler, Raymond R. Sackler, and Felix Martí-Ibáñez, eds. *The Great Physiodynamic Therapies in Psychiatry*. New York: Paul B. Hoeber, Harper and Brothers, 1956.

Simonton, Dean Keith. *Greatness: Who Makes History and Why*. New York: Guilford Press, 1994.

Sneader, Walter. *Drug Discovery: The Evolution of Modern Medicines*. New York: John Wiley, 1985.

Stephens, Trent D., and Rock Brynner. *Dark Remedy: The Impact of Thalidomide and Its Revival as a Vital Medicine*. Cambridge, Mass.: Perseus, 2001.

Storr, Anthony. *The Dynamics of Creation*. New York: Atheneum, 1972.

Swazey, Judith P. *Chlorpromazine in Psychiatry*. Cambridge, Mass.: MIT Press, 1974.

Thagard, Paul. *How Scientists Explain Disease*. Princeton: Princeton University Press, 1999.

Valenstein, Elliot S. *Great and Desperate Cures: The Rise and Decline of Psychosurgery and Other Radical Treatments for Mental Illness*. New York: Basic Books, 1986.

Wallas, Graham. *The Art of Thought*. London: Jonathan Cape, 1926.

Wilson, David. *In Search of Penicillin*. New York: Knopf, 1976.

Wolpert, Lewis, and Alison Richards. *Passionate Minds: The Inner World of Scientists*. Oxford: Oxford University Press, 1997.

Illustration Credits

Index

abdominal aortic aneurysm (AAA), 208–214
abstract art, 7–8
accidental discoveries, xii–xiii, 6
    See also serendipity; *specific discoveries*
acetylcholine, 217–218, 243
acid blockers, 101
"acid-fast" staining technique, 37
actinomycetes, 82–84, 87–88, 335
actinomycin, 84
Act of Creation, The (Koestler), 17
adaptive unconscious, 15
Age of Exploration, 29
Agre, Peter, 3
Albany Medical Center Prize, 366
Albert and Mary Lasker Foundation, 171
Albert Lasker Awards in Medical Research, 171
alcohol abuse, 284–286
Alexander, Albert, 73–74
Alexander, Franz, 100–101
Alexander, Stewart, 119–122
alkaloids, 287
alkylating agents, 122–126

Allison, V. D., 62, 65
Altman, Lawrence K., 294
Alzheimer, Alois, 361
American Cancer Society, 171, 175–176
aminopterin, 129
analogical thinking, 17–19
Anderson, David, 3
aneurysms, 208–214
angina pectoris, 192, 218
angiogenesis, 140–147
angioplasty, 205–207
angiostatin, 146, 152
Anichkov, Nikolai, 226
aniline purple, 48
anomalies, 18–19
Antabuse, 286
anthrax, 33, 34
antianxiety-sedative drugs, 241, 245, 270–272, 281–283, 363–364
antibacterial drugs, 50–57
antibiotics, 55, 245
    from actinomycetes, 87–88
    actinomycin, 84
    overuse of, 90–91

antibiotics (*continued*)
    penicillin, 60–77, 332–334
    streptomycin, 85–91
    streptothricin, 84
    tyrothricin, 84
    workings of, 335
antibodies, 40, 92–94
anticancer drugs, 18
    See also chemotherapy
anticoagulants, 234–238
antidepressants, 241, 245, 273–280
    MAO inhibitors, 274–277, 362
    selective serotonin reuptake
      inhibitors (SSRIs), 278–280
    tricyclic, 277–278, 305
antigen-antibody reactions, 92–94
antigens, 39–40
antimanics, 241
antimetabolites, 129–130, 330, 340
antipsychotics, 241
    See also specific drugs
antiseptics, 61, 62–63
anxiety disorders, 242, 254, 270,
    309
aorta, 199, 208
aortic aneurysms, 208–214
Armstrong, Lance, 136
arteries, clogged, 225–233
arts, similarities between science
    and the, 15–17
aspirin, 49
    anticancer properties of, 148–150
    for blood clots, 237–238
AstraZeneca, 310
atherosclerosis, 225–233
atomic structure, 38
Atoxyl, 44
Aureomycin, 88
Australia antigen, 93–97
Avastin, 146
Avery, Oswald T., 84
Avorn, Jerry, 310
azo dyes, 50–55

bacteria
    discovery of, 30
    H. pylori, 102–113
    resistant strains of, 90–91
    stomach, 102–103
bacterial infections
    streptococcus, 50–55
    during World War I, 61–62
    See also antibiotics
bacteriology, 34–37
Bailar, John C., 177
Baltimore, David, 156
Bari, Italy, 117–126, 339
Barker, Bertha, 193
BAY 43-9006, 163
Beaudette, Fred, 83
Becker, Andrew, 166
Becquerel, Henri, xii
Behring, Emil von, 39, 59
benzodiazepines, 272, 281–283
Berger, Frank, 270–272
Bernard, Claude, 107
Berson, Solomon, 19–20
Beveridge, W. I. B., 45, 103
bewitching, 297
Bing, Richard, 350
Bini, Lucio, 250–251
biomilitarism, 181
bipolar affective disorder, lithium
    for, 260–265
Bishop, J. Michael, 156–158,
    344
bismuth, 105
bisociation, 17
Blackwell, Barry, 362
Blakemore, Arthur, 209–210, 212
Blaser, Martin, 108
blind discovery, 4–6
"Blind Men and the Elephant, The"
    (Saxe), 11–12
blood cells, types of, 165–166
blood clots, 234–238
blood thinners, 234–238

blood vessels
angiogenesis and, 140–147
nitric oxide and, 215–219
triangulation of, 209
Blumberg, Baruch, 22, 92–98
Bohm, David, 18–19
Bohr, Niels, 18
bone marrow cells, 165–166
bone marrow transplants, 165
Borel, Jean-François, 90
Bosch, Hieronymus, 365
Bouck, Noël, 145–146
Bovet, Daniel, 53–54
Boyer, Herbert, 304
brain chemistry
current research on, 298–299
early research on, 242–245
See also psychotropic drugs
brain-imaging studies, 15
brain research, lobotomies and,
252–259
breast cancer, 139
Brinkley, John R., 221–222
Broca, Paul, 253–254
Brown, Alexander Crum, 327
Brown, Michael S., 227–233, 311
Brown-Séquard, Charles-Édouard,
220
Bugie, Elizabeth, 85
Burkitt's lymphoma, 180, 344
Bush, Vannevar, 369

Cade, John, 260–265, 304
Campanario, Juan Miguel, 369
camphor, 249–250
cancer
breast, 139
cervical, 178
colon, 148–150, 342–343
fear of, 169
genetic basis for, 154–164
heavy metals for treatment of,
135–137
leukemia, 127–130, 134, 162
melanoma, 163

metaphors for, 181
metastasization of, xi–xii
mustard gas and, 122–126
prostate, 138–139
statistics on, 177–179
stomach, 111–112
tumors, 140–147, 152–156,
160–161
Vinca alkaloids for, 131–134
cancer-accelerating genes, 154–158,
182
cancer research
conflict over, 170–171
industrialization of, 169–180
serendipity in, 180, 181–183
stem cell research and, 165–168
targeted approach to, 170–178
Cannon, Walter B., 7, 356
cardiac catheterization, 195–203,
350–351
cardiovascular disease
angina pectoris, 192, 218
atherosclerosis, 225–233
coronary artery disease,
200–203
cardiovascular surgery, 208–214
Carliner, Paul, 2
Carrel, Alexis, 209, 352
Catharanthus roseus, 134
catheterization, cardiac, 195–203,
350–351
causation, false attributions of,
13–14
Celebrex, 150
cell membranes, 17
Central Intelligence Agency (CIA),
LSD and, 293–296
Cerletti, Ugo, 250–251
cervical cancer, 178
Chain, Ernst, 70–74, 76, 79, 80,
81, 333
chance
role of, in scientific discoveries,
4–6, 21–22
vs. serendipity, 7

chance (*continued*)
    See also serendipity
Chauveau, Auguste, 195
chemical affinities, 40
chemical properties, 38
chemical warfare, 117–126,
    338–339
chemistry, birth of, 38
chemotherapy, 42, 50, 77
    combination, 130
    ill effects of, 179–180
    for leukemia, 129–130, 134
    nitrogen mustard, 122–126
chloramphenicol, 87–88
chlorpromazine (CPZ), 267–269
cholera, 35
cholesterol research, 225–233
choriocarcinoma, 129–130
chronic myelogenous leukemia
    (CML), 162
Churchill, Winston, 19, 57, 126
CIA, LSD and, 293–296
Cialis, 223–224
Circle Limit IV (Escher), 10
circulatory system, early views on,
    187–188
cisplatin, 136–137
clinically targeted research,
    170–180, 302–303
Close, Chuck, 8–9
clots, blood, 234–238
coal gas, 38–39
coal tar, 38–39
coal-tar dyes, 41–42, 48–55, 327
Cocoanut Grove nightclub, 77
Coghill, Robert, 75
cognitive illusions, 8–13
Cohen, Sidney, 292
Cohen, Stanley, 304
Collip, James, 131
colon cancer, 148–150, 342–343
colonoscopy, 149
coma, induced, 247–249
combination drug therapy, 130
combinatorial chemistry, 312, 368

commercial products, developed
    through serendipity, 39
compound 606, 43–46
Coningham, Arthur, 117
consumption. *See* tuberculosis
Controlled Substances Act (1970),
    295
Cooke, Robert, 141
"corner-of-the-film diagnosis," xii
coronary angioplasty, 205–207
coronary arteries
    bypass surgery, 203
    visualization of, 199–203
coumarin, 236–237
counterstaining technique, 36–37
Cournand, André, 198
COX-1, 150
COX-2, 150
Craven, Lawrence, 237
creative people, characteristics of,
    17
creative thought
    analogies and, 17–19
    anomalies and, 18–19
    pathways of, 14–19
creativity, role of, in scientific dis-
    coveries, 6–7
Crile, George, Jr., 169
crowdsourcing, 316
Curie, Marie, 1
cyanamide, 364
cyclosporin, 90

Dacron grafts, 212–213
Dale, Henry, 68, 244
D'Amato, Robert, 152
Darwin, Charles, 19, 300
Darwin, Horace, 191
Declomycin, 88
deductive reasoning, 6
Delay, Jean, 267
Deniker, Pierre, 267, 305
depression, 242, 273–280, 360–361
diagnostic imaging, xii
diphtheria, 39

direct-to-consumer (DTC) marketing, 308
disulfiram, 284–286
Diuril, 330
Dixon, Laurinda S., 365
DNA, 125, 155, 162,182
dogmatism, 305
Dole, Robert, 213
Domagk, Gerhard, 50–55, 329
dopamine, 18, 167, 244–245, 268–269, 275, 277, 298
Dotter, Charles, 205–206
Dougherty, Thomas, 123–124
Dramamine, 2
Dread Disease, The (Patterson), 169
drug development process, 312–313
drug receptors, 41–42
drug safety tests, 58
drug therapy
   combination, 130
   targeted, 162–164
Druker, Brian, 310
Dublin, Louis, 227
Dubos, René, 84
dye production, 38–39, 48–55

Edelman, Gerald, 14–15
education, changes needed in, 313–315, 317
Effler, Donald, 203
Egas Moniz, António, 252, 254–255, 257, 358
Ehrlich, Paul, 1, 15, 304, 328
   counterstaining technique by, 36–37
   diphtheria work by, 39
   drug receptor research by, 41–42
   immunological research by, 39–40
   malaria work by, 40–41
   syphilis cure by, 42–47
Einstein, Albert, 14, 213–214
Einthoven, Willem, 189–194, 349
Eisenhower, Dwight, 300
Elavil, 291

electrocardiograms (ECGs), 191–192, 193–194
electroconvulsive therapy (ECT), 250–251
electrometers, 188–189
electron microscopes, 105
elements, chemical properties of, 38
Elion, Gertrude, 340
Enders, John, 175
Endicott, Kenneth, 173
endostatin, 146
endothelial cells, 215–219
ENL (erythema nodosum leprosum), 343
Epstein-Barr virus, 344
Equanil, 271
erectile dysfunction (ED), 220–224
ergotism, 296–297
erythromycin, 88
Escher, M. C., 9–10
estrogen therapy, for prostate cancer, 138–139
ethical standards, 97
Ex-Lax, 48–49

fallacies, logical, 13–14
Farber, Sidney, 127–130, 171, 172, 174, 330
Farmer, Frances, 257
Favoloro, René, 203
Feldman, Burton, 89
Ferrara, Napoleone, 144
fever therapy, 246–247
Feynman, Richard, 20–21, 23, 303
fiberoptic endoscopy, 112
Fibiger, Johannes, 13
Fildes, Luke, 69
Fleming, Alexander, 22, 59, 60–69, 71, 78–81, 331
Fleming Myth, 78–79
Florey, Howard, 4, 70–75, 78–81
flumazenil, 364
fluoroscopy, 196
folic acid, 127–130, 330
Folkman, Judah, 140–147, 307

Food and Drug Administration (FDA), 58, 152, 308
Food, Drug, and Cosmetic Act (1938), 58
Forman, David, 112
Forssmann, Werner, 195–198
Fracastoro, Girolamo, 43
Francis, Thomas, Jr., 159
Fraser, Thomas, 327
Freeman, Walter, 255–257
Freud, Sigmund, 357
Friedman, Steven, 212
Furchgott, Robert, 22, 215–219, 353–354

GABA (gamma-aminobutyric acid), 363
Gage, Phineas, 253–254
Galen, 187
Galileo, 326
Galvani, Luigi, 348–349
galvanometer, 190–191, 194
Gardasil, 178
Garrod, A. B., 262
gas gangrene, 61
gastric cancer, 111–112
Gay, Leslie, 2
Gelmo, Paul, 53
gene chips, 162
genes
    cancer-accelerating, 154–158, 182
    cancer-inhibiting, 159–164, 182
Germany
    pharmaceutical industry in, 49–50
    textile industry in, 49
germ theory, 29–34
    See also bacteriology
Gestalt psychology, 8–9
Giardiello, Francis M., 149–150
Gilman, Alfred, 123, 125, 339
Gleevec, 162, 310
Gofman, John, 226
gold salts, 137

Goldstein, Joseph, 227–233, 311, 355
Goodman, Louis S., 123, 339
Gornik, Heather, 177
Gottlieb, Sidney, 293–295
gout, 78
government role, in medical research, 173–175, 300–303
    See also grant application process
Gram, Christian, 331
gram-negative bacteria, 67, 85, 331
gram-positive bacteria, 67, 86, 331, 333–334
Gram's stain, 331
Grant, Cary, 292–293
grant application process
    peer review and, 303–307
    targeted research and, 301–303
Grigg, C. L., 262
Groopman, Jerome, 179–180
groupthink, 301–302
Gruentzig, Andreas, 206–207

H2 blockers, 101
Hald, Jens, 284–286
Haldane, J. B. S., 238
hallucinogens, 287–299
Halsted, William Stewart, 32
Hampton, Caroline, 32
Hardin, Garrett, 305
Hare, Ronald, 63
Harvey, William, 18, 187–188
Hata, Sahachiro, 44–45
HAV (hepatitis A virus) vaccines, 336
HDL (high-density lipoprotein), 226
heart
    electrical current in, 188–189
    recording of, 188–192
heart attacks, 192–194
heart disease
    See also cardiovascular disease
heart research
    on blood thinners, 234–238

on cardiac catheterization,
    195–203
on cholesterol, 225–233
on coronary angioplasty,
    205–207
early work on, 187–194
nitric oxide and, 215–219
on vascular surgery, 208–214
Heatley, Norman, 70–75, 79
Heberden, William, 192
Heffter, Arthur, 364
Heliobacter, 102–113
Heller, John, 173
Helms, Richard, 295
heparin, 235
hepatitis, 92–98
hepatitis A, 94, 97–98, 336
hepatitis B, 94–98, 336
hepatitis C, 336
hepatitis D, 336–337
Herbert, Frank, 13
Herrick, James, 192–193
Hill, W. E., 10
HIV, 180
Hodgkin, Alan, 22, 325
Hodgkin, Dorothy, 333
Hodgkin's disease, 125, 335
Hoffmann, Erich, 43–44
Hofmann, Albert, 287–292, 294,
    296, 298
Hofstadter, Douglas R., 18
Hooke, Robert, 366
Howell, William, 234–235
H. pylori, 102–113
Huant, Ernest, 361
Huggins, Charles, 138–139
human genome, 182
Humphrey, J. H., 23
Hunt, Mary, 75
Huxley, Aldous, 297, 365
hypoglycemic coma, 247–249

ice-pick lobotomy, 256–258
IG Farben, 49–50, 53, 54, 328–329
Ignarro, Louis, 218, 219

illusions, cognitive, 8–13
imagination, as pathway to creative
    thought, 14–15
imipramine, 241, 245, 278, 305
immunology, 39–40, 59–60
immunosuppressants, 90
impotence, 220–224
inductive reasoning, 6
industrial scientific laboratories, 39
infectious jaundice, 94
InnoCentive, 316
innovation, reasons for decline in,
    300–315
insight, 19–21
insulin coma, 247–249
interferon, 180
intuition, as pathway to creative
    thought, 14–15
iproniazid, 275–277
Iressa, 163
isoniazid (INH), 90

Jacobsen, Erik, 284–286
Jacobson, Julius H., II, 204
Jamison, Kay Redfield, 264–265
Janelia Farm Research Campus,
    315–316
Jensen, Carl, 339
Johnston, C. D., 131
Jones, Doris, 85
Jones, Stormie, 231–232

Kandinsky, Wassily, 7–8
Kantorovich, Aharon, 5
Kekulé, Friedrich, 38
Kelsey, Frances O., 152
Kendall, Edward, 3–4
Kennedy, Edward M., 347
Kennedy, Foster, 248
Kennedy, Joseph P., 257
Kennedy, Rose Marie (Rosemary),
    257
Kettering, Charles, 171–172
Kiki (Close), 8–9
Kline, Nathan, 18, 274–277

Koch, Robert, 34–37, 39, 86–87, 108, 137
Koestler, Arthur, 17
Kölliker, Albert von, 188
Krugman, Saul, 96–98
Kuhn, Roland, 277–278, 305
Kuhn, Thomas, 4, 18, 113

Laborit, Henri, 266–267
Laetrile, 143–144
Lakhani, Karim, 316
Langer, Robert, 142–143
Langerhans, Paul, 175, 347
Laqueur, Ernst, 220
Lasker, Albert, 171
Lasker, Mary, 171
Lauterbur, Paul, 306
laxatives, 48–49
LDL (low-density lipoprotein), 226–233
Leary, Timothy, 293, 297
Ledes, Richard, 358
Lee, Tsung-Dao, 13–14
Leeuwenhoek, Antony van, 29–31, 326–327
Lehmann, Heinz E., 268
Lesch, John, 56
leukemia, 127–130, 134, 162
Levi, Primo, 328
Levine, Arnold, 160–161
Levine, Samuel, 193–194
Levitra, 223
Lewis, Thomas, 191–192, 349
Librium, 241, 281–283
lifestyle drugs, 310
Link, Karl Paul, 235–236
Lipitor, 232
Lister, Joseph, 32, 332
lithium, 260–265, 359–360
lobotomies, 252–259, 358
Loewi, Otto, 242–244
logic, 14, 21–23
logical fallacies, 13–14
Loomer, Harry, 276
lovastatin, 232, 335, 355

Lowell, Robert, 264
LSD, 241, 245, 287–299
Ludwig, Bernard, 271
Luria, Salvador, 7
lymphoma
    Burkitt's, 180, 344
    nitrogen mustard and, 122–126
lysergic acid. See LSD
lysozyme, 62–63, 70–71, 80

Macfarlane, Gwyn, 67
"magic bullets" concept, 42, 47
magnetic resonance imaging (MRI), 306
malaria, 40–41, 48
malaria treatment, 246–247
manic-depressive illness, lithium for, 260–265
MAO (monoamine oxidase) inhibitors, 274–277, 362
Marey, Étienne, 195
Marks, John, 295
Marquardt, Martha, 45
Marshall, Barry, 103–113, 306–307
Massengill and Company, 58
Matisse, Henri, 16
Matossian, Mary, 297
May and Baker, 56–57
McCormick, Frank, 163
McCulloch, Ernest, 165–168
McLean, Jay, 234–235
Medawar, Peter, 21, 23–25
medical research
    changing climate for, 300–303
    false impressions of, 1–2
    government role in, 173–175, 300–303
    lack of innovation in current, 300–315
    reality of, 2–4
    targeted, 170–180, 302–303
    See also scientific discoveries
Meduna, Ladislaus von, 249–250, 304
melanoma, 163

Mendeleev, Dmitri, 38
mental illness
    current research on, 298–299
    early treatments for, 246–251
    lobotomies for, 252–259
    See also psychotropic drugs
meprobamate, 245, 270–272
mescaline, 290, 298, 364
metaphors, for disease, 181
Metchnikoff, Ilya, 40, 59
methotrexate, 129–130
methylene blue, 36, 41
"me-too" drugs, 309–310, 317, 368
Metrazol, 249–250
Mevacor (lovastatin), 232, 335, 355
Meyer-Thurow, Georg, 39
microorganisms
    culturing of, 35–37
    discovery of, 29–31
microscopes
    electron, 105
    invention of, 29–31
microvascular surgery, 204
Miller, Mrs. Ogden, 75–76
Millman, Irving, 96
Miltown (meprobamate), 241, 245,
    270–272
Mitchell, S. Weir, 359
MKULTRA, 294–295
molds, 64–69, 71, 332
    See also penicillin
Moniz, António Egas
    See Egas Moniz, António
Monod, Jacques, 323–324
mood-altering drugs
    See psychotropic drugs
Moore, Francis, 175, 181
Morson, Basil, 343
motion sickness cure, 2
Moyer, Andrew J., 75
Mückter, Heinrich, 151
Müller, Heinrich, 188
multiple myeloma, 152–153
Murad, Ferid, 218, 219
mustard gas, 119–126, 338–339

Mustargen, 124–125
myocardial infarctions, 192–194
My Wife and My Mother-in-Law
    (Hill), 10–11

Nabokov, Vladimir, 15
Nardil, 277
National Cancer Act (1971),
    173–175
National Cancer Institute (NCI),
    171, 172–173, 175–177
National Institutes of Health (NIH),
    171, 172, 230, 301–304
National Research Act (1974), 97
Nazis, 55
Ne'eman, Yuval, 5
Neosalvarsan, 46
Nestler, Eric, 280
neurotransmitters, 242–245
Nexium, 309–310
nicotinamide, 361
Nijinsky, Vaslav, 248
nitric oxide (NO), 215–219,
    353–354
nitrogen mustard, 122–126,
    338–339
nitroglycerin, 192, 218, 353
Nitti, Frédéric, 53–54
Nixon, Richard, 173
Nobel, Alfred, 353
Nobel Prize
    acceptance lectures, role of
        chance acknowledged in,
        22, 23, 79, 98, 219
    Nazis and, 55
    recipients. *See* Agre, Peter;
        Baltimore, David;
        Becquerel, Henri; Behring,
        Emil von; Bishop, J.
        Michael; Blumberg, Baruch;
        Bovet, Daniel; Brown,
        Michael; Carrel, Alexis;
        Chain, Ernst; Churchill,
        Winston; Cohen, Stanley;
        Cournand, André; Curie,

Nobel Prize (*continued*)
Marie; Dale, Henry;
Domagk, Gerhard; Egas
Moniz, António; Ehrlich,
Paul; Einstein, Albert;
Einthoven, Willem; Elion,
Gertrude; Enders, John;
Feynman, Richard; Fibiger,
Johannes; Fleming,
Alexander; Florey, Howard;
Forssman, Werner;
Furchgott, Robert;
Goldstein, Joseph;
Hodgkin, Alan; Hodgkin,
Dorothy; Huggins, Charles;
Ignarro, Louis; Kendall,
Edward; Koch, Robert;
Lauterbur, Paul; Lee,
Tsung-Dao; Loewi, Otto;
Luria, Salvador; Marshall,
Barry; Medawar, Peter;
Metchnikoff, Ilya; Monod,
Jacques; Murad, Ferid;
Ossietzky, Carl von; Planck,
Max; Rabi, Isidor; Ramón
y Cajal, Santiago; Richards,
Dickinson; Richet, Charles;
Röntgen, Wilhelm; Rous,
Peyton; Rutherford, Ernest;
Szent-Györgyi, Albert;
Temin, Howard; Vane,
John; Varmus, Harold;
Wagner-Jauregg, Julius;
Waksman, Selman; Warren,
J. Robin; Yalow, Rosalyn;
Yang, Chen Ying
Noble, Clark, 131–132
Noble, Robert, 132–134, 305–306
nonobjective painting, 7–8
nonsteroidal anti-inflammatory
drugs (NSAIDs), 149–150
noradrenaline, 244, 245, 275, 356
norepinephrine, 275
nuclear precession, 21

oncogenes, 157–158, 182
organ transplants, immunosuppressants and, 90
orthodoxy, 305
Ossietzky, Carl von, 55

p53 gene, 160–161
PABA, 329–330
Papanicolaou, George, 178
Pap smear, 178
Park, Robert, 12
PAS (paraaminosalicylic acid), 330
Pasteur, Louis, 7, 31–33
Patterson, James T., 169
Paxil, 279, 309
peer review, 303–307, 318,
366–367
penicillin, 332–334
Fleming's role in discovery of,
60–69
uncovering usefulness of, 70–77
World War II and, 75–77
penicillium, 66–69, 71, 75, 332
Penicillium chrysogenum, 75
Penicillium notatum, 66–69, 71
peptic ulcers. *See* ulcers
periodic table, 38
periwinkle, 131–134
Perkin, William Henry, 48–49
perspective, distortion and, 8–13
Petri, Richard Julius, 35–36
Petri dish, 35–36
pharmaceutical industry, 42
in Germany, 49–50
lack of innovation and, 307–313
marketing by, 308–310, 368
penicillin and, 76–77
regulation of, 317–318
pharmacogenomics, 312, 368
Phenergan (promethazine), 267
phenolphthalein, 48–49
Planck, Max, 5–6
platelets, 165
Plath, Sylvia, 251

platinum, 135–137
polio vaccine, 159–160
polymorphism, 92
polymyxins, 335
post facto logic, 21–23
post hoc, ergo propter hoc, 13–14
postoperative infections, 32
precious metals, 135–137
Prilosec (omeprazole), 310
probenecid, 78
Prontosil, 51–55
Propecia, 311–312
prostaglandins, 238
prostate cancer, 138–139
protozoa, discovery of, 30
Prozac, 278–280, 309
psychiatric treatments
    early, 246–251
    lobotomies, 252–259
    See also psychotropic drugs
psychoactive drugs, 18, 287–299
psychopharmacology, 241–245
psychosomatic medicine, 100–101
psychosurgery, 252–259
psychotomimetic, 292
psychotropic drugs, 18, 241–245
    antidepressants, 273–280
    disulfiram, 284–286
    Librium, 241, 281–283
    lithium, 260–265, 359–360
    LSD, 241, 245, 287–299
    Miltown, 241, 245, 270–272
    Thorazine, 241, 245, 258,
        266–269, 291, 305
    Valium, 241, 245, 281–283, 291,
        363

quartz string, 190–191
questions, asking, 14–15
quinine, 48

Rabi, Isidor, 14
rabies, 33
radioactivity, discovery of, xii
radioimmunoassay (RIA), 19–20

radiology, xii
RAF, 163
Ramón y Cajal, Santiago, 242
randomness, 24
rare diseases, lack of research on,
    310–311
rational drug design, 312, 368
rational thought, 6
reason, as pathway to creative
    thought, 14
red blood cells, 165
researchers
    characteristics of, 6–7
    education of, 313–315
research grants, 301–307, 318
reserpine, 18, 274–275
retrospective falsification, 23
retroviruses, 155–158
reverse transcriptase, 156
Revlimid, 153
rheumatoid arthritis, 137
Rhines, Chester, 83
Rhoads, Cornelius P., 123, 125,
    172, 346–347
Richards, Dickinson, 198
Richet, Charles, 22
RNA, viral, 155–156
Rockefeller Foundation, 301
Rockefeller Institute for Medical
    Research, 301
Rogaine, 311
Rogerson, Gerard, 152
Romazicon (flumazenil), 364
Röntgen, Wilhelm, xii, 22
Roosevelt, Franklin D., 121
Roosevelt, Franklin D., Jr., 54, 58
Root-Bernstein, Robert, 6
Rosenberg, Barnet, 135–137
Rothlin, Ernst, 290
Rothstein, Aser, 17, 22, 24
Rous, Peyton, 154–156
Rutherford, Ernest, 18

sagacity, 6
Sagan, Carl, 369

Sakel, Manfred, 247–249, 259
Salem witches, 297
Salk, Jonas, 159
Salvarsan, 45–47
Sanderson, John Burdon, 332
Sarafem (Prozac), 279, 309
Sargant, William, 255
Saunders, John, 276
Saxe, John Godfrey, 11–12
Schatz, Albert, 84–90
Schaudinn, Fritz, 43
schizophrenia, 242
    electroshock therapy for,
        250–251
    insulin shock treatment for,
        247–249
    Metrazol for, 249–250
    Thorazine for, 266–269
Schou, Mogens, 263
science
    normal vs. revolutionary, 4–6
    similarities between the arts and,
        15–17
scientific discoveries, 12–13
    creativity and, 6–7
    credit for, 276
    portrayal of, by scientific estab-
        lishment, 2
    reality of, 2–4
    reasons for decline in, 300–315
    reporting of, 21–24
    role of chance in, 4–6, 21–22
    through serendipity, xii–xiv, 2–4,
        6–7, 20, 24–25
    unexpected nature of, 170–171
scientific education, 313–315, 317
scientific journals, 318–319
scientific method, 345–346
scientific mind, 6–7
scientific research
    changing climate for, 300–303
    targeted, 170–180, 302–303
Scott, Sir Walter, 16
sedatives, 241, 281–283
Seldinger, Sven-Ivar, 199

selective angiocardiography,
    198–199
selective serotonin reuptake
    inhibitors (SSRIs), 278–280
self-experimentation, 109
Semmelweis, Ignaz, 32
sepsis, 32
Serendip, 6
serendipity
    cultivating, 300–303, 315–320
    role of, in scientific discoveries,
        xii–xiv, 2–4, 6–7, 20,
        24–25
    targeted research and lack of,
        181–182
serotonin, 18, 244–245, 275, 277,
    279, 298
Serpasil, 274–275
sex hormones, 138–139
shark cartilage, 143
Sheskin, Jacob, 343
Shimkin, Michael, 176–177
shock treatment
    electric, 250–251
    insulin, 247–249
sildenafil citrate, 222–223
Sippy, Bertram W., 100
Sippy diet, 100
606, compound, 43–46
sleeping sickness, 35, 43
Sloan, Alfred P., Jr., 171–172
Sloan-Kettering Institute of Cancer
    Research, 172
soil microbiology, 82–83
Sones, F. Mason, Jr., 200–203, 206
Sontag, Susan, 181
Spiegelman, Sol, 174–175
spinal taps, 78
stagnant knowledge, 103
stains and staining techniques,
    36–37, 40–41, 331
Stanley, Leo L., 221
St. Anthony's Fire, 296–297, 365
Starzl, Thomas, 22
Stehelin, Dominique, 344

stem cell research, 3, 165–168, 183
stent-grafts, 213, 352
Sternbach, Leo H., 281–283, 363
Stevens, Jay, 296
St. Mary's Hospital (London), 59–60
Stoll, Arthur, 287
Stoll, Werner, 290
stomach cancer, 111–112
streptococcus, 50–55
streptokinase, 235
streptomycin, 85–91
streptothricin, 84
stress, ulcers and, 100–101
string galvanometer, 190–191, 194
Structure of Scientific Revolutions, The (Kuhn), 4
St. Vitus's Dance, 296
Styron, William, 360–361
sulfadiazine, 57
sulfa drugs, 50–57, 73, 329–330
sulfanilamide, 53–55
sulfapyridine, 56–57
surgery
    bypass, 203
    lobotomies, 252–259
    microvascular, 204
    postoperative infections and, 32
    vascular, 208–214
Svoboda, Gordon, 133–134
sweet clover, 235–237
synesthesia, 292
synthetic dye industry, 48–55
syphilis, 42–47
Szent-Györgyi, Albert, 13, 307
Szmuness, Wolf, 96

Tagamet (cimetidine), 101
tamoxifen, 139
targeted drug therapy, 162–164
targeted research, 170–180, 302–303
Taxol, 177
Temin, Howard, 155–156
Terramycin, 88

tetanus, 61
tetracyclines, 88, 368
textile industry, 49
thalidomide, 151–153, 343
Thomas, Lewis, 69, 345–346
Thorazine, 241, 245, 258, 266–269, 291, 305
thyroxine, 3–4
Till, James, 165–168
Tillet, William, 235
TNP-470, 145
Tofranil, 291
tranquilizers, 18, 241, 245, 270–272, 281–283
Transactions (journal), 326
transorbital lobotomies, 256–258
Tréfouël, Jacques, 53–54
Tréfouël, Thérèse, 53–54
tricyclic antidepressants, 277–278, 305
trypanosome infections, 44
tuberculosis, 36
    cure for, 83–91
    reemergence of, 91
    scourge of, 86–87
tumors, 140–147, 152–153
    p53 gene and, 160–161
    viral cause of, 154–156
Tyndall, John, 332
typhoid fever, 59–60
tyramine, 362
tyrothricin, 84

ulcers, 306–307
    gastric biopsies and, 102–104
    H. pylori and, 102–113
    prevailing opinion on, 99–101
urease, 106

vaccines, 32–33
    hepatitis A (HAV), 336
    hepatitis B, 96
    polio, 159–160
    typhoid, 60
Valenstein, Elliot, 246

Valium (diazepam), 241, 245, 281–283, 291, 363
vancomycin, 88
Vane, John, 237–238
Vaniqa, 312
Varmus, Harold, 156–158, 344
vascular surgery, 208–214
Velcade, 344
Viagra, 222–224, 354
vinblastine, 132, 134
Vinca rosea, 131–134
vincristine, 134
Vinculin, 131
Vinyon grafts, 212
Vioxx, 150
viruses
    retroviruses, 155–158
    spread of tumors and, 154–156
visual imagery, 15
vitamin C, 307
Vogelstein, Bert, 160–161
Voodoo Science (Park), 12
Voorhees, Arthur B., Jr., 209–214

Waddell, William R., 149
Wagner von Jauregg, Julius, 246–247, 357
Waksman, Selman, 82–91
Waller, Augustus D., 188–189
Walpole, Horace, 6, 41
warfarin, 236–237, 356
war on cancer, 173–181
Warren, J. Robin, 102–103, 108–109, 113

Watanabe, Yoshio, 230
Watts, James, 255
Way of an Investigator, The (Cannon), 7
Weinberg, Robert, 164
white blood cells, 59, 122, 127, 129, 132, 152, 165, 339
Wills, Lucy, 127
Wolff, Harold G., 101
wool sorter's disease, 33
World War I
    bacterial infections during, 61–62
    chemical warfare during, 338–339
World War II
    Bari tragedy during, 117–126
    penicillin and, 75–77
    psychosomatic medicine and, 100–101
Wright, Almroth, 59–60, 61, 78, 330

X-rays, discovery of, xii

Yalow, Rosalyn, 19–20, 22, 306, 367
Yang, Chen Ning, 13–14

Zantac (ranitidine), 101
Zoloft (sertraline), 279
Zubrod, C. Gordon, 176
Zyklon B, 329